TRUE VIPERS

TRUE VIPERS

Natural History and Toxinology

of

Old World Vipers

by
David Mallow
David Ludwig
and
Göran Nilson

KRIEGER PUBLISHING COMPANY
Malabar, Florida
2003

Original Edition 2003

Printed and Published by
**KRIEGER PUBLISHING COMPANY
KRIEGER DRIVE
MALABAR, FLORIDA 32950**

Copyright © 2003 by Krieger Publishing Company

All rights reserved. No part of this book may be reproduced in any form or by any means, electronic or mechanical, including information storage and retrieval systems without permission in writing from the publisher.
No liability is assumed with respect to the use of the information contained herein.
Printed in the United States of America.

> **FROM A DECLARATION OF PRINCIPLES JOINTLY ADOPTED BY A COMMITTEE OF THE AMERICAN BAR ASSOCIATION AND A COMMITTEE OF PUBLISHERS:**
> This publication is designed to provide accurate and authoritative information in regard to the subject matter covered. It is sold with the understanding that the publisher is not engaged in rendering legal, accounting, or other professional service. If legal advice or other expert assistance is required, the services of a competent professional person should be sought.

Library of Congress Cataloging-in-Publication Data

Mallow, David, 1955–
  True vipers : natural history and toxinology of Old World vipers / by David Mallow, David Ludwig, and Göran Nilson. — Original ed.
    p. cm.
  Includes bibliographical references.
  ISBN 0-89464-877-2 (alk. paper)
    1. Viperidae. I. Ludwig, David, 1953– II. Nilson, G. (Göran) III. Title.

QL666.O69M34 2003
597.96'3—dc21
                                                            2002043377

Dedication

We would like to dedicate this book to our wives,
Jean Mallow, Cathy Womack, and Margareta Nilson, and our
children Justin and Stephen Mallow; Molly, Jesse, and Colin Ludwig;
and Staffan, Johan, and Peter Nilson. Their love, support,
encouragement, and patience kept us going throughout this project.

Contents

Preface xi

Acknowledgments xiii

Introduction 1

Chapter 1 Azemiopinae 13
   Azemiops 13
      Azemiops feae Boulenger 1888 14

Chapter 2 Causinae 19
   Causus 19
      Causus bilineatus Boulenger 1905 20
      Causus defilippii (Jan 1863) 21
      Causus lichtensteini (Jan 1859) 24
      Causus maculatus (Hallowell 1842) 25
      Causus resimus (Peters 1862) 28
      Causus rhombeatus (Lichtenstein 1823) 30

Chapter 3 Viperinae 35
   Adenorhinos 35
      Adenorhinos barbouri (Loveridge 1930) 35

Chapter 4 *Atheris* 39
      Atheris acuminata Broadley 1998 40
      Atheris broadleyi Lawson 1999 41
      Atheris ceratophora Werner 1895 42
      Atheris chlorechis (Pel 1851) 43
      Atheris desaixi Ashe 1968 45
      Atheris hispida Laurent 1955 47
      Atheris katangensis de Witte 1953 49
      Atheris nitschei Tornier 1902 50
      Atheris rungweensis Bogert 1940 51
      Atheris squamigera (Hallowell 1856) 53
      Atheris subocularis Fischer 1888 57

Chapter 5 *Bitis* 59
   Subgenus *Bitis* Gray 1842 60

Bitis arietans (Merrem 1820) 60
Subgenus *Calechidna* Tschudi 1845 71
    Bitis albanica Hewitt 1937 71
    Bitis armata Smith 1826 73
    Bitis atropos (Linnaeus 1758) 74
    Bitis caudalis (Smith 1839) 79
    Bitis cornuta (Daudin 1803) 86
    Bitis heraldica (Bocage 1889) 89
    Bitis inornata (Smith 1838) 90
    Bitis peringueyi (Boulenger 1888) 92
    Bitis rubida Branch 1997 95
    Bitis schneideri (Boettger 1886) 97
    Bitis xeropaga Haacke 1975 99
Subgenus *Macrocerastes* Reuss 1939 101
    Bitis gabonica (Duméril, Bibron and Duméril 1854) 101
    Bitis nasicornis (Shaw 1802) 115
    Bitis parviocula Böhme 1977 122
Subgenus *Keniabitis* Lenk, Herrmann, Joger, Wink 1999 124
    Bitis worthingtoni Parker 1932 124

Chapter 6 *Cerastes* 127
    Cerastes cerastes (Linnaeus 1758) 128
    Cerastes gasperetti Leviton and Anderson 1967 130
    Cerastes vipera (Linnaeus 1758) 132

Chapter 7 *Daboia* 141
    Daboia palaestinae (Werner 1938) 141
    Daboia russelii (Shaw and Nodder 1797) 150

Chapter 8 *Echis* 161
    Echis carinatus (Schneider 1801) 162
    Echis coloratus (Günther 1878) 164
    Echis hughesi Cherlin 1990 167
    Echis jogeri Cherlin 1990 168
    Echis khosatzkii Cherlin 1990 169
    Echis leucogaster Roman 1972 170
    Echis megalocephala Cherlin 1990 171
    Echis multisquamatus Cherlin 1981 172
    Echis ocellatus Stemmler 1970 173
    Echis pyramidum (Geoffroy and Saint-Hillaire 1827) 175

Chapter 9 *Eristicophis* 189
    Eristicophis macmahonii Alcock and Finn 1897 189

Chapter 10 *Macrovipera* 193

Macrovipera deserti (Anderson 1892) 194
Macrovipera lebetina (Linnaeus 1758) 195
Macrovipera mauritanica (Duméril and Bibron 1848) 200
Macrovipera schweizeri (Werner 1935) 202

Chapter 11 *Montatheris* 205
Montatheris hindii (Boulenger 1910) 205

Chapter 12 *Proatheris* 207
Proatheris superciliaris (Peters 1854) 207

Plates (1.1–14.26) follow page 210

Chapter 13 *Pseudocerastes* 213
Pseudocerastes persicus (Duméril, Bibron and Duméril 1854) ... 213

Chapter 14 *Vipera* 219
Subgenus *Vipera* Laurenti 1768 220
Vipera ammodytes (Linnaeus 1758) 221
Vipera aspis (Linnaeus 1758) 228
Vipera latastei Bosca 1878 236
Vipera monticola Saint-Girons 1954 237
Vipera transcaucasiana Boulenger 1913 239
Subgenus *Pelias* Merrem 1820 241
Vipera barani Böhme and Joger 1984 241
Vipera berus (Linnaeus 1758) 242
Vipera darevskii Vedmederja, Orlov and Tuniyev 1986 252
Vipera dinniki Nikolsky 1913 254
Vipera kaznakovi Nikolsky 1909 256
Vipera nikolskii Vedmederja, Grubant and Rudayeva 1986 258
Vipera pontica Billing, Nilson and Sattler 1990 260
Vipera seoanei Lataste 1879 261
Subgenus *Acridophaga* Reuss 1927 263
Vipera anatolica Eiselt and Baran 1970 266
Vipera ebneri Knoeppfler and Sochurek 1955 267
Vipera eriwanensis (Reuss 1933) 269
Vipera lotievi Nilson, Tuniyev, Orlov, Höggren and Andrén 1995 ... 271
Vipera renardi (Christoph 1861) 272
Vipera ursinii (Bonaparte 1835) 274
Subgenus *Montivipera* Nilson, Tuniyev, Andrén, Orlov,
    Joger and Herrmann 1999 280
Vipera albicornuta Nilson and Andrén 1985 280
Vipera albizona Nilson and Andrén 1990 282
Vipera bornmuelleri Werner 1898 283
Vipera bulgardaghica Nilson and Andrén 1985 285

Vipera latifii Mertens, Darevsky and Klemmer 1967 287
Vipera raddei (Boettger 1890) 290
Vipera wagneri Nilson and Andrén 1984 293
Vipera xanthina (Gray 1849) 295

Bibliography 301

Glossary 351

Preface

The relationship of vipers to man has a long and fascinating history. Even today, vipers evoke fear and excite attention. Improvements in captive care and husbandry have led to a proliferation of private collections and therefore an increase in the need for detailed information on the biology of these snakes.

At the onset it should be emphasized that our objective is to furnish the reader with a summary of taxonomy, general biology, and toxinology for vipers. Within the time it took to complete this book, numerous taxonomic revisions occurred. We understand that full taxonomic agreement is not likely for some time to come. In addition, taxonomic revisions and advances are not always accompanied by advances in general biology and toxinology. A number of taxonomic checklists are available (e.g., Golay et al. 1993; Herprint International's *Catalogue of Valid Species and Synonyms*, 1994; McDiarmid et al. 1999; David and Ineich 1999), and to some extent these are followed, but new and recent information on systematics are also taken into consideration. References are given when necessary.

Published information on biology, ecology and toxinology is not available for many recently described species. Although all species are listed in the book, biological information about these are restricted to the species where informaton is available.

An important lesson can be learned from surveys of published literature. There is a great disparity of information known for this group of snakes. Some species like *Bitis gabonica*, *Vipera berus* and *Echis carinatus* have a great deal of published information about their general biology and toxinology. Other species like *Adenorhinos barbouri*, *Bitis heraldica* and *Bitis worthingtoni* have little more than physical descriptions written about them. Obviously continued research is needed. To help with this, we provide in our bibliography some additional selected references not cited in the text that may be useful to other investigations.

Acknowledgments

This book could not have been written without the assistance of many people contributing in numerous ways. Compiling the vast amount of literature utilized in this book was a formidable task. David Mallow and David Ludwig wish to thank the following: Mary Gazorowski, Wheeling Jesuit University, for her help in running computer literature searches; Maria Joseph, Ohio County Public Library, for arranging interlibrary loans; Dr. Ellen Czensky, Curator of Herpetology, Carnegie Museum, Pittsburgh, for permitting unlimited access to museum files; Dr. Scott Caveney, West Virginia University, for his help in providing access to the WVU literature; Bernard Stole, librarian, Academy of Natural Science, Philadelphia, for his assistance in finding journals; and Dr. Russell A. Mittermeier, Paul Gritis, and Dr. Wolfgang Wuster for their help in providing additional, and often obscure, literature.

A great deal of literature needed to be translated. For this we sincerely thank the following members of the Linsly Language Department: Carol Cook and her student Kristen Agresta, for translating French articles; Dr. Geoffrey Schoolar, for translating German articles; and Dr. Nicoletta Sella, for translating Spanish and Portuguese articles.

Thanks are extended to Michael Chokel, David Hoenig, Chris Gardill and Jim Wissing for their assistance with word processing and computer questions and problems.

We are greatly indebted to R.D. Bartlett and John Tashjian for their gracious donations of slides. The high quality and exquisite detail of these photos greatly enhance the scientific value of this book.

Summarizing the vast quantity of literature into a cogent text required the input of many reviewers. In particular, the assistance and guidance of Dr. Carl Ernst and the late Dr. Sherman Minton were invaluable. Dr. Ernst's input and direction began at the inception of this project and continued unabated throughout this project, and Dr. Minton's advice was instrumental in ensuring that the complex physiological events included in this book were explained correctly and described in a lucid manner. Thanks are extended to our good friends Dr. Douglas Dix and Dr. Bill Hargrove whose critical reviews and advice also helped hone the manuscript into its present form.

Thanks are also extended to Todd Fullerton, Wheeling Hospital, Wheeling, West Virginia, for his help in explaining complex medical phenomena; Dr. Max Hensley, Emeritus Professor of Herpetology, Michigan State University, for his assistance in answering general herpetology questions; Brad McCarthy, snake breeder, Pittsburgh, for providing specimens for observation; and Chris Bernard, for her clerical assistance during certain phases of this book.

Thanks and appreciation are extended to Reno DiOrio and The Linsly School for their financial and professional support during the production of this book.

David Mallow extends special thanks to two former professors and mentors: Dr. Sigurd Nelson, Emeritus Professor of Zoology, SUNY at Oswego, New York, for believing in his zoological potential. Without his faith and efforts, a zoological future would have been very short-lived; and Dr. Richard Snider, Zoology Professor, Michigan State University, under whose tutelage and influence David developed the interest and desire to write. His technical advice and emotional support throughout this project were invaluable.

Several organizations supported this effort, and it is only through their tolerance and patience that it exists. EA Engineering, Science and Technology, through Dr. Loren Jensen and Dr. James J. Gift (a good friend and mentor, as well as boss!), provided time and support. Exponent Environmental Group, via Dr. Marc Lorenzen, Dr. Tom Ginn and Larry Marx, helped enormously. Kimberlee McIntyre was instrumental in finalizing the manuscript and in organizing the assistance of Melisa Leonard and Rachelle Brackett.

Cathy J. Womack, Dave Ludwig's best friend (and, not coincidentally, spouse) assisted not only with patience and love, but with her skilled support as a professional science librarian.

Finally, substantial portions of this manuscript were completed at the Smithsonian Institution's outstanding herpetology library. The kindness, patience and expertise of the staff there is greatfully acknowledged. Among many others, Dr. Bob Reynolds, Dr. George Zug, Ken Tighe and James Poindexter assisted in many ways, including providing access to the collections to allow our original investigation of the head scalation of *Macrovipera lebetina*. We thank these individuals, and the entire institution, for their wonderful facility and fabulous support.

Göran Nilson thanks Claes Andrén, Göteborg, Sweden; Börje Flärdh, Stockholm, Sweden; Nikolai Orlov, Saint Petersburg, Russia; Natalia Ananjeva, Saint Petersburg, Russia; Boris Tuniyev, Sochi, Russia; Keith Corbett, Bornmouth, UK; Ulrich Joger, Darmstadt, Germany; Richard Podlouckey, Hannover, Germany; Anton Stumpel, Arnhem, The Netherlands; Ted Pappenfuss, U.S.A.; Robert Macey, U.S.A.; Victor Khromov, Semipalatinsk, Kazakhstan; Tatjana Dujsebayeva, Alma-Ata, Kazakhstan; Ilya Darevsky, Saint Petersburg, Russia; Aram Agasian, Yerevan, Armenia; Nasrullah Rastegar-Pouyani, Kermānshāh, Iran; Eskandar Rastegar-Pouyani, Sabzevār, Iran; Zoltan Korsos, Budapest, Hungary; Yannis Yioanidis, Athens, Greece; Maria Dimaki, Athens, Greece; and Chris Mattison, London, UK, for brisk discussions in remote viper habitats in Asia, Africa, and Europe.

Introduction

The family Viperidae consists of four closely related subfamilies: Viperinae, Causinae, Azemiopinae, and Crotalinae. The first three subfamilies have traditionally been referred to as "true vipers," and were until recently all included in the subfamily Viperinae. The group of "true vipers" is comprised of approximately 14 genera and 85 species and are all restricted to the Old World. They lack thermoreceptive pits between the nostril and eye (which are characteristic of the closely related Crotalinae, found in both the Old and the New World).

Vipers are generally short, robust snakes ranging in size from *Bitis schneideri*, with a maximum length of 28 cm, to *Bitis gabonica*, which reaches a maximum length over 2 m. The heads of most vipers are spade-shaped and broad to accommodate large venom glands, and are distinct from the neck. Pupils are generally vertically elliptical (Broadley 1990).

Vipers are primarily terrestrial but some species are arboreal, especially *Atheris* spp., the bush vipers. Most are viviparous (ovoviviparous), but egg laying (oviparous) occurs in *Causus* and some other genera.

It is now believed the Viperidae separated from the base of the Colubridae and Elapidae (cobras, kraits, and their allies) lineages early in snake evolution. Based on the fossil record, the oldest viperid fossils are from the early Miocene of western Europe, southeastern Asia and (probably) central Asia (Szyndlar and Rage 1999). True vipers are found throughout Europe, Asia, and Africa. Their absence from Australasia resulted from their appearance after these lands became separated from Pangaea, the ancient supercontinent. While many species are tropical to subtropical, *Vipera berus* range across the Arctic Circle.

Vipers occupy diverse habitats. The heat-retaining heavy body shape and viviparous tendency enable them to live in many habitats over a wide geographic range. Many species are evolved for life in deserts and mountains.

Although vipers lack clearly differentiated heat-sensing pits and associated structures, a nasal sac with sensory function has been described in some species (Lynn 1931). This saclike invagination under the supernasal scale is innervated by the opthalmic branch of the trigeminal nerve in *Pseudocerastes* and *Causus*, and is especially well developed in *Bitis*. Thermal, visual, and chemical cues guide the strike, with warm targets being struck more frequently than cold targets (Breidenbach 1990). Thermal detections may involve scale organs, microscopic organs near the mouth, or highly developed regions of free nerve endings beneath a thinned layer of epidermis (Chiszar et al. 1986; Dickman et al. 1987).

Unique structural traits of vipers include the following: eyes separated from supralabials; keeled dorsal body scales with apical pits; levator anguli oris muscle well developed, distinct, and separate (except in *Azemiops*); pterygoideus glandulae muscle absent; prefrontal bone with anteroventral medial process small or absent, posteroventral medial process absent, dorsomedial process undeveloped; facial carotid artery passes ventral to maxillary and mandibular branch of the trigeminal nerve (Liem et al. 1971; Broadley 1990).

Perhaps the most characteristic traits are the bite and the venom delivery system. Vipers can strike, deliver a lethal dose of venom, and resume a defensive posture very quickly. This ability to kill with minimal contact is advantageous, especially when dealing with large and potentially dangerous prey.

After striking, vipers follow a scent trail to their prey. Some vipers hold specific prey items from initial strike through swallowing, suggesting a discriminatory mechanism. Factors such as prey size, activity and resistance to immobilization determine whether prey are held or released during envenomation.

Venom Apparatus

The venom apparatus includes a pair of retractable folding front fangs. This condition is referred to as solenoglyphous. Alternative fang systems are proteroglyphous (small, fixed front fangs, as in elapids) and opistoglyphous (rear-fanged, as in colubrids). Each viper fang is capable of being moved independently. This can be observed when snakes are yawning or "walking" prey into their mouths.

The skull comprises loosely connected mobile bones. The maxillae are shortened horizontally and are capable of movement on the prefrontal and elongated ectopterygoid bones. The maxillae typically bear a single elongated, hollow fang. This arrangement permits fangs to be rotated around the maxillo-prefrontal joint in a parasagittal plane until they lie against the upper palate when the mouth is closed (Klemmer 1967a, b; Ernst 1992). Thus fangs can grow extremely long, enabling deep bites into prey. Opening the mouth activates a lifting mechanism that puts the fangs in an upright vertical position. The lifting mechanism consists of the downward rotating quadrate bone forcing the relatively long pterygoid forward, which, while reinforced by the ectopterygoid, causes the maxilla to rotate downward. These movements are facilitated by contraction of protractor pterygoidei and levator pterygoidei muscles. As a result, the fangs extend perpendicular or beyond perpendicular relative to the plane of the upper jaw (Kochva 1962, 1987).

Venom Glands and Ducts

The venom gland is the largest and most posterior part of the venom apparatus. Depending on species, it is triangular, pear-shaped, or almond-shaped, and comprises branching tubules arranged around a large central lumen. Structurally and histologically, the venom gland is divided into two distinct regions: posterior and anterior. The posterior region consists of four or five lobes, each containing tubules separated by connective tissue. These tubules, lined with columnar epithelium and basal horizontal cells, open into a large central lumen. The columnar epithelium produces carbohydrate-protein complexes (Gans and Elliott 1968). In the anterior region tubules fuse, becoming fewer in number, and have lumina that increase in diameter. These tubules empty into the central lumen. The narrowed central lumen joins the venom duct (Kochva and Gans 1967, 1970; Gans and Elliott 1968). Enzymes from the main venom gland are grouped into three categories: secretory, nonsecretory, and a combination of both. Approximately 80% of the glandular epithelium is comprised of secretory cells (Brown 1974). Surrounding the venom gland is a muscle layer, the musculi compressor glandulae, which has an outer capsular covering. This muscle surrounds the gland laterally, posteriorly and medially, and inserts on the lateral surface of the lower jaw. It is fixed to several bones by three major ligaments. Its main function is venom gland compression during injection.

Venom glands in several species of *Causus* are unique in being cylindrical and elongate, extending along the sides of the neck (Kochva and Gans 1970; Kochva 1987). Developmental studies in *Causus* show the venom gland primordia and fangs develop from a joint ingrowth of oral epithelium by day 15 of development (postoviposition). At day 19, the venom gland primordium grows posteriorly and differentiates into the main venom gland, accessory gland, and secondary duct. At this time a single fang is present and the musculi compressor glandulae arise dorsally and surround the main venom gland. By day 23, primary and secondary ducts are formed; accessory gland cells form branching tubules, and a few fangs are present in early developmental stages. The accessory gland, lumina of accessory and main glands, and secretory activity appear by day 50 (Shayer-Wollberg and Kochva 1967). Embryonic development in *Causus* is similar in many ways to *Daboia palaestinae* (=*Vipera palaestinae*), an advanced viper. However, the venom gland and associated muscles in *C. rhombeatus* grow faster and hypertrophy, growing beyond the cranium (Haas 1973).

The venom duct extends from the venom gland forward, to below the nostril, just above and exterior to the maxilla. Here it bends down sharply in an S-shaped curve, penetrating the fang sheath at a point anterior to the upper lumen of the fang. The S-shaped curve provides flexibility, absorbing bending movements during fang erection. The venom duct consists of histologically distinct primary and secondary ducts. The primary duct extends from the venom gland to the accessory gland. It consists of cylindrical epithelial cells with limited secretory activity, the specifics of which are poorly known. The secondary venom duct extends from the accessory gland, opening into the venom fang sheath. The cells of this duct are stratified epithelium and produce an acid mucus (Kochva and Gans 1965; 1966, 1970; Gans and Kochva 1965; Gans and Elliott 1968).

The accessory gland includes tubules that enter the primary venom duct, reducing its lumen near the anterior of the gland proper. The accessory gland is differentiated into two discrete secretory systems, anterior and posterior. Posterior accessory gland tubules have a round lumen, and pass into the primary duct in a folded pattern. These tubules are lined with low cuboidal epithelium with round to oval nuclei. Secretions from this gland contain acid mucosus. The anterior accessory gland includes tubules with a narrow folded lumen appearing as an ovoid cleft. These tubules are lined with high cuboidal and lower columnar mucus epithelium. This epithelium contains a cellular network of myoepithelium. Most tubules terminate in the secondary venom duct. Accessory gland function is not fully understood. It is currently thought secretions of this gland enhance toxicity by contributing specific components to the venom (Gans and Kochva 1965; Gans and Elliott 1968; Russell 1983).

A surgical technique for isolating the venom gland from the remainder of the glandular system, rendering the venom apparatus ineffectual, has been developed. Within 2–4 weeks following such isolation the main gland appears full and firm. The gland can remain nonsecretory for indefinite periods without atrophy or necrosis. Postoperative results from over 200 surgeries, monitored for 2 years, found no change in snake appearance, feeding, disposition, or general health. This procedure can enhance the safety of personnel handling snakes, and may be useful for investigating secretory processes (Glenn et al. 1973).

Fangs and Fang Sheath

Available information on fang structure is primarily from American pit vipers (Crotalinae), mostly rattlesnakes. Although some differences exist between fangs of pit vipers and true vipers, many features are similar (Russell 1983).

The fangs are the upper elongated anterior maxillary teeth. They are canaliculated and larger than other teeth. No vestigial exterior groove exists except in *Causus* and certain other juvenile vipers where it appears as a faint line. Fangs are thin, gradually tapering, and elliptical in cross-section. Densely compressed dentine layers provide strength for penetration (Visser and Chapman 1978). For additional strengthening, the fang wall is thickened opposite the major axis. The fang base forms an irregularly shaped buttress, below which is the upper venom canal orifice. This upper orifice is short and wide; it is where venom enters the fang canal. The canal lies deep within the fang except where it surfaces near the tip. The discharge orifice is located on the front edge near the tip. A pulp cavity extends behind the venom canal, terminating below the discharge orifice.

Daboia russelii exhibits a constant ontogenetic growth rate in fang length that correlates with increased head and body length (Ernst 1982). This differs from growth rates in rattlesnakes (*Crotalus, Sisturus*) where fang growth is proportionally faster than head growth in juveniles, constant in young adults, and slower in adults (Klauber 1972; Ernst 1982). Venom orifice length and corresponding fang thickness also increase at a constant rate relative to fang growth. Fang curvature angle increases gradually from juvenile to adult, with adults having greater curvature. Fangs curve in an arc approximating 60–70°, with longer fangs showing greater arcs, since radius does not increase with length. Increases in fang diameter continue after maximum length is reached. Consequently, older snakes have stronger, more heavily walled fangs (Ernst 1982).

Fangs are periodically shed and replaced by reserve fangs. These lie in a double row cluster posterior to and separated by a membrane from the functional fangs. Replacements progress in graduated order of fang maturity. When replacement fangs mature, they move into position while functional fangs loosen. As a result snakes may briefly have 3–4 fangs instead of the usual 2. Old loose fangs

INTRODUCTION

are usually left in prey and eventually swallowed. Replacement is usually symmetrical in young snakes, i.e., left and right sides lose and gain fangs together but asymmetrical replacement may occur in older snakes. Functional fang life-span for an adult rattlesnake is 6–10 weeks, with a functional/reserve overlap of 4–5 days. Age, season, and feeding pattern affect fang replacement frequency. Fang replacement continues through the life of the snake (Visser and Chapman 1978; Russell 1983). Shed fangs may often be observed in and collected from the feces.

The fang sheath is a pinkish-white membrane that encases each fang. Sheaths connect secondary venom ducts and upper fang orifices. The saclike sheaths form pockets which hold venom, ensuring flow into the fang canal. Sheaths grip tightly around the fang just below the upper orifice, preventing leakage and pressure loss. As fangs erect, the sheath rolls up and adheres to the fang base. This brings the venom duct close to the upper fang orifice. As jawbones move forward, sheaths tense, confining venom to the entrance aperture of each fang. Upon fang penetration, sheaths compress, forcing venom into the fangs. During fang replacement, sheaths surround both old and new fangs, permitting venom discharge from two sets of fangs (Visser and Chapman 1978; Russell 1983).

Venom Production and Extraction

Secretory activity is cyclic, with venom synthesis regulated from within the venom glands (Kochva and Gans 1967, 1970). Gland pressure, caused by venom quantity in the lumina, is important in regulating secretory activity. Cellular variations occur during main gland activity cycles (Kochva and Gans 1966, 1967, 1970; Gans and Elliott 1968). During storage phases the epithelium is low and flat. However, within a day of natural feeding or forced venom discharge the epithelia become cuboidal-columnar, indicating the onset of the secretory phase. As venom accumulates there is a concurrent decrease in epithelial height. In most species venom components are produced at different times and rates in different gland/duct regions. However, secretory cells in the main venom gland of *Echis carinatus* produce all venom components (Taylor et al. 1986). The resultant venom is produced as a single "package" by each individual secretory cell. Significant quantities of venom are present within 24 hours after a "severe squeezing," and a milked-out gland will be nearly full within 2 weeks. Repeated milkings keep cells in a secretory phase causing yield to remain fairly constant, even occasionally increasing (Kochva and Gans 1967). The venom gland can store considerable amounts of venom, and milking yields vary according to the snake's age, size, feeding pattern, milking interval, stress, photoperiod, temperature, captive care, geographical area, and season (Latifi 1984).

Venom quantity delivered varies among species. It may range from no venom to full gland contents, but is usually 10–15% of the volume in the gland. (Gans and Elliott 1968; Minton 1974; U.S. Navy 1991). Not all bites result in envenomation. Minton (1974, 1990) reported that 20–60% of bites occurring worldwide resulted in little or no envenomation. Injected venom quantity is under voluntary control. Cues such as prey body mass and activity level determine injection quantity (Visser and Chapman 1978). *Daboia palaestinae* (=*Vipera palaestinae*) discharges equivalent venom quantities in each of a series of bites in small prey (Allon and Kochva 1974). *Agkistrodon piscivorous* inject increasing amounts of venom with increasing prey size (from mice, to rats, to guinea pigs) (Gennaro et al. 1961). Little is known about injected volumes and humans. Estimates are based primarily on symptom severity and laboratory findings. Variables such as snake species, size, length of time the snake holds on, fang and venom gland condition, and extent of fear or anger motivating the strike are important factors (Visser and Chapman 1978).

In nature, venom glands tend to be nearly full, as evidenced by relatively large amounts of venom obtained from first milkings of newly collected snakes. A detailed description of proper milking technique is given by Russell (1983).

Venom Composition and Function

Composition—Snakes possess the most biochemically and pharmacologically complex toxins known. The multicomponent nature of venom has been demonstrated using starchgel and immunoelectrophoresis, which reveals 10–20 protein zones (Master and Rao 1963; Neelin 1963; Boquet 1967a,b). Venom composition is largely genetically determined (Johnson 1968). Variability exists at the family, genus, species, population, individual, and ontogenetic levels. Composition is affected by season, capture area, health, diet, and hibernation status (Chippaux et al. 1991).

Venoms are complex mixtures of peptides, polypeptides, enzymes, glycoproteins, and other low molecular weight substances. Hundreds of protein toxins have been purified from snake venoms (Kini and Evans 1990). Peptides appear to be the more lethal fraction. Inorganic substances such as sodium, calcium, potassium, magnesium, zinc, and small amounts of iron, cobalt, manganese, and nickel are present. In general, snake venoms contain at least 26 enzymes and at least 10 are common to all snake venoms. Viper venom is particularly rich in endopeptidases. For a detailed description of snake venom enzymes, see Zeller (1951), Brown (1974), and Russell (1983).

Venom effects result from three component categories: toxins, enzymes and biologically active amines. Toxins are mostly low molecular weight polypeptides and nonenzymatic proteins. Enzymes are mostly hydrolyses and include proteases, phospholipases, esterases, hyaluronidases, phosphodiesterases and nucleotidases (Kini and Evans 1990). Enzymatic roles in coagulopathy, hemorrhage, necrosis, and hypotension may result from their interaction with other compounds (Minton 1994). The third component, biologically active amines, are compounds such as serotonin and bradykinin. Although all three components may occur in one venom, they vary in their relative degrees of activity. For detailed explanations of venom composition and activity, see Minton (1974), Russell (1983), Kini and Evans (1990), Rosing and Tans (1992) and Ouyang et al. (1992).

Venoms are inherently labile, with composition changing over time and glandular condition, and are not characterized by particular compounds in specific relative proportions. Many components are added to the venom as it moves from the main gland to the fang. Discharged venoms undergo physical and chemical changes, even changing from inactive to active forms. The activation mechanism is unknown but evidence suggests an inhibitor substance is present, necessitating the addition of an activator. Although whole venom is unstable, its components are stable when dried. Snake venoms retain many activities over long periods of time. Desiccated rattlesnake venom retained lethal activity after 26–50 years of storage (Minton 1974, 1990; Russell 1983).

Venom complexity has probably resulted from a spectrum of functional problems including dependence on diverse prey with equally diverse venom tolerances and defensive needs against an array of predators. In some instances prey specificity has led to the evolution of specific venoms. Ontogenetic venom changes may reflect ontogenetic changes in prey choice. Furthermore, juveniles generally have more toxic venom than adults as evidenced by their equilibrating lethality to adult quantity. Venom toxicity and coagulant activity generally decrease with age until maturity, after which toxicity levels off (Minton 1967a, 1975).

Venom lethality is relative and must be expressed in terms of a particular test animal (Gans and Elliott 1968; Russell 1983). Inconsistencies are common in reporting LD_{50} values. Often, experimental units are not uniform, and are reported as "per mouse" or "per gram of mouse," and data pertaining to injection route, i.e. intravenous (i.v.), intraperitoneal (i.p.), or subcutaneous (s.c.), are often omitted. LD_{50} values are usually lowest for i.v. injection compared to i.p. (next highest) or s.c. (highest LD_{50}). In general, humans appear to be more susceptible to snake venoms on a weight-for-weight basis than mice or dogs (Minton 1974).

Venom composition information can be useful in understanding physiological processes and establishing taxonomic relationships among snakes. Comparisons of enzymatic, hemorrhagic, procoagulant, and anticoagulant activities from six vipers (*Bitis, Causus, Cerastes, Echis, Eristicophis, Pseudocerastes*) revealed common characteristics at the family, genus, and species levels. At the family level these venoms exhibited hemorrhagic and arginine ester hydrolase activity, and exhibited low (or no) acetylcholinesterase activity. At the genus level, venoms of *Bitis, Eristicophis* and *Pseudocerastes* either prolonged or did not affect kaolin-cephalin clotting time of platelet-poor rabbit plasma, while *Causus, Cerastes, Echis* and most *Vipera* shortened this time (Jimenez-Porras 1967; Tan and Ponnudurai 1988; 1990a, b, c, 1992; Kini and Evans 1990).

Function—Snake venoms evolved as prey immobilizing substances, and for digestion and defense (Russell 1983). There is a balance between venom lethality and its role in digestion. Lethal effects cannot occur too early or distribution of digestive components would be curtailed. Conversely, if lethal effects are delayed prey may escape. This balance can be modified depending on changes in feeding pattern, predator abundance, and perhaps immunity development by prey (Russell 1983). *Crotalus atrox* venom injected into prey accelerates visceral rupture. This may be important at low temperatures when digestion rates are slowed and bacteria can cause putrefaction and regurgitation of slowly processed prey items. Venom also loosens mammalian hair. On the other hand, individual specimens with venom glands removed feed and survive for years without the digestive action of venom (Thomas and Pugh 1979; Russell 1983).

Envenomation Process

The envenomation process is complex and rapid. The process can be divided into three steps: first, head movement toward prey; second, fang introduction; and third, venom injection. The bite terminates in a stabbing motion, taking as little as one-third of a second. Before striking, vipers usually adopt an S-shaped posture accompanied by slow tongue flicks. During the strike, the mouth opens between 90° and 180°, fangs simultaneously erect and swing down into prey. In rattlesnakes the bottom jaw usually contacts prey first with the upper jaw closing and fangs penetrating rapidly (Gans and Elliott 1968; Kardong 1986a).

There are differences between defensive and predatory strikes. During the latter, the neck arches upward allowing fangs to follow their own curvature for deep penetration. Two types of predatory strikes are recognized: regular and flawed. Flawed strikes are of three types: one fang completely missing; multiple strike attempts before successful fang penetration; and collision with an obstacle (log, rock, cage, etc.). Flawed strikes usually result in poor envenomation and prolonged intoxication, with mice sometimes taking 7 times longer to die than for a regular strike. After a flawed strike, the snake will erect one or both fangs and shift its head toward the prey. These adjustments suggest that vipers use feedback mechanisms for prey location and fang placement. This feedback control compensates for inaccuracies of the initial strike and for evasive action by prey. Regular bites involve fangs embedding one-third to two-thirds of their total length and usually lead to rapid prey death (Kardong 1986a).

In defensive strikes, no neck arching occurs and jaws tend to open maximally. Large numbers of dry bites associated with defensive strikes may be attributed to jaw mechanics (wide jaw angle and unarched neck). Threatened snakes may deliver several rapid stabbing bites in quick succession (Visser and Chapman 1978; Russell 1983; Kardong 1986a).

Following the strike, with fangs embedded, gland muscles contract. Contraction is under active muscular control and can be directed at left, right, both or neither fang. (Gans and Elliott 1968).

Venom Effects

In general, the major effects of snake venom can be broken down into local tissue destruction, hemotoxicity, cardiotoxicity, and neurotoxicity. Snakebite severity depends on factors involving both the snake and victim. Of these, two important and usually unknown factors are injected venom amount and intrinsic toxicity. Controlling variables include venom quantity and composition, gland condition (full or depleted, seasonal variability, proximity to feeding or milking), nature of the bite (direct strike with both fangs or glancing scratch), degree of penetration, degree of fear that caused the snake to bite, number of bites and length of time the snake holds on. For the victim, the following variables are important: general health, size, age, injection site (extremities less dangerous than bites near vital organs, or tissue type, with fat absorbing venom slower than muscle), allergy (susceptibility to protein poisoning, previous history of bites or treatment), psychological condition (panic, increasing heart rate speeds venom distribution) and degree of clothing protection (reduced fang penetration, surface absorption) (Minton 1974; Coppola and Hogan 1992).

Generally, victims of viper bite develop pain, swelling, and discoloration at the bite site within minutes. Edema advances centripetally, often crossing the trunk, spreading up and down the opposite side. Swollen skin reddens, especially around lymph vessels, and watery or bloody blisters may appear. Tenderness may occur at the bite site. Cytolysis, ischemia, induration, and blood extravasation may also occur. Vasculitis, subfascial edemic pressure, direct proteolytic activity and poor treatment may result in necrosis. Vomiting, abdominal pain and diarrhea may also occur. Often, hypotension, restlessness or borderline consciousness are exhibited. If left untreated, swelling may continue and anaphylactoid reactions passing to secondary shock from fluid loss and bleeding into soft tissues may ensue. The numerous toxins and diverse action of snake venoms ensure that nearly all organ systems will be affected (Minton 1974; Jena and Sarangi 1993; Spawls and Branch 1995).

Elevated fever may indicate secondary infection. Putrefactive gangrene may affect deep tissue, especially where necrosis occurs (Minton 1974; Jena and Sarangi 1993).

The primary action of viper venom is on blood and endothelial cells lining blood vessels (Tu et al. 1969). Minton (1994) stated that the most common clinical manifestation of viper bite is incoagulable blood, usually the result of defibrination. This may cause prolonged oozing of blood from fang punctures and hemorrhage from the mouth, nose, stomach, intestines, and other organs. It appears snake venoms alter numerous coagulation mechanisms, often with more than one effect present in any given venom.

There are six pathway sites where venom procoagulants interact (Russell 1983; Minton 1994), and several procoagulants have been isolated and identified (Kornalik and Hladovec 1975; Markland and Pirkle 1977). Four prothrombin activator groups have been classified and these encompass diverse enzymes with varying structural and functional properties. Group I activators convert prothrombin to meizothrombin, a thrombinlike molecule with less than 5% the activity of thrombin. Groups II and III activators convert prothrombin to thrombin by cleaving appropriate peptides. Group IV activators also cleave peptides from prothrombin, but the product is not enzymatically active and functions as a thrombin precursor (Rosing and Tans 1992).

Broad-spectrum proteases cause coagulant, anticoagulant, and fibrinolytic activities. There is substantial proteolytic variation within viper genera (Tu et al. 1966; Russell 1983).

For the victim, venom excretion occurs primarily through the kidneys, with the intestines playing a minor role. However, venom effects on kidneys and associated blood vessels make excretory metabolism studies difficult to assess (Efrati and Reif 1953; Efrati 1969).

Treatment and Antivenins

Epidemiology—Snakebite kills 30,000–40,000 people worldwide each year (Russell 1983). Assuming a nominal 15% mortality rate, over 300,000 people are bitten each year. This figure is probably underestimated since most envenomations occur in rural areas and are therefore not reported. The actual number of bite cases may be closer to 1 million (Minton and Minton 1969). The majority of snakebite cases occur in Asia, with India experiencing the highest number, 10,000–15,000 deaths per year (Spawls and Branch 1995). Jena and Sarangi (1993) provide additional epidemiological statistics.

Diagnosis—Initial diagnosis is based on visual identification of the snake, nature of the wound, local pain, swelling, discoloration and, if possible, enzyme-linked immunosorbent assay (ELISA) verification. Following diagnosis, bite severity is assessed to support decisions regarding conservative or aggressive management. An initial coagulation profile is often obtained on severely envenomated hospitalized patients, with follow-up tests every 6–8 h (Minton 1990). In remote areas, simple clotting observation in freshly drawn blood is a useful diagnostic tool (Reid 1967).

Proper identification of the offending snake is important. Unfortunately, species diagnoses based on patient physical signs are unreliable. Thus, optimal patient treatment is not always possible. Aspirates from bite sites or venom on skin or clothing may be useful for species identification using immunodiagnosis (Minton 1990). This process involves identifying species specific venoms in fluids from sera, blisters, wound aspirates, and urine. Although only 40% successful, use of more specific antisera and improved assay techniques could enhance this procedure. Recent advances with immunoassay techniques enable clinicians to differentiate snake venom from other venoms relatively quickly. Nanogram quantities of snake venom from tissues, body fluids, serum, and urine can be detected and determined within 40–60 minutes. Bite severity can be estimated from venom antigen levels in plasma.

ELISA offers the potential for epidemiological studies of snakebite incidence by detecting venom antibodies in a population. Using ELISA, Pugh and Theakston (1980) determined that 58% of snakebite victims in the Benue Valley in Nigeria were bitten by *Echis,* with remaining bites due primarily to *Bitis arietans*. However, Minton (1987) cautioned that nonspecific reactivity of serum samples and cross-reactions between snake venoms currently limit the use of this approach.

Antivenin Usage—Antivenins are the only specific and effective therapy for life-threatening envenomation (Minton 1994; Spawls and Branch 1995). However, decisions regarding antivenin use must be made with caution (Benbassat and Shalev 1993). The benefits of preventing complications versus the risk of anaphylaxis and serum sickness make management decisions difficult.

Variation in antivenin effectiveness involves many factors, of which titer, degree of purification, concentration, and antivenin specificity are most important. High antivenin concentration and purification reduce chances of side effects such as anaphylactoid reactions (Russell and Lauritzen 1966).

Antivenin Specificity—Venom neutralization is proportional to antivenin specificity. Comparisons between polyvalent and monovalent antivenins show monovalent antivenins produce more effective species-specific neutralization (Mohamed,

Fawzia et al. 1977). New World polyvalent pit viper antivenin (from three species of *Crotalus* and one species of *Bothrops*) neutralize some Old World viper venoms to a moderate degree. The neutralized Old World venoms failed to show any clear taxonomic relationships. For example, *Vipera ammodytes, Daboia palaestinae* (=*Vipera* palaestinae) and *Bitis gabonica* were neutralized, but not *Daboia russelii, B. arietans* or either species of *Echis* (Minton 1976).

Geographic variability affects antivenin effectiveness. For example, antivenin from Asian *Echis* venom is ineffective in treating West African *Echis* bites. Symptom diversity in bite victims from the same species has alerted physicians to the need for more specific antivenins (Chippaux et al. 1991). Fortunately, similarities in some toxic fractions among members of the same genus, or even the same family, enable antivenins prepared from a single species to neutralize venoms from a variety of species. Polyvalent antivenin from four species of pit vipers (three *Crotalus* spp., one *Bothrops* sp.) neutralize venoms from 17 pit viper species and is used in treatment for the bites of 65 species of pit vipers from North, Central, and South America (Russell and Lauritzen 1966).

Antivenin Administration—Antivenin is most effective delivered i.v., at room temperature via slow saline drip. Effective doses vary with severity of envenomation and potency of antivenin. Antivenins are limited in their effect on local edema and necrosis, but are effective in reversing hemostatic defect and hypovolemic shock. Although antivenin is often implemented in severe cases, awareness of patient hypersensitivity and possible anaphylactic response is important (Minton 1994). For a thorough review of antivenin use and availability, Theakston and Warrell (1991) and Spawls and Branch (1995) should be consulted.

Additional Treatment—In addition to antivenin therapy, supportive measures for shock, respiratory distress, electrolyte imbalance, blood fluid loss, and renal failure are available (Minton 1994). Ashley and Burchfield (1968) review safety and health issues arising from management of captive snake collections.

Chapter 1
Azemiopinae

Azemiops feae is nowadays considered to belong to a separate subfamily at the base of Crotalinae (Cadle 1988, 1992; Knight and Mindell 1993; Heise et al. 1995; Underwood 1999).

Azemiops

The monospecific genus Azemiops is commonly known as Fea's viper. The first specimens were collected by the European explorer M.L. Fea and the genus was officially described by Boulenger in 1888. Only two specimens were known until 1935. Between 1935 and 1985, a few more specimens were collected but these did not survive long in captivity. These snakes inhabit remote mountainous areas of southeast Asia and are rare. Little is known about them (Marx and Olechowski 1970; Mehrtens 1987).

Fea's viper has a unique combination of morphological features. Superficially, it resembles a colubrid. Based on similarity of microdermatoglyphic patterns, Price (1987) suggested that *Azemiops* is phylogenetically related to *Causus* despite their differing in zoogeography, body form, and pupil shape. Analysis of mitochondrial ribosomal DNA sequences supports an *Azemiops*/pit viper clade (Knight and Mindell 1993; Heise et al. 1995).

Fea's viper's status as a "primitive" snake is rarely contested and is based upon various skeletal and integumental characteristics. In a detailed analysis of morphological character-states, Liem et al. (1971) found 41 primitive traits. Of these, the medial or nasal wing of the prefrontal clearly distinguishes it from the other vipers. It also has the fewest derived external character-states (one), making it the least derived species in the entire family. *Azemiops feae* also shares some holdover colubrid features such as supralabials incorporated into external border of orbits, crown with 9 scutes, and dorsal scales without carination.

Despite these unusual traits there are sound reasons for including Fea's viper in the Viperidae, but as a separate subfamily. These include overall cephalic anatomy, protraction of the palatomaxillary unit, lack of the pterygoideus glandulae muscles, and histology of venom and accessory glands (Liem et al. 1971).

Azemiops feae
Boulenger 1888

Recognition
(Plate 1.1)

Head—Flattened, elliptical, not triangulate like typical vipers; distinct from the neck; snout broad and short; frontal slightly broader than long; parietals long as their distance from end of snout, pointed posteriorly, bordered laterally by anterior temporal and 3 scales; prefrontals subequal; no scales present between subocular and supralabials; 6 supralabials, the third entering part of the orbit; 7–8 sublabials; 2–3 preoculars; 1–2 postoculars; pair of short chin shields separated from ventral shield by 3 transverse series of scales; palatine and pterygoid teeth numerous (Boulenger 1888). Unlike other vipers, the fangs possess a ridge at the tip lateral to the discharge orifice, and a bladelike structure on the ventral surface seen only in some opistoglyphous and atractaspid snakes (Mebs et al. 1994).

Body—Cylindrical, moderately slender with short tail; dorsals smooth, in 17 nonoblique rows at midbody; ventrals rounded, 180–189; subcaudals paired, 42–53, with possibly some single ones anteriorly; anal single (Boulenger 1888).

Size—*Azemiops feae* measures less than 1 m in total length. Liem et al. (1971) reported a maximum length of 77 cm. Orlov (1997) reported 78 cm for a female and 72 cm for a male.

Pattern—Head is orange to slightly yellow, with a distinct cross-pattern outlined in gray; eye yellowish with vertical pupil; body coloration shiny blue-black, with a series of widely spaced, thin (1–2 scales), white-orange transverse bands. These bands arise on the venter and may or may not meet mid-dorsally; venter evenly cream-gray with a slight pinkish tinge (Kardong 1986b; Mehrtens 1987; Mara 1993).

Taxonomy and Distribution

Azemiops feae was first described by Boulenger (1888) from a specimen collected in Burma. No subspecies are recognized.

Distribution records are from southeast Asia, China and southeastern Tibet, and Vietnam (Leviton 1967; Kardong 1986b; Marx and Olechowski 1970).

Habitat

Found in mountainous regions up to about 1000 m. It prefers cooler climates, with temperatures averaging 20–25°C (Marx and Olechowski 1970, Mehrtens 1987, and Mara 1993). It is sometimes found on roadsides, in rice fields, within straw and grass, and even around and within homes (Zhao and Zhao 1981). In Vietnam, Orlov (1997) describes the habitat as bamboo-tree fern forest with open light sites, and with the forest floor covered with soft layers of decidious leaves, decomposed trunks of tree ferns, and vigorous outcrops of the carst formation, permanently permeated by numerous open and subterranean streams. The species is nocturnal, and Orlov (1997) gives 18–20°C on soil surface as preferred night temperatures. *Azemiops* seem to live in very moisture substrata and always stay under shelters with absolutely wet substrates.

Food and Feeding

A common gray shrew, *Crocidura attenuata,* was found in the stomach of a captured immature female (Marx and Olechowski 1970). A captive specimen was fed small prekilled mice (Mehrtens 1987). Kardong (1986b) reported these snakes as reluctant feeders. However, when they did feed it was on newborn mice, and only during the night. During three observed feedings, prey were not released after being struck. Amphibians, fish and lizards were all refused.

Behavior

These snakes are crepuscular in the wild and are active from early March to late November (Zhao and Zhao 1981).

Azemiops feae displays a distinct defensive behavior when threatened. The body flattens, becoming wider. The normally ovoid head becomes triangular by the outward flaring of the posterior jaw. Some captive specimens vibrate their tails. If threatened at close range they strike. Fangs may or may not be erect during strikes. A 63-cm specimen reportedly reached over 18 cm during a vigorous strike.

Kardong (1986b) observed these snakes pushing potting soil with their snouts, but not actually burrowing.

Reproduction and Development

Azemiops feae is believed to be oviparous, but information on clutch size and hatchlings is lacking (Mehrtens 1987). Five weeks after acquiring two females, Kardong (1986b) observed three and five ovarian follicles in the left ovaries, the largest measuring 1.5×8 mm. However, neither the ovaries nor reproductive tracts revealed whether eggs are laid. Histological examination of one testis from each of five males revealed the seminiferous tubules to be up to three cells thick, with conspicuous lumina. Advanced spermatids and spermatozoa were present, although in low numbers.

Orlov (1995, 1997) observed *Azemiops* in the field and in captivity, and describes the mating behaviour as similar with other vipers. The male and female move parallel and the male is twitching his head along the female body. Copulation lasts about 10 minutes. Mating was observed during the period from 5 to 15 July.

Bite and Venom

Epidemiology—No records of envenomation are known.

Yield—Of seven adult specimens milked, five delivered only trace amounts of venom despite vigorous biting. The other two yielded 1.75 mg/snake (Vest 1986).

Toxicity—An LD_{50} of 0.52 mg/kg was determined for i.v. injected mice weighing between 18 and 22 g (Vest 1986).

Content—Venom is similar in color and viscosity to other vipers, and is a clear golden yellow. Immunological and electrophoretic characteristics and enzyme activity are generally similar to other viper venoms (Mebs et al. 1994). Electrophoretic analysis revealed 22 visible proteinaceous bands with molecular weights ranging from 10,000 to 80,000 (Vest 1986).

Symptoms and Physiological Effect—Venom causes pronounced paralytic symptoms in mice, without eliciting local and/or hemorrhagic syndromes. Mice injected intravenously (i.v.) with doses of 0.50–0.60 mg/kg, exhibited minor to moderate vasodilation of ear vessels within 10 minutes. After 20–40 minutes they became sluggish with slowed breathing. This was followed by clonic convulsions leading to flaccid paralysis 90–110 minutes postinjection. Despite nearly complete paralysis, mice retained slight ability to move their legs and were able to right themselves until the terminal stage of poisoning. Respirations reached a critical minimum of 35 per minute with paroxysm and death occurring between 86 and 150 minutes postinjection. Vest (1986) described the envenomation progression as resembling elapid venoms that contain postsynaptic neurotoxic peptides. Subcutaneous injected doses of 0.50–0.60 mg/kg resulted in death, while mice receiving s.c. doses 0.40 mg/kg or less survived. All mice surviving 180 minutes postinjection survived. Local tissue response and degradation were relatively minor (Vest 1986).

Hematological Effects—Hemorrhagic activity is slight. Intracutaneous injection of venom in rabbits failed to produce evidence of hemorrhage (Vest 1986).

Treatment—Significant precipitation lines developed between *A. feae* venom and antivenins from tiger snakes (*Notechis*), death adders (*Acanthophis*), mamba (*Dendroaspis*) and Iranian cobra (*Naja oxiana*) (Vest 1986). Strong multiple precipitation lines developed against Serum Europe, Serum North Africa, FitzSimon's Polyvalent, and Wyeth Polyvalent antivenins. Wyeth Polyvalent antiserum was the only crotalid preparation reacting against *A. feae* venom. Immunodiffusion reactions of *A. feae* venom against multigenus antivenins demonstrate the presence of antigenetically related components between *A. feae* and all major groups of terrestrial venomous snakes.

Remarks

Fea's vipers are reportedly difficult to maintain in captivity (Mara 1993). Specimens kept by Kardong (1986b) were housed individually or in pairs in terraria with 2–3 cm of potting soil on the bottom. Temperature gradients were maintained between 15 and 33°C. Hide boxes and loose paper were provided, and water was constantly available. However, despite several observations of drinking, the snakes remained susceptible to dehydration, evidenced by a 10% weight loss in 30 days in some specimens. Lighting was cycled every 12 hours upon arrival of new snakes and after 2 weeks was reduced to 8 hours. Hide boxes with their cooler 15–20°C temperatures received much use (Marx and Olechowski 1970; Mehrtens 1987; Mara 1993).

Orlov succeeded in keeping a pair alive for several years by keeping them in a more or less aquatic substrate. Their skin must always be kept wet to avoid dehydration (Orlov pers. obs.).

Chapter 2
Causinae

The group of species in the genus *Causus* together form the subfamily Causinae.

Causus

The genus *Causus* contains snakes commonly known as night adders. Night adders look like nonvenomous snakes or elapids, having heads only slightly distinct from the neck and eyes with round pupils. As the common name indicates, they are primarily nocturnal (Broadley 1990).

This genus is regarded as the most primitive of the Viperidae based on head shields, oviparity and venom apparatus.

Fangs are relatively short compared to other viperids. There is a fine line, or suture, on the fang surface from base to tip which represents the vestigial edge where groove lips, from incomplete fang canal closure, meet. Some species have elongated venom glands (50–75 cm in adults) which extend along each side of the neck (Underwood 1967; Parker 1977).

Venom is weak and tends to dribble from fangs, resulting in only a small injected amount. Envenomation usually causes only localized pain and swelling. Fatalities are rare (Sweeney 1961; Turner 1972).

Recognition

Head—Moderate in size; eye moderate in size, with round pupil; 9 large symmetrical head shields; prefrontal immovable; no hinge action where prefrontal bone engages frontal, but maxilla rotation almost as great as other viperines; rostral broad, sometimes pointed or upturned; nostril located between 2 nasals and internasal; frontal and supraoculars long; loreal present, separating nasal and preoculars; suboculars separated from labials; mandible with splenial and angular elements (Pitman 1938; Marx and Rabb 1965; Broadley 1990; Branch 1992).

Body—Cylindrical or slightly depressed; moderately slender; dorsal scales smooth or weakly keeled, with apical pits; ventral scales rounded; subcaudals single or paired; anal entire; tail short.

Other characteristics include unusually long kidneys, well-developed tracheal lung with two tracheal arteries; liver overlaps the tip of the heart (Underwood 1967).

Taxonomy

This genus, established from *Sepedon rhombeata,* currently contains six species. Cladistic analysis by Ashe and Marx (1988) revealed this genus to be monophyletic and the most basally derived viperid.

Causus bilineatus
Boulenger 1905

Commonly known as the lined night adder.

Recognition

Head—Slightly distinct from neck; snout fairly long and tapering.

Body—Dorsal scales weakly keeled, 15–18 rows at midbody, soft and velvety in appearance; ventrals 128–144 in females, 122–141 in males; subcaudals 18–30.

Size—Averages 30–50 cm with a maximum reported length of 65 cm (Spawls and Branch 1995).

Pattern—Head with dorsal V-shaped mark, with vertex on the frontal; each side marked black from the posterior corner of eye to angle of maxilla; dorsal coloration ash to auburn to brown; numerous irregular or vaguely rectangular black patches dorsally, lying within 2 distinct, narrow, pale stripes which run the length of the body. Ventral coloration dark to dark cream (Spawls and Branch 1995).

Taxonomy and Distribution

Described as *Causus rhombeatus* from specimens collected between Benguella and Bihe, Angola.

This species occurs in south central Africa, from southern Zaire to northern Namibia, west to eastern Angola. In Zambia, it is recorded from the northwestern and western provinces with two specimens collected from a swamp south of Lake Bangweulu (Broadley 1971; Mehrtens 1987; Spawls and Branch 1995).

Habitat

Favors moist savanna, swampy areas, and forest-savanna landscapes. Presence of clawed frogs, *Xenopus,* in the guts of wild-caught specimens suggests more aquatic behavior than other night adders (Spawls and Branch 1995).

Food and Feeding

Feeds primarily on frogs and toads, especially the clawed frog, *Xenopus* (Spawls and Branch 1995).

Behavior

Spawls and Branch (1995) suggested that general behavior is "probably similar to other night adders."

Reproduction and Development

Oviparous.

Bite and Venom

Nothing is known regarding venom composition and toxicity. No case histories are reported.

Causus defilippii
(Jan 1863)

Commonly known as the snouted night adder
because of its pointed, upturned snout.

Recognition

Head—Short, broad, and wide; snout prominent, pointed, and upturned; eyes medium sized with round pupil; rostral large; circumorbital ring with 1–2 preoculars and postoculars; 1–2 suboculars separate orbit from supralabials, which number 6–7; sublabials 7–10, the first 3–4 contacting the short anterior chin shields; posterior chin shields very small, indistinguishable from other posterior scales; temporals 2+3, (occasionally 2+4, exceptionally 1+2).

Body—Dorsal scales 16–18 rows at midbody, weakly keeled, velvety in appearance; ventral scales 108–128, seldom exceeding 117 in males or less than 118 in females; subcaudals 10–19, paired, usually not less than 14 in males or over 15 in females; anal entire.

Size—Adults average 20–35 cm in length, rarely exceeding 50 cm in length; tail short (Spawls and Branch 1995).

Pattern—Head dorsally marked with characteristic V-shaped marking, the apex of which is on the frontal; oblique dark streak behind eye; body coloration from light brown, pinkish brown to gray or grayish green; 20–30 crescent-shaped dark markings on dorsal surface (may be indistinct); venter yellowish white, uniform or with scattered small grayish brown spots; juveniles usually glossy black or gray (Sweeney 1961; Broadley 1990).

Taxonomy and Distribution

Described as *Heterodon de filippii*. No subspecies are recognized.

Causus defilippii is an East African species, from the southern coast of Kenya, south to Tanzania, west to Malawi, eastern Zambia, south on the Mozambique plain to Zimbabwe to northern and eastern Transvaal and northern Natal (Spawls and Branch 1995). It is common in the plain and lowveld areas of Zimbabwe. In the Transvaal, it completely replaces *C. rhombeatus* (Broadley 1971, 1990). It is widely distributed, but nowhere common (Sweeney 1961).

Habitat

Snouted night adders favor moist environments but have been found in arid areas on rocky hillsides and escarpments (Sweeney 1961; Phelps 1989; Broadley and Cock 1993). Although terrestrial, it occasionally climbs into low vegetation in pursuit of frogs (Spawls and Branch 1995).

Food and Feeding

Snouted night adders feed primarily on amphibians (Sweeney 1961; Broadley 1990). Mice and other small rodents are also eaten (Pienaar 1966). It has a "voracious" appetite in captivity. Steehouder (1989) fed road-killed toads and recommended making rats smell like toads for feeding. In captivity it drinks frequently (Sweeney 1961).

Behavior

Snouted night adders are not entirely nocturnal. Sweeney (1961) stated that all the specimens he collected were taken during the day in cool weather, during or after rain. He also reported that these snakes are good swimmers. When not basking, individuals hide in ground cover, brush piles and in holes (Spawls and Branch 1995).

When angered it inflates its body and hisses. In captivity, it readily becomes tame and unwilling to strike (Sweeney 1961).

Reproduction and Development

Males engage in combat during courtship periods. A newly captured male introduced into a cage containing another male resulted in the two snakes rearing up, entwining and wrestling until the larger snake forced its opponent to the ground, whereupon it fled (Broadley 1990).

From 3 to 9 eggs averaging 20–25 × 12–16 mm in size are laid during the summer. A gravid female collected by Haagner (1986a) in early October laid 5 eggs on 29 November. Eggs measured 22.0–30 × 10–13 mm (mean 26 × 12 mm), and weighed 6 g in total (24% of the postpartum weight of female). A 44-cm female collected by Loveridge (1959) in Tanganyika on 29 December contained 15 eggs, each measuring approximately 13 × 21 mm. Incubation takes about 3.5 months and neonates are approximately 10 cm in length (Broadley 1990; Broadley and Cock 1993). A gravid female caught in Natal in December deposited 9 eggs. Oviposition began on 8 December at 1600 h and by 1720 h, 7 eggs were laid. Egg number 9 was laid at 1815 hours, but was soft and dented. Eggs were incubated individually at room temperature (22–30°C) on damp paper towels in a glass bowl, sealed in a plastic bag. Fungal spots were scraped off weekly. Egg number 9 collapsed by day 8 and 2 other eggs failed to hatch. On 4 February, after 54 days incubation, the 6 remaining eggs hatched, all within 1 hour (Botha 1984).

Bite and Venom

Venom glands in *C. defilippii* are short and do not extend into the neck. Nothing is known regarding venom composition or toxicity.

Marais (1981) reported on a 23-year-old 70-kg male bitten on the left forefinger by a 21-cm specimen. The bite occurred at 1600 h while the victim was handling the snake. Penetration by both fangs resulted in bleeding punctures. Swelling and mild throbbing pain occurred within minutes. After 20 minutes, the patient began sneezing, his eyes were blood-shot and watering, and swelling progressed to adjacent fingers. Despite swelling, some finger movement was possible. Pain rapidly progressed to the armpit, at which time respiratory difficulty occurred. In addition, a rash appeared on both arms, the neck, lower back and legs. After 30 minutes, pain progressed to the legs and feet and numbness developed around the chest, contributing to breathing difficulty. Nasal blockage made breathing through the nostrils impossible. After 45 minutes, the victim was hospitalized with reddened sensitive legs. The victim complained of pain due to contact with sheets. At this time, blood pressure was 130/70 mm Hg, pulse 120, and temperature 36°C. After 50 minutes, the patient was given a saline drip, a 200-mg cortisone i.v., and 1 mg of antihistamine i.m. followed by 50 mg i.v. Atherex. The arm rash soon disappeared. Between 3.5 and 8 hours postbite, the hand, even while elevated, swelled to a degree that made movement impossible. Fourteen hours postbite, swelling progressed past the wrist, limiting movement to 45°. At 18 hours postbite, the saline drip was removed, swelling was subsiding, but arms remained sensitive. The patient was discharged after 3 days with swelling greatly reduced but with the affected limb still sensitive and impaired. By the fifth day, swelling was further reduced, permitting opening and closing of the hand. Full recovery took 10 days. This patient was bitten three previous times by *C. rhombeatus,* suffering only minor effects each time. Minton (1997, pers. comm.) suggested that many symptoms may have resulted from anaphylaxis due to past sensitization from prior bites.

Another case history (Creighton and Haagner 1986), involved an individual bitten on the hand. There was immediate bleeding from the puncture site, and swelling and throbbing pain appeared within minutes. After 2 hours, despite application of ice, the hand swelled to "tennis ball size" and it was impossible to bend the fingers. Glands in the armpit were extremely painful. By 4 hours postbite, pain was intense, swelling reached the elbow and knuckles were blue. At this time, the patient was hospitalized and given 25 mg Phenergan and 500 mg Solu-Cortef. Two days following the bite, the hand was still swollen, bruised and painful. By day 3, swelling was subsiding and the hand returned to normal after 5–7 days.

Causus lichtensteinii
(Jan 1859)

Commonly known as Lichtenstein's night adder.

Recognition

Head—Not greatly broadened; snout obtuse; supralabials 6; sublabials 9; circumorbital ring 5–7; temporals 2+3 (occasionally 2+2), with first and second upper temporals as long together as first lower one; loreals 1+1 (except for one specimen that had 2 on one side).

Body—Dorsal scales weakly keeled, 15 at midbody; ventral scales 128–152; subcaudals in males 18–22, females 17–19; anal scale entire.

Size—This is a small species, averaging 30–55 cm with a maximum reported length of 70 cm; body generally slim (Spawls and Branch 1995).

Pattern—Neck with distinctive white V-shape; throat with black and yellow bands; body coloration greenish or olive, "velvety" in texture; dark, narrow, backward-pointing chevrons down dorsal surface; chevrons may be indistinct or completely developed into rhombic markings.

Juveniles usually dark brown; with white line extending from rostral, over the eye, across temporals to angle of jaw; a second white line joins the first on the labial border; nape with prominent white V anteriorly edged with dark brown with apex on parietal suture; 18 dark backward-pointing chevrons, especially distinct at midbody; venter with dark crossbands anteriorly; lighter interspaces form bands extending laterally, 3 of which are distinct and subequally spaced; tail base with white band 5 dorsal scales in width, and another, two scales wide at the tip. Of all juvenile characteristics, the white V is the most persistent, although it may be absent in adults (Schmidt 1923; Pitman 1938; Roux-Esteve 1965; Manacas 1981–1982; Phelps 1989).

Taxonomy and Distribution

Described as *Aspidelaps lichtensteini* from specimens collected from the Gold Coast. No subspecies are recognized.

Distributed through western and central Africa; in Sierra Leone, Ghana, across the Dahomey Gap in Nigeria, western Kenya, Uganda, and northern Angola. Isolated collections have been made from southeastern Sudan, southwestern Zaire and northern Zambia (Phelps 1989; Golay et al. 1993; Spawls and Branch 1995).

Habitat

Unlike other members of the genus, *C. lichtensteinii* is found primarily in undisturbed rain forests with low-intensity light. It prefers swampy areas near water. Specimens have been collected as high as 670 m in the Atewa Range Forest Reserve in Ghana (Leston 1970b).

Food and Feeding

Feeds almost exclusively on frogs and toads (Leston 1970b).

Behavior

Diurnal and primarily terrestrial, but it can swim well (has colonized islands in Lake Victoria). Inactive periods spent in brush piles, holes, tree root clusters and other ground cover. Disturbance elicits hissing, puffing and inflating responses similar to other members of the genus (Pitman 1938; Spawls and Branch 1995).

Reproduction and Development

Ova were found in a female in mid-September (Pitman 1938). Spawls and Branch (1995) reported 4–8 eggs are produced.

Bite and Venom

Venom glands do not extend down the neck (Haas 1973). Nothing is known regarding venom composition and toxicity; case histories are not available. No antivenin is available. (Spawls and Branch 1995).

Causus maculatus (Hallowell 1842)

Known as the forest rhombic night adder or West African night adder. Formerly considered a subspecies of *C. rhombeatus*.

Recognition

(Plate 2.1)

Head—Snout obtuse; rostral rounded; single loreal; circumorbital ring with 2–3 preoculars, 1–2 postoculars, and 1–2 suboculars; supralabials 6; sublabials 9–10, 4 in contact with the sublinguals; temporals 2–3.

Body—Dorsal scales 17–19 at midbody; ventrals 118–137 in females, 124–144 in males (a clinal increase occurs in ventral counts from south to north and east to west); subcaudals in females 14–23, males 15–26 (Hughes 1977).

Size—West African night adders are small and stout, averaging 30–60 cm. Maximum size is 70 cm or slightly longer (Spawls and Branch 1995).

Pattern—Head with forward directed V-shaped mark; space between parallel lines of V in juveniles solid black, which disappears in most adults; variably developed short black line behind the eye; body with black rhombic areas dorsally, least distinct on neck, most pronounced posteriorly; individual rhomboids not encircled in white unless skin is stretched, revealing small white specks; laterally with irregular pattern of black specks (absent in some specimens); individuals from Zaire usually uniformly brown (Hughes 1977).

Taxonomy and Distribution

Described as *Distichurus maculatus* from specimens collected from Liberia. No subspecies are recognized.

These snakes occur in west and central Africa, from Senegal east to Chad, southeast to Zaire, northeast into southeastern Sudan. Also found in river gorges and low country of southwestern Ethiopia, southwest to northern Angola, and Zaire (Golay et al. 1993, Spawls and Branch 1995).

Collection of 767 *C. maculatus* in Haute-Volta represented 13.4% of the total snakes collected. Of the *C. maculatus,* 438 were collected in summer from areas averaging greater than 90 cm of rainfall, 314 from areas averaging between 60 and 90 cm of rain, and 15 were from areas averaging less than 60 cm of rain (Roman 1980).

Habitat

Occupies a wide array of habitats including forest, savanna and semidesert (Hughes 1977; Spawls and Branch 1995). In Zaire, is confined to forest areas (Hughes 1977).

Food and Feeding

Feeds primarily on frogs and toads (Spawls and Branch 1995).

Behavior

Primarily terrestrial but occasionally climbs into low bushes after prey. Active night and day, basks in sunlight. Hides in holes, brush piles and ground cover when inactive. Most active during the rainy season (March–October), estivates during the dry season (Spawls and Branch 1995).

In Tafo, a cacao farm locality in Ghana, specimens were regularly observed moving and feeding in open plantations. This behavior may result from plantation lighting which attracts insects and, hence, toads (Leston and Hughes 1968).

When threatened, exhibits hissing, puffing and inflating responses similar to other *Causus* (Spawls and Branch 1995).

Reproduction and Development

In West Africa, oviposition occurs from February–April. Lays 6–20 eggs averaging 26 × 16 mm. Hatching occurs in May–July with hatchlings 13–16 cm (Spawls and Branch 1995).

Oviducts in many specimens collected during the rainy season contained eggs (Leston and Hughes 1968). One specimen had 3 eggs in each oviduct, averaging 17 × 8 mm; another had 3 eggs total, averaging 14 mm long; and a third specimen had 16 eggs, 11 on the left and 5 on the right, averaging 11 × 5 mm.

Bite and Venom

Causus maculatus has elongated venom glands, with an exceptionally long 120-mm venom gland recorded (Roman 1980). No information is available regarding epidemiology. Spawls and Branch (1995) reported an average venom yield of 100 mg and an i.v. "toxicity" of 10 mg/kg. It is unclear what the toxic endpoint or test species were in this report.

Symptoms in humans are generally mild, characterized by pain, moderate swelling, local lymphadenopathy and mild fever. Symptoms usually resolve over 2–3 days (Spawls and Branch 1995).

Causus maculatus bites have been studied in Nigeria (Warrell, Ormerod et al. 1976). Of ten cases, nine were bitten on the toes, foot or ankle. The tenth victim was bitten on the finger. Single fang marks were found in two victims. Two bites occurred in the morning (0630 and 0900 hrs). All patients complained of local pain, starting within 30 minutes of the bite. The extent of swelling was variable, from a few centimeters around the bite to the entire limb. Maximal swelling occurred within 1 day of the bite. Blistering, necrosis and hemorrhage were not observed and elevated temperatures occurred in only three patients. One patient showed mild neutrophil leucocytosis, but results of all other blood tests (clotting times, clot quality, PCV, platelet count, fibrinogen concentration and FDP) were within normal limits.

Of the ten cases, only one indicated severe envenomation. The patient, a 5-year-old boy, was bitten on the right foot by a 60-cm *C. maculatus*. The child was admitted to a medical facility 5.5 hours later. At that time, swelling involved the foot and ankle. The child was drowsy but rousable, his muscles were flaccid and tendon reflexes were sluggish. There was no respiratory distress, blood pressure was 64/38 mm Hg with a pulse of 120 per minute, and clotting was normal. During the next 2 hours, his blood pressure rose to 80/40 mm Hg with a pulse of 90 per minute. He remained drowsy until 12 hours postbite, when blood pressure was 100/50 mm Hg. By the next day his leg was swollen to the knee, a 12% increase of his normal limb circumference. Neither necrosis nor blistering occurred. Recovery was rapid, with swelling virtually gone within 3 days.

Using monovalent *C. maculatus* antivenins, Janssen et al. (1990) found low cross-reaction with other heterologous viperid venoms. This suggests that *C. maculatus* venom has little antigenic affinity to other viperid venoms. However, antigenic studies involving *Causus* venom as an indicator of phylogenetic relationships have revealed contradictory results. *Causus* venom cross-reacted with most crotalid and viperid antivenins, presumably reflecting the primitive nature of this genus (Minton 1968). However, Detrait and Saint-Girons (1979) found cross-reactivity markedly lower than for any other viperid venom tested. Janssen et al. (1990) concluded that phylogenetic relationships should not be speculated based upon cross-reactivity outcomes alone; rather, they should be used in conjunction with other systematic structural criteria.

Remarks

Nematode and cestode infestations of the gut can be heavy, with nematodes commonly observed in the feces (Leston and Hughes 1968).

Causus resimus
(Peters 1862)

Commonly known as the green night adder.

Recognition

(Plate 2.2)

Head—Short, slightly distinct from the neck; snout upturned; orbit with 2 preoculars, 2 postoculars, and 1 or 2 subocular scales; temporals 2+3 (rarely 2+4); sublabial 6–7.

Body—Noticeably stout; dorsal scales smooth or faintly keeled, 19–22; ventrals 131–155; subcaudals paired, 16–27; anal entire.

Size—Average size 30–60 cm, maximum 75 cm (Spawls and Branch 1995).

Pattern—Head dorsally covered with black scales forming V-shaped outline, distinct in juveniles; body coloration various shades of green (olive, grass green, bright green) with a velvety appearance; dorsum with a series of dark, inverted, chevronlike crossbars similar to *C. rhombeatus* and *C. defilippii;* chin and throat yellow; venter yellowish, cream or pearly (Corkill 1935; Pitman 1938; Manacas 1981–1982; Spawls and Branch 1995).

Taxonomy and Distribution

Described as *Heterophis resimus* from specimens collected from Jebel Ghule, Sennar, Sudan. No subspecies are recognized, although Spawls and Branch (1995) believe there may be grounds for separating races or even separate species.

Causus resimus is distributed in isolated populations across tropical Africa. Records are available from Angola, around Lake Victoria, Kenya, Somalia, central and southwestern Sudan, southeastern Ethiopia, northern Zaire, eastern Cameroon, Uganda, Tanganyika, northern Mozambique, Ruanda-Urundi, and the Congo. Other occurrences are recorded in Chad and Nigeria. In the Arab Emirates, Sudan, Uganda, and Kenya, it is sympatric with *C. rhombeatus*. However, in Somalia they are segregated, where *C. resimus* is found only in the southern lowlands (Broadley 1967; Hughes 1977; Spawls and Branch 1995).

Habitat

Warm, low-lying, moist savanna, high grassland, wooded hills, in riparian ecosystems of rivers running through semideserts, swamps, coastal scrubland and rocky river gorges. They have been taken from man-made habitats such as sugar cane plantations, abandoned quarries, and shallow "borrow pit pools" near roadsides (Spawls and Branch 1995).

Food and Feeding

Feeds primarily on frogs and toads (Spawls and Branch 1995).

Behavior

Slow-moving, but can strike quickly. When angered, the body is inflated, accompanied by ferocious hissing and puffing. The anterior is raised into a coil, with swiping strikes that tend to "lash" rather than "stab."

Primarily terrestrial, but they will climb reeds and sedges in pursuit of prey. They swim well. Despite their name, they are active during the day and often bask. When inactive, they hide under ground cover, brush piles and in holes (Spawls and Branch 1995).

Reproduction and Development

Male combat occurs during courtship (Curry-Lindahl 1956). Loveridge (1933) reported a clutch of 9 eggs from Kenya, each measuring 12 × 5 mm, and 4 other eggs measuring 19 × 9 mm. No breeding season is discernible in captive specimens, which have produced clutches at 2-month intervals (Spawls and Branch 1995).

Bite and Venom

Nothing is known regarding venom composition or toxicity. No case histories are recorded. Venom glands are elongated. No antivenin is known to be effective (Spawls and Branch 1995).

Causus rhombeatus
(Lichtenstein 1823)
The rhombic night adder.

Recognition
(Plates 2.3 and 2.4)

Head—Snout obtuse, moderately prominent, more rounded than other species in the genus; nostrils laterally positioned; circumorbital ring with 2–3 preoculars, 1–2 postoculars; 1–2 suboculars separate orbit from the supralabials; temporals typically 2+3, occasionally 2+4 (exceptionally 2+2 or 3+3); 6 supralabials (exceptionally 7); sublabials usually 7 or 10 (rarely 8, exceptionally 11, 12, or 13); first 3–4 sublabials in contact with anterior chin shields; posterior chin shields small, often indistinguishable from throat scales.

Body—Dorsal body scales 15–21 rows at midbody, moderately keeled with satiny texture; ventrals 120–166; subcaudals 15–36.

Size—Largest species in the genus, averaging 60 cm; a record 93-cm male was collected in eastern Zimbabwe (Barton 1966). Males with tail proportionately longer than females. Girth comparable to that of "an adult man's thumb" (Loveridge 1933). Harper (1963) described *C. rhombeatus* as appearing several sizes too small for its skin, even after a large meal or when swollen in anger.

Pattern—Head dorsally marked with dark brown to black "arrowhead," the apex of which rests on the frontal shield between the eyes; oblique dark streaks on either side of head behind eyes; labials usually dark-edged; body dorsally covered with varying shades of light to dark gray to olive, or light to pinkish brown; 20–30 dark brown or black rhomboidlike markings or chevrons down the back (markings sometimes lacking in northern populations); lateral markings fuse with dorsal blotches in West African specimens; in central Africa, lateral blotches alternate with the dorsal markings; venter uniformly pearly yellowish, dirty white to grayish, or pinkish gray. Greenish specimens can adjust color slightly from grass green to gray-green to match their background. Ontogenetic pattern changes may occur, with pattern becoming less distinct, with large adults often unicolored (Sweeney 1961; Hughes 1977; Greene 1988; Broadley and Cock 1993).

Taxonomy and Distribution

Described as *Sepedon rhombeatus*. No subspecies are recognized.

Rhombic night adders occur throughout sub-Saharan Africa. Its northern limit extends from Gambia, Sierra Leone, and northern and eastern Nigeria to the Nile region of Sudan and Somalia. In southern Africa, it is confined to the eastern coastal regions, except in the extreme south where it extends westward along the coast. Its range extends through Natal to Mozambique, Transvaal, Zimbabwe and northern Botswana. It is widely distributed in Angola in the west and Tanzania in the east (Broadley 1990). Other records are from Malawi, Liberia, Uganda, Cameroon, Ghana, Ivory Coast, Kenya, Namibia and Zambia (Berger-Dell'Mour 1987; Broadley 1988; U.S. Navy 1991). Loveridge (1933) and Harper (1963) reported it the most common snake in Ghana and Nairobi.

Widely distributed and commonly encountered. It comprised 22% of snakes collected in western Ghana during a 2-year survey (Swiecicki 1965).

Habitat

Especially fond of damp environments from coastal plains to altitudes of 1500 m (Sweeney 1961; Phelps 1989; Broadley 1990). Open woodlands, grasslands, savannas near streams, swamps and marshes are favored habitats (Pitman 1938; Mehrtens 1987). It is absent from rain forests. Often encountered in or near farmsteads and houses where they find abundant prey and hideouts under floors or in stone walls (Broadley 1990). During collection forays in which bounties were paid for specimens, the majority of night adders were collected around buildings and in workers' compounds (Leston 1970a). Rubbish heaps, piles of stones and the litter of "outbuildings" are favored (Pitman 1938).

Food and Feeding

Feeds primarily on frogs, (*Breviceps*), toads (*Bufo regularis* and *B. lemairei*), and occasionally small rodents. Small turtles are also eaten (Sweeney 1961).

Vision is poor and prey is detected primarily by smell. Hunting is usually at night. Prey is seized, held tightly, then swallowed headfirst if large. Small prey are swallowed in any orientation while still alive and struggling. When swallowing, the upper jaw moves forward with fangs erect, but they are folded back against the roof of the mouth before teeth hook into prey (Lambiris 1966).

They are voracious feeders that will eat several frogs or toads in succession. Digestion is fairly rapid, usually 7–10 days. Pitman (1938) reported that during 1 week a specimen (of undescribed size) ate a "large frog, three small toads and nine very small toads." One week later, its stomach was examined and all were digested (except the frog's feet). One toad per week was adequate for captive adult specimens (Harper 1963). In captivity they become tame and feed freely, even from the hand (Broadley 1990; Broadley and Cock 1993).

Behavior

An active snake, often moving relatively rapidly. Crawling speed has been estimated at 92 cm per second (Harper 1963). Most often encountered on the ground, but can climb and swim effectively (Sweeney 1961).

Although primarily nocturnal, it is often encountered basking in early morning or late afternoon. A dozen specimens collected by Harper (1963) were all active during the heat of day. Usually basks near old termite hills, animal holes, stone heaps or rubbish piles, into which it can readily escape. It hunts on dull rainy days, in the cool of evening or early morning. During the day, it rests in undergrowth, under stones or logs, or in termite mounds (Sweeney 1961).

While some individuals are temperamental, most are inoffensive and docile, seldom attempting to bite except under extreme provocation (Broadley 1990). According to FitzSimons, quoted in Pitman (1938), *C. rhombeatus* in captivity "become so tame that you may allow them to creep, climb, and slither round your neck and inside your garments." When seriously threatened, it assumes a "ferocious" attitude, coiling up, flattening its anterior, hissing loudly and inflating its body making its dark markings more conspicuous (Phelps 1989; Broadley 1990).

Causus rhombeatus is reportedly one of the best "jumping" snakes. Forceful inflation and deflation of the anterior one-third to one-half of the body while striking frequently lifts the entire length off the ground (Pitman 1961; FitzSimons 1980).

Seasonal activity patterns vary. Bounties paid for *C. rhombeatus* in a cacao plantation in Tafo, Ghana, during a 13-month period revealed peak activity in July and September, corresponding with peaks in rainfall and the abundance of the toad, *Bufo regularis* (Leston 1970a).

Reproduction and Development

Mating usually occurs in early spring, with oviposition in summer. Specimens from Ghana oviposit from May to early November. Eggs number 7–26, measure 25–37 × 14–20 mm. There is little or no prenatal development. Incubation requires 70–85 days. In captivity, females often incubate the eggs. Normally 1 clutch is produced per year, however, in northern populations where seasonal change is less marked, several small clutches may be produced at intervals of a month or 2 throughout the year. Hatchlings are 10–13 cm long and feed on small frogs and toads (Sweeney 1961; Cansdale 1961; Spawls and Branch 1995).

A female, isolated for at least 5 months, laid 4 clutches of fertile eggs before a sterile one was produced (Woodward 1933; Broadley 1990).

Molts frequently as an adult, usually at monthly intervals, reflecting a relationship between shedding and the high metabolic rate of this small active viper (Harper 1963). Reported to live 10 years (Sweeney 1961).

Bite and Venom

Epidemiology—Most bites are inflicted at night on feet or ankles. In Zimbabwe, this species is responsible for the majority of bites in built-up areas (Broadley and Cock 1993).

Yield—Venom glands are long. Yield varies from 20 or 30 mg to 300 mg of venom (Branch 1992; Spawls and Branch 1995, respectively).

Toxicity—LD_{50} values vary: i.v. 10.8, 14.6, and >16.0 mg/kg (Mebs 1978); s.c. 15.0 mg/kg (Minton 1974); 185 µg per 16–18 g mouse (Christensen 1967b). Although *Causus* venom is not very toxic to humans, frogs and toads are very susceptible and usually succumb rapidly when bitten (Broadley 1990). Pitman (1938) reported the venom is "particularly deadly when injected into a rat, far more so than the venom of either the puff adder, *Bitis arietans*, or the gaboon viper, *B. gabonica*."

Content—Little is known regarding composition. Peptidase activity hydrolyzes most di- and several tripeptides (Tu and Toom 1968). However, tripeptide linkages were not all equally cleaved. Dried venom is distinctive in appearance, consisting of dull spongy flakes (Pitman 1938). Spawls and Branch (1995) described it as "waxy" in appearance.

Symptoms and Physiological Effects—Bites in humans cause pain and local swelling. Tachycardia, swelling of the lymph nodes and local necrosis, possibly requiring skin graft, may also occur (Broadley and Cock 1975, 1993). A hemotoxic effect is reported. An antithrombin III inactivating enzyme, "CR-serpinase," has been isolated (Janssen et al. 1992). This compound inhibits serine proteinases involved in blood coagulation. A clot-promoting factor is demonstrable when the powerful anticoagulant factor is eliminated (Chapman 1967). Fatalities are rare.

A 5-year-old boy bitten on the toe by an adult night adder experienced swelling, hemorrhage from the bite, slight fever and mild hematuria. He recovered after 11 days (Corkill 1935). Brown (1973) reported only one of eight envenomation cases had "severe" effects. Two zoo workers carelessly handling night adders were bitten and became "dangerously ill" (Cansdale 1961).

Remarks

The egg-eating snake, *Dasypeltis*, is often confused with the rhombic night adder. Geographic variations in pattern of *C. rhombeatus* are paralleled by *Dasypeltis* (Broadley 1990).

Causus rhombeatus does well in captivity if supplied with sufficient frogs and toads. They require dry, well ventilated caging with hiding places, drinking water, and a basking spot that is kept slightly warmer (26–27°C) than the substrate temperature. They will accept prekilled rodents and are often voracious feeders. Feeding should be monitored to prevent obesity (Mehrtens 1987).

Four were kept in a 2-cubic foot cage with a "sanded" floor. A dish of water was provided and individuals would often immerse themselves for long periods (Harper 1963).

Interspecific interactions are poorly known. The file snake, *Mehelya* spp. is a major predator (Pitman 1938). Broadley (1974a) reported a case of ophiophagy involving *C. rhombeatus*, the Cape File snake, *Mehelya capensis* and the herald snake, *Crotaphopeltis hotamboeia*. A 450-mm *C. rhombeatus* swallowed tail-first was found inside a 450-mm herald snake. No prior record of ophiophagy by

C. hotamboeia exists. In this case, the herald snake itself was found inside a Cape File snake, with just its tail extending from the mouth.

Studies on *C. rhombeatus* kidneys by Fox (1976) revealed a large number of renal arteries (20). This runs contrary to the contention that primitive snakes have fewer renal arteries than advanced snakes (Underwood 1967).

Chapter 3
Viperinae

Viperinae is the biggest group of vipers in the Old World. All European, west Asian and African vipers (except *Causus*) belong to Viperinae. In Central and East Asia, *Azemiops* and pitvipers, Crotalinae, share the space with vipers of the subfamily Viperinae. Viperinae contains 12 genera here divided into 78 species. However, newly described species, as well as subspecies raised to full species, will increase this number.

Adenorhinos

The genus *Adenorhinos* contains a single species, *barbouri,* commonly known as worm-eating viper, Uzungwe viper, Barbour's short-headed viper or Barbour's viper. It was originally described as an *Atheris* but later placed in a separate genus, *Adenorhinos*. It is morphologically divergent from all other members of the genus *Atheris,* but recent studies (Lenk et al. 2001) have suggested that it is closely related to and forming a sister group with the sympatric *Atheris ceratophora*. Future studies will show wheather *Adenorhinos barbouri* should be returned to *Atheris* (which thus in its present composition is paraphyletic) or if it together with *Atheris ceratophora* should form a separate clade.

Adenorhinos barbouri
(Loveridge 1930)

Recognition

Head—Moderately broad, distinct from neck; snout short and rounded; eyes prominent, approximately 1.5 times the distance to the mouth; pupils vertically elliptical; crown covered with small, strongly keeled, imbricate scales; nostril in an extreme anterior position, part of a single nasal; nasal in contact with preocular; single row of scales separates suboculars from supralabials, with some broadly in contact with 3 supralabials; supralabials smooth, 5–6; 5 smooth sublabials; 2 pairs of smooth chin shields, with a median groove; gulars smooth, with the anterior-most contacting posterior-most chin shields; anterior and posterior temporals single, enlarged, and smooth; a subcutaneous nasal gland and concave depression are in the posterior area of the nasal scale.

Body—Moderately slender; tail relatively short; dorsal scales strongly keeled except for outermost row, 20–23 rows at midbody; lateral body scales not serrated;

ventrals smooth and rounded, 116–122; subcaudals smooth, single, 19–23; anal single (U.S. Navy 1991; Spawls and Branch 1995).

Size—Small, reaching approximately 40 cm in total length (Spawls and Branch 1995).

Pattern—Ground color brown to blackish olive; pair of straw-colored, zigzag dorsolateral stripes extending from back of head to end of tail; stripes may form an irregular chain of darker rhombic blotches along the back; tail sometimes with faint black checkering; venter greenish white to olive (Spawls and Branch 1995).

Taxonomy and Distribution

Described from the type specimen *Atheris barbouri* collected at Dabaga, Uzungwe Mountains, Tanzania. No subspecies are recognized.

Worm-eating vipers are morphologically distinct and are meanwhile considered the most basally derived monophyletic genus in the Viperinae (but see above). Based upon detailed cladisitic analysis of 66 character states, Ashe and Marx (1988) found 8 of 17 derived character states unique to the *Adenorhinos* lineage. These include: presence of nasal shield depression; anterior position of nostril in the nasal shield; length of anterior portion of the skull; ectopterygoid bone centrally broad without a spine; premaxilla with posterior process on lateral arms; relatively large eyes; absence of loreal shield; relative length of longest maxillary tooth. The first four character-states represent transformation of structural features not found in any other viper. Character-states five and six represent transformation series unique to vipers as a whole, and numbers seven and eight are unique end points of transformation series found in less highly derived forms.

According to Ashe and Marx (1988), the large number of highly derived character-states confined to *Adenorhinos* reflects the unique evolutionary pathway taken by this lineage.

Adenorhinus barbouri has a very limited distribution. Specimens have been collected from the Udzungwa and Ukinga mountains in western Tanzania (Spawls and Branch 1995).

Habitat

Terrestrial. Found in brush and other undergrowth on mountain slopes at altitudes of 1800 m (Spawls and Branch 1995). Preferred habitat seems to be moist forest. It has been found in thick bush and bamboo undergrowth, but also in gardens of tea farms (Spawls et al. 2002).

Food and Feeding

Thought to be a specialist feeder on slugs, earthworms and other soft-bodied invertebrates (Spawls and Branch 1995) or frogs (Spawls et al. 2002).

Behavior

Subterranean habits have been suggested but are unlikely due to lack of morphological adaptations for even a semifossorial life (Spawls and Branch 1995).

Reproduction and Development

The species appear to be oviparous. Three females collected in February 1930 each contained 10 eggs, the biggest measuring 1 × 0.6 cm (Spawls et al. 2002).

Bite and Venom

No cases of envenomation are known. Its restricted distribution makes bites unlikely. Nothing is known regarding toxicity, composition or effects of venom.

Chapter 4
Atheris

The genus *Atheris* was established by Cope in 1862 from specimens previously included in *Chlorechis*. Species are highly arboreal and fully adapted to life in trees, shrubs and brush.

Many members of this genus have isolated fragmented distributions. They are confined to rain forest areas, which locally is under development pressure throughout the range, and remnant parcels of habitat support relict populations of *Atheris*.

Because of their isolated and dense habitats and arboreal habits, these species are not of great medical importance. Their venoms are not well studied and antivenins are not available.

Recognition

Head—Broad, sharply distinct from narrow neck; no enlarged dorsal scales; canthus distinct; snout broad; crown with small imbricate keeled or smooth scales; eyes large with vertical pupils, separated from supralabials by 1–3 scale rows and nasal by 2–3 scales.

Body—All species are small to medium sized (adult maxima 60–80 cm); dorsal scales are heavily keeled and overlapping; bodies are slender, tapering, slightly compressed, and cryptically colored; subcaudals are single; tail prehensile; dorsal scales strongly keeled with apical pits. Lateral scales smaller than dorsals, not serrated, in 14–36 oblique rows at midbody; ventrals rounded, 133–175; subcaudals 38–67, single (U.S. Navy 1991; Spawls et al. 2002).

Taxonomy and Distribution

The genus *Atheris* currently includes 12 species: *A.acuminata, A. broadleyi, A. ceratophora, A. chlorechis, A. desaixi, A. hispida, A. katangensis, A. nitschei, A. squamigera* and *A. subocularis*. In addition, the former subspecies *A. nitschei rungweensis* is nowadays treated as a full species. The genus is confined in distribution to central Africa, with one species (*A. chlorechis*) found as far west as Guinea. However, recently a new species, *Atheris hirsuta,* Ernst and Rödel (2002) was described from the Taï National Park, Ivory Coast.

Atheris acuminata
Broadley 1998

Called the acuminate bush viper, the specific epithet refers to the sharply pointed and elongated (acuminate) dorsal scales on the back of the head and anterior section of the body.

Recognition

Head—Flat and subtriangular, with a rounded snout, distinct from the neck, covered with pointed and sharply elongated and keeled scales. Pentagonal rostral shield in contact with two large suprarostrals; 6 supralabials. Eyes large.

Body—Strongly keeled, the sharply pointed and elongated body scales extend to mid-body (in *A. hispida* the upturned spiny scales do not extend to posterior part of back). Scale in 14 midbody rows. Ventrals 160; subcaudals 54; anal entire.

Size—A small and slender bushviper of 44 cm total length; tail 18% of total length.

Pattern—This species is yellowish green with dark H-shaped markings on head and dark markings on the tail. Many green scales on top of head (Spawls et al. 2002).

Taxonomy and Distribution

This species is closely related to *Atheris hispida* and differs primarily in head and body scalation. Almost nothing is known about its distribution. The type is at present the single specimen known—a male from a forest near Nsere Lodge, Kyambura Game Reserve, just south of lake George, Ankole District, western Uganda, at an altitude of 950 m (Spawls et al. 2002).

Habitat

Very little is known about this exceptionally rare species. The single known locality is a riverine gallery forest near a small lake, surrounded by the savannah of the Kyambura Game Reserve (Spawls et al. 2002).

Food and Feeding

No information is available.

Behavior

No information is available.

Reproduction and Development

No information is available.

Bite and Venom

Nothing is known of the toxinology.

Atheris broadleyi
Lawson 1999

This species is called Broadley's bush viper.

Recognition

Head—Flat and triangular, distinct from the neck, covered with keeled scales; rostral scale 3.5–4 times broader than high; 3–8 interorbital scales across; 9–12 upper labials, in contact with suboculars; 9–12 lower labials, 2–3 lower labials in contact with the chin shield; 12–16 circumoculars.

Body—Keeled, but lateral scales without serrated keel. 17–23 midbody dorsal scale rows, ventrals 157–169, subcaudals 45–61.

Size—768 mm total length.

Pattern—Citrine to greenish olive or olive brown above with pale yellowish crossbands that may be very faint. A broad black postocular stripe runs from behind the eye to the corner of mouth. Throat whitish or pale yellow below, and ventral side is pale whitish or light bluish, and heavily mottled with black. Unlike *Atheris squamigera,* the color and pattern of this species appears to be remarkably consistent (M.A. Jacobi www.kingsnake.com/atheris/; Lawson et al. 2001).

Taxonomy and Distribution

Atheris broadleyi has previously been confused with the widespread *Atheris squamigera,* but it is distinguished from this species and others by differences in scalation and color/pattern. The species is distributed throughout southern Cameroon, Congo, Central African Republic and Gabon.

Habitat

It inhabits moist evergreen and semideciduous transition forest (Jacobs file).

Food and Feeding

No information is available.

Behavior

No information is available.

Reproduction and Development

No information is available.

Bite and Venom

Nothing is known of the toxinology.

Atheris ceratophora
Werner 1895

A common name is the Usambara Mountain bush viper, after its original locality.

Recognition
(Plate 4.1)

Head—Head triangular, well set off from neck, covered with fragmented keeled scales; rostral little more than 2 times and less than 2.5 times as long as broad; 8–9 scales across interorbital; 3–5 enlarged hornlike superciliary scales; 9 upper labials; 3 lower labials in contact with the chin shield.

Body—21–25 midbody dorsal scale rows, ventrals 142–152, subcaudals 41–56.

Size—Largest male 42 cm total length, of which 8 cm is tail, maximum length of females (slightly larger than males) 54 cm.

Pattern—Dorsum yellowish green, olive, gray or black; variable markings, some individuals are unmarked, others are patterned with irregular black spots or cross bars, sometimes lined with yellow or white spots. Venter orange to nearly black, sometimes with dark speckles (Barbour and Loveridge 1928; Spawls and Branch 1995).

Taxonomy and Distribution

Described by Werner in 1895 from a single male specimen, there is only the nominate form which is found in the Usambara, Udzungwa, and possibly the Uluguru mountains of Tanzania.

Habitat

Atheris ceratophora occupies grass or low bushes about 1 m from the ground (Spawls and Branch 1995) in forests and woodlands from 700 to above 2000 m altitude.

Food and Feeding

The only feeding recorded for this species is by Barbour and Loveridge (1928) who found one specimen containing an arboreal frog, *Hyperolius* sp.

Behavior

Probably crepuscular or nocturnal (Spawls and Branch 1995).

Reproduction and Development

Some indications suggest that the breeding period is in September and October; young are born in March and April (Spawls et al. 2002).

Bite and Venom

Very little is known about the venom of this species. A large female (55 cm total length) bit a 43-year-old male weighing 87 kg. The victim was previously bitten by *Atractaspis bibronii;* no antivenin was used in that case. The *A. ceratophora* bite occurred when the snake fell from a branch onto which it was being placed. One fang penetrated the victim's left shoulder. After 10 minutes, there was a black spot 10 mm in diameter at the bite. At hour 2 pain increased and spread over the shoulder, which became especially sensitive to movement. The victim took 16 mg of chlorpheniramine maleate and slept peacefully on the opposite shoulder. Pain and discoloration disappeared by day 5; symptoms were gone by day 6. As this is the only bite record for this species, it is unknown if all are this mild (Emmrich 1993).

Remarks

Described as a peaceful and decorative captive, it is easy to keep if provided with climbing and hiding places. Feeds readily on mice and young sparrows (Vogel 1964). Emmrich (1997) reviews the biology and ecology of this species.

Atheris chlorechis
(Pel 1851)

Called the western bush viper.

Recognition
(Plate 4.2)

Head — Broad and flat, very distinct from the narrow neck, head as broad as broadest part of body; covered with numerous tiny, overlapping, strongly keeled scales.

Body — Dorsal scales small and very strongly keeled, in 25–36 midbody rows.

Size — Relatively slender, adults average 50 cm with a maximum of 70 cm; tail long and fully prehensile.

Pattern — Adults uniform pale green, darkening on sides and toward tail; more or less paired golden spots about 2.5 cm apart along upper surface; venter pale green; color reversed in young specimens (less than 25 cm), being yellowish with green spots (Cansdale 1961; Spawls and Branch 1995).

Taxonomy and Distribution

Described by Pel in 1851 on three syntypes. Type locality is Butre, Ghana.

The distribution is restricted to forest habitats of West Africa (Guinea-Bissau, Guinea, Sierra Leone, Liberia, Ivory Coast, Ghana, Togo, Benin and isolated localities in Nigeria, Cameroon, Equatorial Guinea and Gabon (McDiarmid et al. 1999).

Habitat

An arboreal forest snake, it lives in dense foliage 1–2 m from the ground.

Food and Feeding

Reported to eat rodents, tree frogs and lizards (Cansdale 1961).

Behavior

No information is available.

Reproduction and Development

Neonates measure 13–15 cm and are born in March and April. There are 6–9 babies per litter (Spawls and Branch 1995).

Bite and Venom

Its fangs are long, thin and sharp, but the animal is generally inoffensive and there are no reported bites to humans. Venom yield is up to 100 mg (wet weight per individual). Subcutaneous LD_{50} (presumably for laboratory mice weighing approximately 20 g) is reported to be 8 mg/kg (Cansdale 1961; Spawls and Branch 1995).

Atheris desaixi
Ashe 1968
Called the Mount Kenya bush viper.

Recognition
(Plate 4.3)

Head—Broad and triangular, distinct from neck, covered with small keeled scales; eye set well forward in the head; 14–17 circumorbital scales; 8–11 interorbital scales; 2–3 scales between eye and nasal; 2 between eye and supralabials; 10–12 upper labials; 11–14 sublabials; rostral broader than deep; upper margin of rostral highest at the center, with an even number of suprarostrals; superciliaries not enlarged; nasal circular and entire or partially divided; scales in front of nasals around eyes, and chin shields smooth; other head scales short and strongly keeled; gulars strongly keeled; 9 gulars from sublabials to chin shields, chin shields in pattern 1 small, 1 large, 4 small.

Body—Dorsals short, heavily keeled; keels on upper dorsals end before the end of the scales (vs. *A. chlorechis* in which the keels end in swellings at the posterior of each scale), keels on the lower dorsum serrated; 24–31 middorsal scale rows; 160–174 ventrals; 41–54 subcaudals in females, 53 in one male specimen (Ashe 1968).

Size—Adults reported in the original description (Ashe 1968) ranged from 49–68 cm total length, head and body length ranged from 43–59 cm, tail length from 6–9 cm. In a recent report, the average adult size is listed as 40–60 cm, with 70 cm maximum, and hatchlings 17–22 cm (Spawls and Branch 1995).

Pattern—In adults, the body color is charcoal black, the tip of each dorsal scale mustard yellow, with festoons or loops of the same yellow color along the sides of the body, indistinct forward and becoming clearer and more contrasted posteriorly, becoming indistinct again on the tail, sometimes with the yellow markings fusing into zigzags. The venter is yellowish in the forward half, becoming suffused with and dominated by purple to purplish black at the rear, there may be sexual dimorphism of the venter, with males exhibiting more and more frequent darker pigmentation, undersurface of tail purplish black except for the very posterior which is marked or blotched with yellow. Hatchlings are generally yellow with a white tail tip, they darken with age and reach adult color and pattern at around 30 cm (Ashe 1968; Spawls and Branch 1995; Hedges 1983; Spawls et al. 1998)

Taxonomy and Distribution

Described by Ashe in 1968 from a female holotype and several additional individuals collected in 1967. A number of specimens have been collected subsequently.

This is the only *Atheris* species currently known to occur east of the Great Rift Valley in Kenya. There are two known locations, one in Igembe in the northern Nyambeni range of central Kenya, and one on southeastern Mount Kenya at Chuka. It is likely that the discontinuous nature of the distribution is due to loss of the forested habitat required to support this species. It has been hypothesized that other locations still supporting remnant forest on Mount Kenya, in the Nyambeni range, or in the region around Meru may support populations (Spawls and Branch 1995; Spawls et al. 1998).

Habitat

This is a forest snake. The original specimens were collected in clearings and along pathways in dense rain forest at an altitude of 1600 m. Of the three original specimens for which habitat notes are available, one was taken about 2 m up in a tree growing through the dense undergrowth of a clearing, one was captured at a similar height in a bush densely covered with creepers, and the third, also taken about 2 m up, was caught in the canopy of a small tree in a densely vegetated area. All three specimens were captured within a 1.6 km radius and the forest at this location was very humid (Ashe 1968).

Food and Feeding

Reported to be a generalist feeder, eating small birds, rodents and amphibians (Hedges 1983). Spawls and Branch (1995) report that *A. desaixi* "appears to feed on small mammals." This report is reinforced by (or is perhaps based upon) notes included with the description (Ashe 1968) which indicate that captives fed readily on white mice.

Behavior

Very little is known regarding the behavior of this species. It is unclear at this time whether they are diurnal, nocturnal or crepuscular (Spawls and Branch 1995). In captivity, they seem to feed equally readily both day and night (Ashe 1968).

In typical fashion for arboreal *Atheris*, *A. desaixi* is cryptically colored and moves slowly and deliberately in the branches that are its preferred microhabitat. When captured, individuals strike readily and struggle with great energy to escape. However, they take readily to captivity and quickly become calm (Spawls and Branch 1995). The threat display includes a scale-rubbing demonstration with coils counterlooped to produce a loud hissing noise. This habit is lost as individuals settle into captivity and become used to handling and disturbance (Ashe 1968).

Reproduction and Development

A wild-caught female from the Nyambeni range of Kenya gave birth to 13 young. Young were born in August and ranged from 17–21 cm in total length (Spawls and Branch 1995).

Bite and Venom

There is apparently a single reported bite. A Kenyan collector was bitten on the index finger of the right hand, with penetration by a single fang. There was considerable pain and swelling. The individual was treated with unspecified antivenin and recovered fully. Available antivenins are of unknown efficacy. The relatively large size and potentially powerful anticoagulant properties of the venom suggest that bites should be treated as serious (Spawls and Branch 1995). Bites in the wild are unlikely due to the rarity, arboreality and general inconspicuousness of individuals of this species.

Atheris hispida
Laurent 1955

Called the rough scaled bush viper, spiny bush viper, or hairy bush viper.

Recognition
(Plate 4.4)

Head—Triangular, with snout short, shorter in males than females, distinct from slender neck; 9–16 circumorbital scales; 7–9 interorbital scales; nostril is a vertical slit 3 scales from eye; 1 scale row between eye and supralabials, which number 7–10, the fourth of which is enlarged; 9–10 sublabials, 3 in contact with chin shield; posterior labials keeled; rostral much wider than deep, not visible from above, surmounted by 3 large scales; 2 scales separate nasal and eye, the anterior one is keeled, sole pair chin shields followed by 3 or 4 pairs gulars or 4–5 pairs gulars and no chin shield, gulars mostly strongly keeled.

Body—Adults are shaggy, almost bristly in appearance; narrow elongate scales; exaggerated enlarged keels, each scale prolonged backwards and incurved to become a spine on the neck and anterior body, this development decreasing progressively posteriorly; hatchlings lack this prolongation of scale form, midbody rows 15–19; ventrals 149–166; subcaudals 35–64; anal entire.

Size—males recorded to 73 cm (58-cm body, 15-cm tail), females to 58 cm, male remarkably slender and elongate relative to female.

Pattern—Female olive to blackish with almost indistinct to clearly distinct black occipital chevron, lighter anteriorly with apex dorsal scales even lighter, yellowish or russet, also blackish outlining dorsal spots which disappear toward rear body, tail uniformly blackish, or ringed or streaked black or dull green. Male vivid yellow green, occipital chevron and temporal band, black, sharply contrasting general light coloration, numerous dorsal spots in a roughly zigzag pattern, belly light green anterior, darker posterior, anterior edge ventrals black, black spots below posterior, dominant on tail which is ringed. One authority (Laurent 1960) reports no sexual variation, that both males and females are dull green or shades of vivid green to blackish green, occipital chevrons often reduced to black blotches or spots, below green or greenish yellow, bluish posteriorly more and more dotted black until subcaudals become entirely black (Laurent 1955; Pitman 1974; Spawls and Branch 1995).

Taxonomy and Distribution

Described by Laurent in 1955 as *Atheris hispida*. Found in Kivo and Orientale provinces, Zaire, Uganda and the Kakamega Forest in Kenya (Broadley 1968; Pitman 1974; Spawls and Branch 1995).

Habitat

Rain forest and rain forest wetlands, particularly those with emergent reedbeds and in papyrus marshes adjacent to lakes and rivers. Requires dense foliage. Found to approximately 1800 m elevation (Pitman 1974; Spawls and Branch 1995).

Food and Feeding

Poorly known. Diet presumed by Pitman (1974) to be similar to other *Atheris*. A snail was found in one stomach. May forage on the ground for small mammals. Frogs, lizards and occasionally birds are also eaten (Spawls and Branch 1995).

Behavior

Active nocturnally. Moves easily through foliage with prehensile tail, found in dense thick bushes 1–2 m above ground. Often climbs reeds and stalks to bask on flowers or terminal leaves.

Reproduction and Development

There are 5–12 young per litter, measuring approximately 17 cm.

Bite and Venom

Unknown. Presumed to be of little threat to humans and venom is supposed to be of low toxicity. Spawls and Branch (1995) state that available antivenins are ineffective.

Remarks

In captivity they may not drink from a bowl, in which case they must be misted with warm water several times weekly. They feed readily on small prekilled mice.

Atheris katangensis de Witte 1953

Called the Shaba bush viper.

Recognition

Head—Flat and triangular, with a rounded snout, distinct from the neck, covered with small keeled scales.

Body—Strongly keeled, in 24–31 dorsal rows at midbody, tail is short, with 38–42 subcaudals in females and 45–59 in males.

Size—40 cm maximum total length.

Pattern—Dorsum purple-brown or yellow-brown, with paired dorsolateral lines in contrasting shade running from head to tail, lines may break into zigzag pattern. Venter and tail tip yellowish (Spawls and Branch 1995).

Taxonomy and Distribution

Described from specimens taken in Shaba Province, Zaire. To date, found only in a restricted area of eastern Zaire (Spawls and Branch 1995).

Habitat

Collection location is in canopied forest adjacent to rivers at an altitude of 1200–1500 m (Spawls and Branch 1995).

Food and Feeding

No information is available.

Behavior

No information is available.

Reproduction and Development

No information is available.

Bite and Venom

Nothing is known of the toxinology.

Atheris nitschei
Tornier 1902

Known as the Great Lakes bush viper and Nitsche's bush viper.

Recognition
(Plates 4.5 and 4.6)

Head—Large, triangular, flattened, distinct from neck, densely covered with small keeled scales, last 4 upper labials keeled, scales below head to end of gape not keeled, 12–19 scales around eye, 8–12 across top of head between eyes, usually 3 scales between nasal and eye (rarely 2 or 4), and 4–8 scales separate mental from first ventral.

Body—Middorsal scale rows 23–33, ventrals 141–164, undivided subcaudals 35–59.

Size—Large, stout and heavy-bodied, average length 63 cm, maximum to at least 73 cm (62-cm body, 11-cm tail); female larger than male; tail of male 5.85–6.3 cm in total length, tail of females 6.2–7.7.

Pattern—Dorsum in green shades with broad black zigzag broadening into inverted V at top of head, bright yellowish green, sometimes olive green above and greenish, pale green or yellowish below, with or without interstitial black, broad zigzag may coalesce into wide irregular stripe and inverted V, narrow or almost solid centrally on head but not on all specimens, keels on scales normally black, anterior head scales margined black, often lateral series black spots, at least 10 mm of the tail broadly tipped with black, ivory white or pinkish in hatchlings which are dull grayish green and sometimes lacking typical adult markings (Bogert 1940; Vesey-Fitzgerald 1958; Sweeney 1961; Hedges 1983).

Taxonomy and Distribution

Described as *A. nitschei* by Tornier from specimens taken in Mporo Swamp in central Africa (Bogert 1940).

Habitat

Found in mountain forests, in association with wetland and meadow areas, along small streams, in papyrus reed and elephant grass marshes, occasionally in scrub and bush in elevated valleys and montane forest up to the bamboo zone at

1600–2800 m of altitude. Abundant in papyrus fringing small lakes (Sweeney 1961; Broadley 1971; Pitman 1974; Spawls and Branch 1995).

Food and Feeding

Feeds largely on cricket frogs (*Phrynobatrachus*), tree frogs (*Hyperolius*), other frogs, and young mammals (Sweeney 1961). Pitman (1974) stated the diet is varied, with a preference for mammals including pygmy mice (*Leggada grata*), shrews, amphibians, and lizards. It may also eat birds and rob nests of eggs and hatchlings (Hedges 1983).

Behavior

Largely arboreal, although in eastern Zaire it is usually found on the ground at high altitudes. Feeds at night but basks during the day. Climbs to the top of papyrus stems to bask coiled in the inflorescence. Very quick when disturbed, often releasing and falling free from the basking flower. The strongly prehensile tail enables easy climbing and it swims well (Spawls and Branch 1995).

Not generally aggressive but fierce when disturbed. Some settle into captivity, most retain irascibility and object to handling, remaining generally bad-tempered. Hatchlings are particularly "vicious."

Juveniles employ caudal luring, waving the yellow tail tip to lure prey into striking range (Pitman 1974).

Reproduction and Development

Poorly known. In one collection made in late October to mid-November, 17 females contained developing ova or full term hatchlings. Ova are spherical, the largest about "marble-size," hatchlings just emerging averaged 16 cm total length (Pitman 1974).

Bite and Venom

Unknown. Presumed by most authors to be of little medical significance. Spawls and Branch (1995) state that available antivenins do not neutralize the venom.

Remarks

Adults are usually heavily infested with cestodes and nematodes. They are eaten by birds of prey.

Atheris rungweensis Bogert 1940

Called the Rungwe tree viper. The name refers to Rungwe Mountain, Tanzania, the type locality.

Recognition

Head—Flat and triangular, distinct from the neck, covered with keeled scales.

Body—A large bush viper. 22–33 midbody scale rows; ventrals 150–165; subcaudals 46–58.

Size—65 cm maximum total length.

Pattern—Bright green to green and black above, often with a yellow pattern on the back of the head and a pair of yellow dorsolateral zigzag lines and/or a row of yellow lateral spots where the sides meet the ventral scales. Venter yellow to grayish green (Spawls et al. 2002).

Taxonomy and Distribution

This species was formerly considered to be a subspecies of *Atheris nitschei*. The species differ in scalation and coloration. While *A. nitschei* is typically black and green, its southern relative, *A. rungweensis* is predominantly green (or black) and yellow and has smaller cephalic scales that are more strongly keeled anteriorly. It also has keeled gulars, whereas *A. nitschei* has smooth gulars. Due to the morphological differences and the fact that these two taxa do not appear to intergrade, *Atheris rungweensis* is now considered to be a full species (Spawls et al. 2002). Can be distinguished from *A. nitschei* by its normal color pattern of green and yellow, rather than green and black. Its head scales are also smaller and more strongly keeled anteriorly.

The distribution of the species is south of the range of *A. nitschei* in western Tanzania and northwestern Zambia. It also occurs on the Misuku Hills of northern Malawi. This distribution represents the southernmost range of the genus *Atheris*.

Habitat

Found in low bushes along streams and at the edges of montane forest at altitudes of 800–2000 meters. Sporadically observed in moist savanna, woodland and hill forest habitats (Spawls et al. 2002).

Food and Feeding

No information is available.

Behavior

No information is available.

Reproduction and Development

No information is available.

Bite and Venom

Nothing is known of the toxinology.

Atheris squamigera (Hallowell 1856 [dated 1854/55])

Generally called the green bush viper.

Recognition
(Plates 4.7 and 4.8)

Head—Large, broad, flat and distinct from neck, mouth with enormous gape; head thickly covered with keeled imbricate scales; 7–9 interorbital scales; 9–12 supralabials, 9–12 sublabials with anterior 2 or 3 touching chin shields; 2 scales between eye and nasal; rostral invisible from above; small scale above rostral flanked by very large scale on either side; nostrils lateral; 10–18 circumorbitals; 2 (rarely 1 or more than 2) rows of small scales between eye and labials; gular scales keeled; pupil vertically elliptical.

Body—Dorsal scale rows at midbody 15–23, fewer (11–17) posteriorly; ventrals 152–175, subcaudals undivided 45–67; possibly variation in morphometric characters related to habitat: averages of 17 midbody dorsal rows in southern forest, 21 in northern grasslands, subcaudals 52 in forest, 58 in grassland, ventrals 171 in forest, 168 in grassland.

Size—Possibly the largest bush viper species, average total length 46–50 cm, maximum to at least 78 cm, mature females generally larger than males. In a Nigerian population maximum male length was 67.0 cm, and maximum female length was 70.5 cm (Luiselli et al. 2000).

Pattern—Coloration consistent in some populations, variable in others. Dorsal color varies from sage green or light green to green, dark green, bluish, olive or dark olive brown, rare individuals are yellow, reddish or slate gray. Scales with light-colored keels, scales may be yellow-tipped forming series of 30 or more light crossbands or chevrons, 10–19 chevrons on tail, not always clearly defined though usually present, ventral edge of dorsum with light spots in pairs, interstitial black color apparent only when skin is stretched. Venter dull or pale olive or yellow, heavily mottled with blackish or uniformly yellowish or greenish, throat sometimes yellow. Tail with conspicuous 7–12-mm ivory white tip extending back over 10 subcaudals. Neonates are dark olive with wavy bars, paler olive or yellowish olive with fine dark olive margins, bars at 5-mm intervals, venter paler greenish olive, adult coloration develops after 3–4 months (Thireau 1967; Pitman 1974; Stucki-Stern 1979; Phelps 1981; U.S. Navy 1991; Spawls and Branch 1995).

Taxonomy and Distribution

Originally described as *Echis squamigera* (Hallowell 1856) and then placed in *Atheris* as *Atheris squamigera* by Peters (1864) and Boulenger in 1896. Type locality is "near the river Gaboon."

Several subspecies have been erected (Golay et al. 1993). These include: *A. s. squamigera* (Ghana to Cameroon, Zaire, Uganda, western Kenya and Angola), *anisolepis* (southern Congo to northern Angola) and *robustus* from the Ituri Forest in Province Orientale. Spawls and Branch (1995) suggested that yellow-colored specimens from the Congo Republic may be treated as a separate species (*A. laeviceps*). Laurent (1958) reported on *robustus* from Ituri. He stated that the dorsals for this race are 21–24 with two rows of scales between upper labials and eye, and the subcaudals may be fewer (Spawls and Branch 1995). Lawson et al (2001) treated it as a "putative" subspecies. In a recent study Lawson and Ustach (2000) compared the type of *A. squamigera* with topotypes of *A. anisolepis,* and considered them to be synonymous. Also *A. laeviceps* is considered to be synonymous as the single character, yellow color is not a geographical characteristic, but occurs in most populations and even within clutches. At present, two subspecies can possibly be defined:

Atheris s. squamigera, and
Atheris s. robustus.

Widely distributed in western and central Africa (Phelps 1981). Found from Ghana to Cameroon and the Congo Republic, through Zaire and Angola, east to Uganda and Kenya. Two fragmentary decomposed male specimens were found in Tanzania, in the Rumanyika Game Reserve at Karagwe District (Broadley 1995a). These are the first records from Tanzania and the locality is a few kilometers east of Mpororo Swamp, the type locality for *A. nitschei.*

Habitat

This is primarily a rain forest species (Spawls and Branch 1995), although it is also found in thin forests associated with grasslands (Stucki-Stern 1979). In primary forest and structured vegetation it is highly arboreal. In remnant forest patches and at the fringe of primary forest areas it inhabits bushes and tall grass. It may persist in areas from which forest has been stripped by logging or other development (Pitman 1974). Detailed studies of habitat preferences of African snakes record this species only in densely vegetated and closed canopy forest areas (Villiers 1975; Butler and Reid 1986).

In Nigeria it mainly occurs in secondary forest patches (both dry and flooded) and in shrublands, and frequently also in primary forest patches (Luiselli et al. 2000).

Food and Feeding

Small mammals, frogs, lizards and snakes generally define the diet (Pitman 1974; Villiers 1975; Phelps 1981; Spawls and Branch 1995). Loveridge (1933) examined 50 stomachs and found 44 empty, 4 with small rodent remains (*Dendromus, Mastomys, Leggado*), and much unidentifiable rodent fur. One small tree frog was also recorded. From Kinshasa in the Congo Republic, 5 specimens contained mammal remains.

This species feeds at night by descending through the foliage to hang just above the ground. From this position it strikes small mammals passing underneath. It also manages to drink while hanging from vegetation by ingesting condensing water running downward on the body (Spawls and Branch 1995).

In a Nigerian population, the main food item is small rodents that are caught by vipers hanging from branches near the forest floor in the night (Luiselli et al 2000).

Behavior

Active nocturnally, during the day it basks in the sun above foliage. A very alert and irritable snake, it strikes violently when molested. It is not, however, aggressive when not disturbed. Deliberate in its climbing movements it is nonetheless highly (if not totally) arboreal. When disturbed in vegetation it freezes. If the disturbance continues it may drop in a "free fall" through the foliage (Stucki-Stern 1979; Spawls and Branch 1995).

Reproduction and Development

In the wild (Uganda) mating occurs in October and November. Young are born in March–April, and number 5–9 per litter (Pitman 1974; Mehrtens 1987; Spawls and Branch 1995).

In captivity they require very high humidity to breed. Sexes in one case were kept separate from January to the end of November. Only one mating was observed but two females were pregnant. Each female produced 8 young, of which most were green and a few were yellow. There was one stillborn green individual in each brood. Some neonates readily ate frogs, others required force-feeding of pink mice. After a few months all fed independently (Hagstrom 1994).

Bite and Venom

The venom has been characterized as "not very active" and the species as "doubtfully dangerous to humans" (Pitman 1974). The latter is likely the case simply because humans have little contact with this highly arboreal snake of dense forest habitats.

Full specifics of venom yield, toxicity and content are not known. Intravenous injections of venom samples are reported to be very toxic to mice (Pitman 1974; Spawls and Branch 1995). The i.v. LD_{50} is 11 µg for mice weighing 16–18 g (Christensen 1967a). Phospholipases are present in the venom. It is only slightly neutralized by polyvalent antivenin, barely detectably by trivalent *Dendroaspis* antiserum, and fairly well neutralized by *Echis carinatus* monovalent antivenin (Spawls and Branch 1995). Several bite case histories are available, one of which resulted in the death of the victim. In one case, a "well-built, 14-stone" caucasian male was bitten with one fang penetrating the left-hand little finger by a 60-cm individual. There was immediate sharp pain. Further effects were "obscured by the astonishing variety—mostly unnecessary—of drastic remedies applied which resulted in the victim's temporary transfer to a hospital." Antivenins were applied and it was concluded that *Echis carinatus* antivenin was more effective than polyvalent. This victim recovered with no permanent aftereffects (Pitman 1974).

Knoepffler (1965) described a bite which he experienced himself. His forearm was bitten by a captive animal identified as "probably" *A. squamigera*. The snake was small, about 30 cm. There was severe pain, swelling of the forearm, dizziness, chills, severe nausea, regional swelling of lymph nodes, interference with vision and pain on breathing. His rectal temperature peaked after 12 hours at 38.5°C. No specific therapy was employed, aspirin was taken for pain and recovery was complete.

Lanoie and Branch (1991) documented a fatal bite. The victim, a "robust" 37-year-old male, was bitten on the right shin by a large 71-cm specimen while walking in a forest. After 10 h, there was massive swelling of the right leg and thigh. The victim's blood refused to clot. On day 2 postbite, 40 ml of polyvalent antiserum was administered intravenously. No further antivenin was given after it was discovered that the snake was *A. squamigera*. Symptoms were treated with intravenous saline, antacids, acetaminophen, vitamin K and prescribed rest. On day 3, blood was still not clotting, and on day 5, the victim began vomiting blood. Matching donors were located, and whole blood was administered after the patient became hypotensive and went into shock. Coagulation did not improve, neither did the hypotension. The patient died on day 6.

Remarks

The green bush viper is preyed upon by snake-eating file snakes, *Mehelya* spp., and the forest cobra, *Naja melanoleuca*. Recorded parasites include *ticks* (*Aponomma falsolaeve*) and trematodes (*Mesocoelium monodi*) (Pitman 1974).

In captivity, *A. squamigera* is generally peaceful and responsive. It is said to be particularly well suited to vivarium life, having no special temperature or space requirements, and it readily feeds on mice and young sparrows. It needs only something to climb on and a hiding place (Vogel 1964).

Atheris subocularis
Fischer 1888

The name southwest Cameroon bush viper has been suggested.

Recognition

The scientific name refers to the head scalation—where the eye is in contact with the supralabials. In the related *A. squamigera* the eye is separated by suboculars (males) or suboculars and a variable number of interoculabials (females).

Head—Large, broad, flat and distinct from neck, mouth with big gape; head thickly covered with keeled imbricate scales; 6–7 interorbital scales; 8–10 supralabials, 8–9 sublabials with anterior 2–4 touching chin shields; 2 scales between eye and nasal; rostral invisible from above; small scale above rostral flanked by very large scale on either side; nostrils lateral; 11–14 circumorbitals; the fourth and/or fifth supralabial scale is in contact with the orbit, or separated by greatly reduced suboculars; gular scales keeled; pupil vertically elliptical.

Body—Dorsal scale rows at midbody 14–16, fewer (12–13) posteriorly; ventrals 152–163, subcaudals undivided 58–65 (males);

Size—Total length 491 mm (males).

Pattern—The head and body are greenish olive or yellowish olive-green above and the top of the head is marked with an incomplete black chevron or several black spots. The interstitial skin is black, olive-green or citrine beneath the crossbands and black interstitial skin is clearly visible around the margins of the ventral scales. The body has 30 or more faint olive-yellow crossbands, which become even less distinct posteriorly and on the tail, and these crossbands are often bordered anteriorly by a darker band. The ventral surfaces are a uniform dull lime-green that is sparsely smudged with black and are slightly darker towards the tail. The eye is yellowish-green and the tongue is red with a black tip (Lawson et al. 2001).

Taxonomy and Distribution

Lawson et al. (2001) resurrected this species from synonymy with *Atheris squamigera* based on specimens from southwestern Cameroon and subsequent DNA and character analysis.

Known only from the Southwest Province of Cameroon west of the Cameroon Highlands. Occasionally it can be found in extreme eastern Nigeria.

Habitat

It is a forest species. Specimens have been observed at an elevation below 300 m along the western base of the Cameroon Highlands. According to Lawson et al. (2001) the species may be more common at moderate elevations in the Bakossi Mountains and Rumbi Hills.

Food and Feeding

No information is available.

Behavior

No information is available.

Reproduction and Development

No information is available.

Bite and Venom

No information is available.

Chapter 5
Bitis

Bitis contains snakes commonly known as puff adders or African adders. This genus includes the largest and smallest vipers in the world, and many species are small to moderate in size. Members of the genus are characterized by their behavior of inflating and deflating their bodies in a loud hissing or puffing threat display. All species are terrestrial and stocky. They are sluggish but can strike with great speed. All are dangerous, some extremely so. The entire genus is viviparous and some species produce large numbers of offspring.

Recognition

Head—Large, flattened, triangular, broader than neck; canthus distinct; without enlarged plates on crown, instead covered with small, keeled, imbricate scales; snout well rounded, with some species having enlarged erect scales; eye relatively small with vertically elliptical pupil, separated from supralabials by 2–5 rows of small keeled scales; some species with enlarged erect scales above eye; nostrils large, directed upward and/or outward; rostral separated from nasal by 0–6 rows of small scales; skull with well developed supranasal sac; maxillary much shortened anteriorly, bearing only 1 pair of recurved fangs, which in *Bitis gabonica* are among the largest in the world; palatines and pterygoid with small recurved teeth; anterior mandibular teeth large, decreasing in size posteriorly; mandible without splenial element; postorbital bone very large, in contact with the ectopterygoid, which has an outer, hook-shaped flange (FitzSimons 1980; Broadley 1990; U.S. Navy 1991).

Body—Somewhat depressed, moderately to extremely stout; scales all keeled and imbricate, with apical pits; dorsal scales 21–46 at midbody; lateral scales may be nonoblique to slightly oblique; ventral scales large, rounded, or with faint lateral keels, as broad as body; ventrals 112–153; subcaudals 16–37, paired, laterally keeled in some species; anal plate entire; tail short (U.S. Navy 1991).

Taxonomy and Distribution

The genus *Bitis* was established from *Vipera* (*Echidna*) *arietans* Merrem, 1820 (=*Bitis arietans* (Merrem 1820)) by Gray 1842. The genus is distributed throughout Africa extending into Saudi Arabia. Currently 16 species are recognized: *B. arietans* from the southwestern section of the Arabian peninsula (southwestern Saudi Arabia and Yemen) and most of Africa except Egypt, Libya and Tunisia; *B. atropos, B. caudalis, B. inornata,* all from South Africa; *B. cornuta* from Namibia; *B. gabonica* and *B. nasicornis,* both from western, eastern and central Africa;

B. heraldica from Angola; *B. parviocula* from southwestern Ethiopia; *B. peringueyi* from Angola and Namibia; *B. schneideri* from the Orange River and South Africa; *B. worthingtoni* from Kenya; *B. albanica, B. armata, B. rubida* and *B. xeropaga* from localized areas in southern Africa. In addition the two subspecies *Bitis g. gabonica* and *B. g. rhinoceros* are separated on species level by Lenk et al. (2001).

The genus is divided into subgenera *Bitis, Macrocerastes, Calechidna* and *Keniabitis* by Lenk et al. (1999) based on molecular studies.

Subgenus *Bitis*
Gray 1842

Bitis arietans is the single species in the subgenus. The subgenus is characterized by nasal separated from first supralabial by one to three scales, from rostral by one to two scales, no hornlike scales on head, and a different cytochrome b (Lenk et al. 1999).

Bitis arietans
(Merrem 1820)

The puff adder is one of the most well-known vipers.

Recognition
(Plate 5.1)

Head—Subtriangular; snout blunt and rounded; rostral small; circumorbital ring with 10–16 scales; interoculars 7–11; 3–4 scales between the suborbital and supralabials which number 12–17; sublabials 13–19 of which the first 3–4 (rarely 5) contact chin shields.

Body—Dorsal scales 29–41 rows at midbody, strongly keeled except for outermost row on either side, ventrals 123–147; subcaudals 14–38, with females not exceeding 24; anal plate entire.

Size—One of the largest vipers, averaging approximately 1 m in total length. Characteristically very stout. Large specimens, over 190 cm, weighing more than 6.0 kg with a girth of 40 cm are reported. Saudi Arabian individuals are typically smaller reaching 80 cm total length. Males are typically longer than females, readily distinguished by proportionally longer tails.

Pattern—Colors and patterns vary geographically. Head with 2 dark bands, 1 on crown, the other between eyes; 2 oblique bars from eye to lip; head ventrally yellowish white with scattered black blotches; eye varies from gold to silvery gray; dorsal body coloration from straw yellow, light brown to orange or reddish brown;

a backwardly directed pattern of chevron-shaped dark brown to black bars or bands extends down back and tail; some populations heavily speckled with brown or black, often obscuring other coloration, giving snake a dusty brown or blackish effect; neonates with golden head markings with pinkish to reddish ventral plates toward the lateral edges (Sweeney 1961; FitzSimons 1980; Mehrtens 1987; Gasperetti 1988; Broadley 1990; Spawls and Branch 1995).

Branch and Farrell (1988) described an unusual puff adder from Summer Pride, East London, in South Africa, that was striped. The marking consisted of a narrow (1 scale wide) pale yellow stripe extending from head crown to tail tip.

Taxonomy and Distribution

Described by Laurenti as *Cobra lachesis* in 1768 and as *Vipera arietans* by Merrem 1820. Type locality is given as the Cape of Good Hope. The name *lachesis* was officially suppressed by ICZN 1945, Opinion 188, by use of the Plenary Powers and *Bitis arietans* was designated as the type species for the genus *Bitis*. In addition, *Bitis arietans*, proposed by Boulenger in 1896, was added to the official list of Species Names in Zoology by ICZN, 1954, Direction 1, Opinions Decls. 3(30): 401–416. As a result no other names can be used for *Bitis arietans* (Peters 1967; McDarmid et al. 1999).

Puff adders are probably the most common and widespread snake in Africa. They are distributed throughout Africa, from southern Morocco east to Saudi Arabia and Yemen and south to Cape Province (Joger 1984; Mehrtens 1987; Phelps 1989; Broadley 1990).

Two subspecies are recognized:
B. arietans arietans and
B. arietans somalica.

The former is the most widespread. The latter, the Somali puff adder, is restricted to the Somali Republic and northern Kenya. This subspecies is distinguished by keeled subcaudals (Parker 1949; Mehrtens 1987; Branch 1992).

Habitat

Adequate cover, presence of permanent water and suitable temperature are the most important environmental factors controlling distribution (Visser and Chapman 1978). Puff adders are commonly found in semiarid savanna habitats and rocky grassland. Rain forests and extreme deserts are avoided. Agricultural and residential areas are frequented due to the abundance of rats and chickens. In human environs they seek undisturbed overgrown sites. Instances of puff adders entering huts and storage areas are reported. Often lie on roads at night to absorb heat during cooler weather. Puff adders are found from sea level to 3000 m (Pitman 1938; Mehrtens 1987; Cloudsley-Thompson 1988a; Broadley 1990; U.S. Navy 1991).

Generally terrestrial but occasionally found in water where they swim easily. In water, they rest periodically with the anterior two-thirds of the body at the surface, the remainder submerged, hanging at a 90° angle. They have been observed submerging for up to 10 minutes (Sweeney 1961). They may even swim in swift-flowing rivers. Mackay (1980) observed a puff adder make a "deliberate" crossing of the Galana River in Kenya which had a flow rate of 0.57 m/sec. The author followed an adder track from a grassy slope to the river. After crossing 6 m of sand and mud the track disappeared at the river's edge. It reappeared again on a small mud island 25 m downstream. The track continued across 5 m of mud before disappearing again at the water's edge. Water is apparently not a barrier to puff adder movement.

Food and Feeding

Although rodents (*Rattus, Mastomys, Rhabdomys, Arvicanthis*) are preferred, *B. arietans* are opportunistic feeders, taking birds, lizards, toads and frogs if mammalian prey are unavailable. Other prey include button quail, duiker deer (Phelps 1989) and a small leopard tortoise (Branch 1977). Young readily take small toads and orthopterous insects (cockroaches, grasshoppers, crickets) (Akester 1989).

Puff adders are sit-and-wait predators, often lying camouflaged along rodent runs. Birds, especially nightjars (Caprimulgidae) frequenting open roads in search of insects, often become prey (Cansdale 1961; Sweeney 1961). In captivity they will take previously killed food items (Broadley 1990).

Puff adders have long (12–18 mm) fangs, which are rotated via the pterygoid rather than the quadrate (Boltt and Ewer 1964; Broadley 1990). Following envenomation, *B. arietans* makes a "leisurely" advance, carefully inspecting its prey with its tongue. Prey are usually swallowed headfirst, but larger specimens take prey head or tail first. The whole process from strike through swallowing takes about a half hour for a large rat (Broadley 1990). Frogs and toads may be swallowed alive without venom injection. Instances of live intact prey being removed from the stomach soon after being swallowed have been reported (Sweeney 1961).

Ophiophagy has been observed involving a puff adder that died swallowing a 50-cm vine snake (*Thelotornis capensis oatesi*). Although the vine snake was packed into the puff adder's digestive tract in tight horizontal S-bends, an additional 40 cm of the body extended from the mouth. Food was not found in the vine snake's mouth so prey dispute was not the initiating factor (Broadley 1974a).

During winter, puff adders hibernate and do not feed. During warm months, they feed ravenously to build fat reserves to last through hibernation (Pienaar 1978; Broadley 1990). Death of a young puff adder from gluttony is reported by Haagner (1988). The young snake ate three mice weighing 13.8 g total. Postmortem weight of the snake alone was 14.2 g. Suffocation apparently resulted from the snake eating 97% of its own body weight. Captive-raised specimens sometimes refuse food for intervals ranging from 38–166 days (Jacobsen 1986a).

Behavior

Puff adders are generally sluggish, relying on cryptic coloration to avoid notice. Movement is primarily rectilinear through ventral scale action aided by heavy body weight that provides broad contact with the ground. On hot surfaces, they may wriggle violently from side to side to escape the discomfort. However, this movement does not increase forward progress. When agitated, *B. arietans* resorts to typical serpentine movement for acceleration (Sweeney 1961; Broadley 1990).

Although primarily ground dwelling, they may climb into bushes or low shrubs to bask, particularly gravid females (Broadley 1990). One specimen was found 4.6 m up a highly branched tree (Sweeney 1961). Puff adders have been found in recesses of stone walls 0.5 m or more above ground (Visser and Chapman 1978), and in the roofs of huts (Pitman 1938).

Puff adders are active primarily at night, when they forage (Broadley and Cock 1975; Broadley 1990). Sweeney (1961) found them active during rainy days, and Branch (1977) reported them basking during daylight.

Hibernation may be interrupted with emergence of individuals to bask in the midday sun or in individuals with insufficient body fat needing to find food (Broadley 1990). During the dry season, *B. arietans* estivates up to 3 months (Sweeney 1961).

The strike occurs at very high velocity. They can strike forward or sideways and quickly return to a defensive position (Pienaar 1978). During the strike fangs can extend past 90° and penetrate deeply, often causing death of a prey item from physical trauma alone. Broadley and Cock (1975) reported fangs can penetrate soft leather. Striking distance is approximating one-third the body length. However, young individuals launch their entire body forward when striking (Turner 1972). *Bitis arietans* seldom grips its victim, releasing quickly to resume the "ready position" should another strike be necessary (Broadley and Cock 1975; Broadley 1990).

Reproduction and Development

During the breeding season combat occurs between males. One male moves over the body of another while flicking his tongue. When the male reaches about one-third up the other's body, the lower male raises his head and starts pushing against the aggressor. The snakes intertwine their necks and push against each other. They move while intertwined, switching top and bottom positions. Hissing sometimes accompanies combat (Thomas 1972; Darlington 1983; Haagner 1990a).

Courtship occurs from October to December. It is initiated by a male moving in a "jerky" motion over a female while flicking his tongue. As his head approaches hers, he curls his tail around hers. A receptive female remains still while he passes his tail underneath, to insert one of his hemipenes. If nonreceptive, she may move away or flatten her body to prevent the male tail from passing under. Successful copulation may depend on voluntary opening of the cloaca by the female (Gillingham et al. 1977; Broadley 1990; Haagner 1990a).

An early October courtship of two puff adders on an asphalt road in Naivasha, Kenya, involved a male attempting to crawl on top of an outstretched female that was trying to find cover along the roadside. Bending his body in a zigzag along her length, he tried to coil his neck around hers, which was raised off the ground. The female moved away slowly with the male pursuing. When the male eventually crawled on top, the female began butting him with the side of her head trying to knock him off. The entire series of events was repeated numerous times before they disappeared into the brush. At no time did the female indicate she might bite the male (Clapp 1977).

A pair of copulating puff adders was reportedly so engrossed that the two did not attempt to part, even when observed at close range. The pair was killed and the copulatory position maintained in preservation (Maurer 1975).

During gestation females typically do not feed, absorbing fat from the abdominal fat bodies. They tend to seek warmth, spending considerable time basking. Gravid females become increasingly aggressive (Haagner 1990a).

Puff adders are viviparous, with young sometimes freeing themselves from the retaining membranes while inside the mother. This has yielded to the myth that young eat their way out of their mother's body (Pienaar 1978; Broadley 1990). Gestation requires approximately 5–6 months. Immediately prior to each birth, a large swelling can be seen moving toward the vent, displacing the tail. During birth, the female's entire posterior undergoes muscular contractions. Delivery of 13 young in 3 hours is reported by McDonald (1962).

Young are born during summer and early autumn (December to April) (Broadley 1990). Litter size is typically 20–40, varying with the size of the mother (Boycott 1978). Exceptional litters of 147 from Kenya and 156 from a Czechoslovakian zoo have been recorded (Spawls 1979, Branch 1977, respectively).

Newborns typically measure 15–25 cm total length and weigh 8–16 g (Haagner 1990a). Neonates are active immediately and are often vigorously aggressive, exhibiting considerable puffing and hissing. Yawning, involving flexing of the maxilla from side to side, is often observed during the first few minutes after birth. Newborns do not feed until after the first molt, which is usually within 2–12 h after birth, although it may occur within the first 10–12 minutes. Individuals will kill small mice almost immediately. When prey is unavailable, baby puff adders can live for up to 3 months without eating. In captivity neonates should be separated into individual cages with a water dish and hide box. Feeding one mouse per week the first year results in very rapid growth under these conditions (Haagner 1990a).

Coulson and Riddell (1988) provided a detailed growth record of a young (1–4-week-old) puff adder captured in Zimbabwe. During 301 days in captivity, the snake increased from 8 g and 16 cm to 170 g and 55 cm. Thirty-seven meals were eaten totaling 391 g, consisting of 23 amphibians, one lizard, four shrews and nine rodents. The largest meal was a 19 g rodent eaten when the snake weighed 31 g. After approximately 4 months a change in diet occurred with rodents becoming a definite preference. Feeding intervals were irregular, averaging 10 days, with 29 days the longest time between meals. Mean daily food consumption, mass increase, and length increase were 1.30 g, 0.54 g, and 1.3 mm respectively. Molting occurred on four occasions; 14 April (76 days after capture) at 27 g; 31 July at 55 g; 10 September at 111 g; and 5 November at 211 g.

A detailed 32-month study of growth in four juvenile puff adders was undertaken by Jacobsen (1986a). Snakes were fed mice weekly (mean food weight 59 g per meal). Average daily consumption was between 2.48 g and 3.18 g, resulting in growth rates 0.49–0.66 mm/day respectively. Every 100 g of food consumed yielded 18.4 mm average growth. Growth slows after sexual maturity due to energy demands of gonadal maturation and other costs associated with reproduction. Sexual maturity is achieved, on average, at 45 months (range 37–53 months). After 7 years, growth continues at a slower rate as the animal ages.

Bite and Venom

Epidemiology—The wide distribution, common occurrence, and potent venom of *Bitis arietans* makes this species responsible for more fatalities than any other African snake (Minton 1967b; Broadley 1968; U.S. Navy 1991). Puff adder bites cause over 60% of serious bites in southern Africa and 75% of all serious snakebites in Zimbabwe and Rhodesia (Broadley and Cock 1975, 1993). In a 7-year period in Natal, South Africa, 210 of 219 (96%) bites were confirmed *B. arietans* bites. Of these, 67 were considered serious and resulted in 11 deaths (Chapman 1967).

Puff adders are easily trod upon because of their cryptic coloration. Consequently, most bites occur at ankle height or below. In addition, their reliance on camouflage makes them reluctant to flee potential threats and they will often stand their ground rather than retreat. This habit results in the deaths of many grazing animals. Curiously, members of the pig family (Suidae) and guinea pigs (Caviidae) have high resistance to *B. arietans* venom. This may be due in part to relatively large insulating layers of fat below the skin in these animals (Broadley 1990).

Yield—Venom discharge is typically between 100–350 mg (Delpierre et al. 1971, Kochva 1978, Broadley 1990). An average venom yield of 71 mg (165 mg maximum) taken from 17 milked puff adders was recorded by Freyvogel (1965). Cloudsley-Thompson (1988a) reported a maximum venom yield of 750 mg.

Puff adders often have paired fangs on each maxilla which for a time are both functional (Lake and Trevor-Jones 1987). Lake et al. (1988) provide a detailed histological description of the venom apparatus.

Toxicity—Puff adder venom is one of the most toxic of any viper. LD_{50}s in mice vary: 0.35 mg/kg, to 0.42–2.0 mg/kg, 0.5–1.1 μg (Freyvogel 1965; Minton 1974; Mebs 1978); 1.0–7.75 mg/kg s.c. (Freyvogel 1965; Minton 1974; Mebs 1978); and 0.50–3.68 mg/kg i.p.(Minton 1974; Mebs 1978).

LD_{99} values and average time to death for mice and dogs is reported as 0.99 mg/kg, 154 minutes and 0.5 mg/kg, 480 minutes, respectively (Vick et al. 1966). Death in i.v.-injected mice occurs in less than 2 minutes (Christensen 1967b). Injection of 100 mg is usually sufficient to kill a healthy adult human male, with death taking 24 hours or more (Minton 1967a). An average size puff adder contains enough venom to kill 4–5 men (U.S. Navy 1991).

Content—Electrophoretic studies of I^{125}-labelled *B. arietans* venom showed at least 10 protein zones (Gumaa et al. 1974). Gelfiltration chromatography yielded four separate specific amino acid esterase activities (Delpierre et al. 1971) Copper, sodium, and trace elements Ca, Mg, P, Fe, Al, Cu and Ag are present (Boquet 1967a; Christensen 1967a,b; Minton 1974).

Three major and two minor proteases have been confirmed (van der Walt and Joubert 1971). Protease A is a metalloprotease with proteolytic activity (Tu and Toom 1968; Strydom et al. 1986). The venom hydrolyzed 24 of 29 dipeptides and 6 of 6 tripeptides. High proteolytic activity without marked specificity for peptide bond hydrolysis was described by van der Walt and Joubert (1972a, b) and Steyn and Delpierre (1973). In addition to proteases, Christensen (1967a, b) reported other enzymes including L-amino acid oxidase, phosphodiesterase, phospholipase A, hyaluronidase, ATPase, 5'-nucleotidase and cholinesterase. *Bitis arietans* venom rapidly hydrolyzed serum phospholipids (Condrea et al. 1962).

Successive milkings alters venom composition. A 63% decrease in total protein concentration was found after 4 consecutive days of milkings. Differences in composition among individuals disappears following consecutive milkings, and a constant electrophoretic pattern is obtained. This latent uniformity may result from early milkings yielding accumulated stored venom which is altered during storage. Consecutive milkings extracts fresh venom from the glands, yielding greater uniformity (Willemse, Hattingh, Karlsson et al. 1979).

Symptoms and Physiological Effects—In rats, radioactively tagged venom is distributed largely to the thyroid and kidneys (Gumaa et al. 1974).

In dogs, puff adder venom produces a transient fall in arterial blood pressure and heart rate, with decreases progressing over the 15–30-minute interval preceding death (Vick et al. 1966). Sporadic irregular body movements followed by complete respiratory cessation were observed approximately 15 minutes postinjection. Profound bradycardia and EKG changes which rapidly progress to cardiac arrest occur shortly after envenomation. In addition, cortical electrical activity decreased sharply at approximately 3–5 minutes.

Five-mg doses of venom cause severe shock and rapid death in adult baboons (*Papio ursinus*). The most prominent effects are hypofibrinogenemia and transient thrombocytopenia. These effects are dose-dependent. All test animals receiving 5-mg venom doses showed decreases in plasma fibrinogen and dramatic thrombocytopenia within 5 minutes, lasting up to 24 hours. At a 50-mg dose, fibrinolysis occurs and blood becomes incoagulable. Death is due to extensive hemorrhage, particularly of gastrointestinal mucosa (Brink and Steytler 1974).

Puff adder envenomation in humans produces severe local and systemic symptoms. Bites can be divided into two symptomatic categories based on the degree and type of local effect: bites with little or no surface extravasation and bites with hemorrhages evident as ecchymosis, bleeding and swelling. Bites in the second category cause widespread superficial or deep necrosis. Severe pain and tenderness occurs in both bite categories. In serious cases, limbs become immovably flexed indicating significant hemorrhage or coagulation in muscles. However, residual induration is rare and these areas usually completely resolve (Chapman 1967).

Other bite symptoms occurring in humans are as follows: shock; pain which rapidly becomes intense; watery blood oozing from fang punctures; nausea and vomiting; red to purple discoloration around the wound, which changes to blue-black; pain and enlargement of regional lymph nodes and profound swelling which is hot and painful to the touch. After several days, swelling usually subsides except for the area immediately around the wound. Large blisters may also appear in the area of the bite (Broadley and Cock 1975).

At the cellular level, *B. arietans* venom causes mitochondrial swelling and strong inhibition of mitochondrial respiration (Taub and Elliott 1964). Concurrent with mitochondrial effects are biochemical changes associated with blood glucose and liver and muscle glycogen. Lethal and sublethal venom doses caused blood glucose levels to rise 1–2 hours postinjection, with a steady decline to initial levels after 6 hours. A steady decline in muscle and liver glycogen occurred with lethal and sublethal doses, the former having a more pronounced effect (Mohamed, Fouad, Abbas et al. 1980). *Bitis arietans* venom lysed bovine spleen and mouse embryo cells *in vitro*. The rapid and complete degradation of cells, which may occur up to 24 hours, is the primary cause of the localized necrosis that often occurs in serious envenomations.

Bitis arietans venom exhibited the highest collagenolytic activity of all viperids tested (Kaiser and Raab 1967). However, the venom is only weakly myonecrotic (Tu et al. 1969). Necrotic skin, subcutaneous tissue and muscle separate from healthy tissue and eventually slough with serous exudate. The slough may be superficial or deep, sometimes exposing bone. Gangrene is common and may result in appendage loss. Secondary infection may further complicate healing. Despite prompt treatment, complete healing of the sloughed region may take an extended period and some victims are permanently disabled with stiffness (Broadley and Cock 1975; Buys and Buys 1980).

In human victims, hypotension with accompanying dizziness, weakness and periods of semi- or unconsciousness is reported. Bradycardia and EKG changes occur shortly after envenomation (Brown 1973). Venom can disrupt the beta-globulin enzyme system, inducing complimentary inactivation of the C3-C9 sequence (Minton 1974). If untreated, death may ensue within several days due to cerebral hemorrhage leading to convulsions, kidney failure, and complications caused by extensive swelling (Broadley and Cock 1975; Branch 1992).

Neurotoxic properties have been detected. Venom decreases action potentials in giant squid axons up to 75% (Rosenberg 1965). This effect may cause rapid death in small prey (Broadley 1990).

Hematological Effects—Puff adder venom has strong hemotoxic activity, causing hemolysis and internal bleeding from mucus membranes. The primary mode of action is extensive hemorrhage due to blood and capillary breakdown, causing suffusion of blood into tissues. Blood is sometimes found in the sputum, urine, feces and vomitus after 18 hours and small petechial hemorrhages may be observed under mucus membranes and skin.

Blood tests reveal definite disruption of the coagulation mechanism, with procoagulant and anticoagulant activity occurring. Human platelets demonstrate extreme susceptibility to puff adder venom *in vitro,* having a dose-related irreversible aggregation response. In human and rat blood, platelet count was depressed (Brink and Steytler 1974). In humans, prothrombin time and low activated partial thromboplastin time was prolonged for up to 7 days. Partial thromboplastin time was accelerated in human blood but prolonged in rat blood. Fibrin monomers were not detected and platelet count and fibrinogen levels were borderline (Phillips et al. 1973a).

In vitro procoagulant activity is found in rat blood at low venom concentrations (10 μg/ml plasma) (Phillips, Weiss and Christy 1973). Activation of the contact system (factor XI) was indicated by accelerated partial thromboplastin times. Fibrin monomers formed rapidly, and factors V and VIII gradually decreased. Prothrombin and thrombin times changed little. At high venom concentrations (50 μg/ml plasma) thrombin times increased markedly, a result of fibrinogen/fibrin degradation products from fibrinolysis. Recalcification times also were prolonged from 132 seconds (without venom) to over 1 hour (with venom). Clotting occurred at 10 and 8 minutes at venom concentrations of 500 and 1000 μg/ml respectively. Anticoagulant effects on dog and human plasma result from reduced plasma prothrombin, a direct result of lytic action on factors VII and X (Rosenfeld et al. 1967).

Coagulation defect is not due to enzymatic reactions alone. Immunological reactions between blood coagulating factors and venom also affect coagulation (Willemse, Hatting, Karlsson et al. 1979).

Case Histories—Warrell et al. (1975) reported on 10 patients bitten by puff adders in Nigeria over a 3-year period. Patients ranged in age from 7 to 55 years old. Four were bitten on the calf, two on the ankle. The remaining four were bitten on the toe, heel, thumb and hand. Fang punctures 1.5–5 cm apart were found in four cases. All reported local pain and swelling within 20 minutes. Three vomited (one from an emetic). Local blistering occurred in five cases, necrosis and spontaneous systemic bleeding in three. Two cases had nose bleeds and one bled from otherwise healthy gums for 4 hours. Other clinical symptoms and number of cases exhibiting them were ecchymoses (2), fever (5), drowsiness (4), bradycardia (1) and hypotension (1).

Laboratory investigations revealed neutrophil leucocytosis in five of six patients tested. Anemia, possibly from systemic bleeding and microangiopathic hemolysis, developed in three cases, as did thrombocytopenia. Blood was coagulable in all nine cases tested. Clotting factor analysis revealed little effect on levels of fibrinogen; factors II, V, VII, VIII, X; fibrinogen degradation products; and plasma electrolytes and urea.

No blood was found in vomitus, stools or urine. Swelling maximized in 1–2 days and resolved in 5–21 days. Antivenin was given in seven cases, in two of which (it was reported in retrospect) it was not warranted. Four patients showed signs of severe envenomation and were given 40–80 ml polyvalent antivenin (Beringwerke, *Bitis, Echis, Naja*). Another patient admitted almost 4 hours postbite exhibited hypotension and bradycardia. After receiving 80 ml of antivenin, blood pressure rose by 60 mm Hg, and pulse increased from 52 to 80 bpm.

Of the above cases, two resulted in death. One patient received 40 ml antivenin 2.5 hours postbite. Despite treatment, blood pressure fell to 85/50 mm Hg after 16 hours, with death 1 hour later. Necropsy failed to show cause of death. An 18-year-old man bitten in the calf was the second fatality. The patient was hospitalized after 3 hours with a very tender leg, swollen to the knee. No antivenin was available. After 18 hours, the patient complained of abdominal pain, was cold and sweaty, had a bp of 100/70 mm Hg, and a pulse of 132. A dextrose-saline i.v. was given which normalized blood pressure. After 25 hours, swelling extended to the groin and blisters coalesced from ankle to midcalf. Two days later, the leg was cold and numb and lacked circulation. By day 4 the patient was severely anemic and had two episodes of nosebleeds. Blood film with poikilocytes and schistocytes suggested microangiopathic hemolysis. One unit of blood was transfused. The leg became gangrenous and was amputated on day 23. Postoperative paralytic ileus developed and plasma potassium and urea concentrations rose. The patient died in ventricular fibrillation 24 days postbite.

Takahashi and Tu (1970) described a case history of a 30-year-old student bitten on the right index finger near the middle joint by a pet puff adder. Two fang marks, one slightly larger than the other, were present and bleeding. Hospital admission was within 45 minutes. One vial of crotalid polyvalent antiserum was given i.m. and s.c. After 2 hours, severe swelling from wrist to fingers occurred and another vial of antiserum was given. Edema progressed and the patient experienced painful swollen lymph nodes in the right armpit. Ten hours postbite, arm and hand were slightly cyanotic, painful and throbbing, and the shoulder was tender. Swelling extended past the elbow. Twenty-one hours postbite, 10 cc South African Institute for Medical Research (SAIMR) polyvalent antivenin for *B. arietans, B. gabonica, Naja nivea,* and *Hemachatus haemachatus* was s.c.- and i.m.-injected into the forearm. Four hours later, another 20 cm^3 of polyvalent antivenin was administered intravenously. One day following antiserum injection, swelling and pain subsided. By the day 3, edema, pain and tenderness were minimal. Four days after antivenin treatment, there was only slight swelling in the upper forearm and slight pain in the fingers.

Phillips, Weiss and Christy (1973) and Phillips, Weiss, Pessar and Christy (1973) reported on a 22-year-old male bitten on the medial bone of his right index finger by a 9-inch pet specimen. The patient immediately applied a tourniquet and incised the wound. He arrived at the hospital 2 hours after envenomation and received 30 ml of SAIMR tropical polyvalent antivenin, 0.5 ml tetanus toxoid and a rapid glucose infusion intravenously. The bitten finger was cyanotic and edema extended to the elbow. Vital signs were stable and no generalized bleeding was observed. The only systemic signs of envenomation were a 38.8°C fever and edema to the armpit with painful axillary lymphadenopathy. By day 3, all these symptoms disappeared. The patient was released after 7 relatively uneventful days.

Another case history involving a finger punctured by both fangs was described by Visser and Chapman (1978). Within 20 minutes, the finger swelled with intense pain. A ligature was applied to the base of the finger and 7 ml of antivenin was injected into the dorsum of the hand. The ligature was removed after 5 minutes. After 20 minutes, the patient was hospitalized. The entire hand was painful and throbbing with swelling reaching the elbow. Ten ml of antivenin was injected i.m. followed by a second injection 15 minutes later. By 36 hours, the elbow and armpit were discolored and swelling reached the shoulder. The fingers had large blisters and were hemorrhaging. Blood blisters were incised and allowed to drain. Saline and two pints of blood were administered. By day 4, hemorrhaging stopped. However, pain continued and swelling gradually increased, rupturing the skin and muscle at the bite site. Arterial hemorrhage ensued. By day 5, gangrene occurred at the fang punctures and the finger was amputated at the base. Incisions were made on the wrist, palm and dorsum of the hand to relieve edema. At day 7, an urticarial serum rash was reported. The patient was released on day 10. Antibiotics were given on day 27 to treat the hand, swollen by secondary infection.

Treatment and Mortality—In a clinical accounting of nine fatalities due to puff adder envenomation (Visser and Chapman 1978), extravasation leading to oligemic shock was the cause of death. For this reason, prompt fluid replacement, especially blood and plasma, has been reported to be important. Other reported treatments included surgical procedures to remove dead tissue, antibacterial therapy to combat putrefactive organisms, and aspirin and/or codeine and cold compresses to alleviate pain. Heparin is not effective in preventing thrombocytopenia or hemorrhagic death (Phillips, Weiss and Christy 1973; Brink and Steytler 1974). However, at low venom doses (1 mg or less) heparin has been reported to be beneficial in preventing platelet aggregation. It has also been used in treating mild cases where absorption into blood is slow, where no antivenin is available or in serum-sensitive patients (Brink and Steytler 1974).

It has been stated (Broadley and Cock 1975; Visser and Chapman 1978) that antivenin use is indicated only if major local or systemic signs develop. Mohamed and coworkers (Mohamed, Saleh, Ahmed and El-Maghraby 1977; Mohamed, Saleh, Ahmed, El-Maghraby and Allam 1977; Mohamed, Fawzia et al. 1977; Mohamed, Abdel-Baset et al. 1980) reported that 35 and 40 mouse LD_{50}s were neutralized by 1 ml of monovalent and polyvalent antivenin respectively. Visser and Chapman (1978) state that antivenin should not be administered around the wound site, as this may increase local pressure and tissue damage. Large muscle mass areas such as the upper outer quadrant of the buttock or deltoid muscle should be used as i.m. injection sites. Elevation and immobilization of the affected limb, cold water compresses and treatment for shock are reported to be effective (Turner 1972).

Despite numerous accounts of death due to puff adder envenomation, mortality rates are proportionally low (Broadley 1990).

Remarks

Under suitable conditions puff adders do well in captivity, living upwards of 14 years with a record longevity of 15 years, 10 months (Russell 1983). They must be provided with well ventilated housing, a hiding place, and plenty of water, warmth and sunlight. Haagner (1990a) successfully housed and bred puff adders in fiberglass cages measuring $1.0 \times 0.3 \times 3.0$ m with no provision for basking. They feed well at temperatures of 26–29°C with a slightly cooler temperature at night (Mehrtens 1987).

Murphy and Joy (1973) reported on a defective fang mechanism in a captive puff adder. Five years after raising a young specimen which fed well and grew appreciably, it started producing a yellow exudate between the labials. The cause was replacement fangs lodged within the vagina dentalis. Eight replacement fangs, a quantity of yellow exudate which had spread to the pterygoid teeth, and a considerable amount of necrotic tissue were surgically removed from the fang sheath. A small syringe inserted into the fang sheath infused the area with hydrogen peroxide, and antibiotic ointment was packed into the sheath. This temporarily remedied the problem, but 2 months later the same fang impaction occurred. Surgical removal of the entire defective fang mechanism prevented further difficulties.

Puff adders are susceptible to mouth diseases such as cankers, which rot away soft parts and make swallowing impossible (Broadley 1990). Wild-caught puff adders are often infested with worms, pentastomids, linguatilids, ticks (*Amblyomma* and *Rhipicephalus*) and mites. Newly imported specimens should be quarantined and treated for parasite infection (Gasperetti 1988).

A puff adder should not be noosed during capture. The weight of the snake can, with one convulsive jerk, dislocate its neck. The large head in relation to its neck makes picking it up by the neck easy. However, its tendency to suddenly give convulsive jerks to free its head makes it a dangerous snake to handle (Visser and Chapman 1978).

In parts of Africa, *B. arietans* are sometimes tethered to a game trail either to catch prey or to kill tribal enemies (Sweeney 1961).

Warthogs, wildcats, genets, honey badgers, foxes, mongooses and birds of prey are known predators of puff adders (Gasperetti 1988). Buzzards, eagles, the secretary bird and ground hornbill (*Bucorvus leadbeateri*) are its principal avian enemies (Pienaar 1978). Cattle egrets are reported to have mobbed a swimming puff adder (Chenaux-Repond 1974). Man is probably their chief enemy. During winter, their stored fat is sought by native herbalists as a cure for rheumatism. In addition, individuals near human habitations are killed and natural habitat is being converted to agriculture throughout the range (Broadley 1990).

Subgenus *Calechidna* Tschudi 1845

The South African small *Bitis* are members of this subgenus, which is characterized by a series of anatomical features (no anterolateral pocket in the lung; a gap present between heart and liver; angular and splenial bones united into a single, much reduced bone and lacking close approach to the dentary) (Groombridge 1980; Lenk et al. 1999).

Bitis albanica Hewitt 1937

Known as the Albany adder.

Recognition

Head — Subtriangular with a rounded snout; supraorbital region swollen with some slightly elongated scales; neck distinct and narrow. Crown with finely keeled scales.

Body — Stout, cylindrical in cross-section; dorsals keeled, 27 at midbody; tail short. Ventrals 126–130, subcaudals 21–27.

Size — Largest recorded specimen is 30.2 cm. It is a small species, and most specimens are around 24–26 cm (Branch 1999).

Pattern — Top of head light gray, with a narrow blunt-tipped arrow shape between eyes; temporal region is light gray; side of head with a dark, narrow to broad stripe running from behind the eye to the upper lip (three supralabials wide); chin and throat are light cream to yellowish with dark brown infusons on the lower labials; body ground color light gray brown dorsally, with 15–20 prominent dark brown, dorsolateral blotches which are usually paired or irregular on the latter half of the body, the blotches have occasional vestiges of white lateral and posterior borders; the belly is uniformly light gray; all dorsal scales are heavily speckled with fine dark spots (Branch 1999).

Taxonomy and Distribution

The description of this species is based on eight specimens collected in the Eastern Cape Province. The species belong to the *Bitis cornuta-Bitis inornata* complex.

The Albany adder is known only from a few localities in the Eastern Cape Province in South Africa. It has an altitudinal range from 50 m to 500 m (Branch 1999).

Habitat

The eight specimens known were found in dry habitats like xeric succulent thicket and open mosaic of grassland and thicket underlain by limestone (Branch 1999).

Food and Feeding

Stomach contents have been cycloid skink (*Mabuya*) scales and mammal hair.

Behavior

No information is available.

Reproduction and Development

No information is available.

Bite and Venom

Nothing is known about toxicity or composition of the venom. It is assumed that in nature and severity envenomation will be similar to that of other dwarf *Bitis* (except *B. atropos*).

Bitis armata
Smith 1826

Known as the southern adder.

Recognition

Head—Subtriangular with a rounded snout; supraorbital region swollen with a group of 6–7 elongated scales; neck distinct and narrow. Crown with finelly keeled scales.

Body—Stout, cylindrical in cross section; dorsals keeled, 27 at midbody; tail short. Ventrals 115–128, subcaudals 19–31.

Size—Largest recorded female is 41.4 cm, and largest male 37.2 cm. It is a small species (Branch 1999).

Pattern—Top of head with a dark, irregularly shaped, forward pointing arrow mark. Two dark bands are running from the eye down to supralabials 4–6 and the second from eye to supralabials 8–10; the upper temporal region is gray-brown and the throat is yellowish cream with dark brown infusons on the lower labials; body ground color is gray brown dorsally, with 22–28 paired rectangular black, dorsolateral blotches which each has a pale centre; the belly is white (Branch 1999).

Taxonomy and Distribution

Description is based on specimens from south-western Cape in South Africa, and it is a member of the *Bitis cornuta-Bitis inornata* complex.
  It has an altitudinal range below 200m (Branch 1999).

Habitat

The species occurs in low-lying coastal flats with winter rainfall. Mean daily temperature range from 12 to 22°C, with an average of 17°C (Branch 1999).

Food and Feeding

No information is available.

Behavior

No information is available.

Reproduction and Development

The female type contained seven ova in her oviducts.

Bite and Venom

Nothing is known about toxicity or composition of the venom.

Bitis atropos
(Linnaeus 1758)

Commonly known as the berg adder, Cape mountain adder
or simply the mountain adder.

Recognition
(Plate 5.2)

Head—Typically elongate but still broader than neck; scales on the back of head elongate; 11–16 interocular scales; 10–16 circumorbital scales; 1–3 (exceptionally 4) scales between subocular and supralabials, which number 9–13 (rarely 14); 10–16 sublabials, the first 3–4 in contact with the chin shields.

Body—Scales are keeled except for outermost rows; dorsal scales in 27–33 rows at midbody, ventrals are 118–144, subcaudals smooth, 15–31, males usually more than 24 and females have fewer than 25 (FitzSimons 1980; Broadley 1990; Spawls and Branch 1995).

Size—Adults average 30–40 cm, with some females exceeding 50–60 cm. Tail length into total length is 8.5–10 times in males and 10–13 times in females. Older specimens tend to become heavy, even obese, reaching a girth of 9–10 cm (Fitz-Simons 1980; Spawls and Branch 1995).

Pattern—Crown of head with an arrowhead-shaped dark blotch with two pale streaks on either side; dorsal body coloration typically dark brown with dorsolateral series of subtriangular to semicircular black markings on each side; markings edged below by white to yellowish line, below which is another series of similarly shaped but smaller, pale-edged, dark spots; between upper and lower rows of dark markings is a series of Y-shaped dark blotches; chin and throat usually pink to yellowish, spotted with black, especially toward the sides; venter dirty white with dusky infusions, or slate gray to black (FitzSimons 1980; Spawls and Branch 1995).

Taxonomy and Distribution

Bitis atropos was first described from the Cape of Good Hope as *Coluber atropos*. Its current range is known from about a half-dozen disjunct populations in eastern mountainous regions of South Africa, Lesotho and eastern Zimbabwe. Haagner and Hurter (1988) believe the distribution is continuous and much broader, including a wide distribution along the escarpment of Swaziland far south to Piet Retief. Populations are found from the Cape Peninsula eastward along the coastal mountains to the eastern Cape Province, northward along the Drakensberg escarpment to the northeastern Transvaal and in the Chimanimani Mountains and Inyanga district along the eastern border of Rhodesia and Zimbabwe (Spawls and Branch 1995).

Habitat

The berg adder occupies a diversity of habitats but prefers areas with cool weather and high rainfall. It is usually found on rocky hillsides and mountain slopes up to an altitude of about 3000 m. In Zimbabwe, it is not found below 1500 m (Broadley and Cock 1975). In the eastern Cape Province, it is found in small rock outcrops at sea level where it experiences cold wet winters and warm dry summers (FitzSimons 1980; Spawls and Branch 1995). Grassland with patches of bushes and shrubs are frequented (Broadley 1990).

Food and Feeding

Diet is varied and includes small birds, rodents (*Mus minutoides*), shrews, amphibians (Cape mountain toads, *Capensibufo tradouwi*, raucus toad, *Bufo rangeri*, rain frog, *Breviceps*), lizards and other snakes (Branch 1977; Ellis 1979; FitzSimons 1980; Broadley 1990; Spawls and Branch 1995).

Behavior

These snakes can be bad tempered if disturbed, hissing loudly, striking viciously and "throwing themselves around in a frenzy." This temperament makes them unsuitable for captivity (Ellis 1979; FitzSimons 1980; Spawls and Branch 1995).

At high altitudes, berg adders are diurnal, spending their time foraging for prey or basking on rock ledges (Spawls and Branch 1995).

Reproduction and Development

From 5–16 live young are produced in a litter (FitzSimons 1980). Young are approximately 14 cm long (Branch 1977). Haagner and Hurter (1988) collected 14 gravid females which gave birth mostly from November to December. Litter size ranged from 5–12 (mean 7.8). Neonates ranged in size from 9.4 to 12.5 cm (mean 11 cm) and 1.1–3.5 g (mean 2.1 g). A female received in mid-April gave birth to 4 young in early September. A gestation period in excess of 130 days was estimated. During its time in captivity, no males were present, suggesting sperm retention and delayed fertilization (Haagner 1990b).

Bite and Venom

Epidemiology—The remote habits of *Bitis atropos* on mountain ledges at altitudes above 1300 m makes encounters rare and bite incidence low. Only one bite out of 1067 reported in Natal between 1957 and 1963 was from a berg adder. Eighty-five percent of known Berg adder bites occurred on feet and ankles (Ellis 1979).

Yield—A total of 536 mg dry venom weight was obtained from 20 extractions, with a mean yield of 26.8 mg. All specimens were adults, ranging from 38 to 55 cm in length, with an average fang length of 7 mm (Rivers and Koenig 1981).

Toxicity—No specific data regarding LD_{50} values are available. Spawls and Branch (1995) reported toxicity levels to be equivalent to horned vipers, *Cerastes*.

Content—*Bitis atropos* venom is unusual, containing a neurotoxic fraction, possibly evolved as a defensive mechanism (Broadley and Cock 1975; Ellis 1979; FitzSimons 1980). Although not fully characterized, it is heat stable, nondialyzable and contains phospholipase and an L-amino acid oxidase.

Protein content was spectrophotometrically determined to be 68.5 mg/ml, with a protein nitrogen content of 10.9 mg/ml. Electrophoresis revealed five distinct protein bands, four of which were negatively charged (Rivers and Koenig 1981).

Symptoms and Physiological Effects—Initial local symptoms include rapidly spreading swelling and pain (Spawls and Branch 1995). Neurologic disturbance may not always occur (Paget and Cock 1979), but if present it occurs in approximately 1 hour (Ellis 1979) causing eye muscle paralysis (ophthalmoplegia), severe bilateral ptosis and widely fixed dilated pupils. Nausea, vomiting, vertigo and impairment of vision, taste and smell occur (Spawls and Branch 1995). Neurotoxic effects in mice are indistinguishable from effects associated with elapid venoms (Christensen 1967b). Tissue damage is often absent, never severe and may result from cytotoxic action (Broadley and Cock 1975; Ellis 1979). Hematological disturbance is minimal with blood changes and hemorrhage often absent.

Case Histories—Hurwitz and Hull (1971) reported on a 16-year-old boy bitten on the left thumb by a 20-cm berg adder he was carrying in his pocket. Both fangs penetrated the finger. Within 30 minutes, the boy was weak, the wounds were discolored, and the hand and forearm were tender and swollen. After 45 minutes, vision was blurred and eyelids felt heavy. One hour postbite there was nausea, vomiting, impaired balance, generalized muscular weakness and paresthesia (numbness) of the tongue and mouth. Eyes were heavy, causing difficulty in movement and focusing. Between 2 and 3 hours postbite, the senses of smell and taste were lost, nausea, vomiting and general weakness continued. The eyes were incapable of opening, moving or focusing. After 12 hours, hearing returned to normal but the patient was still drowsy. The patient was admitted to the hospital 24 hours after being bitten. At this time the entire left hand and forearm were swollen and tender. There was complete bilateral ptosis and ophthalmoplegia, no pupil response to light and no corneal reflex. Taste remained lost, balance impaired, speech slurred and the gag response was absent. Pulse, blood pressure, temperature and hearing were normal. There was no sign of cerebellar dysfunction although cranial nerves III, IV and VI were noticeably affected. Clinical examination of blood showed prothrombin consumption time to be depressed (9%), one-stage prothrombin time slightly elevated (13.3 sec), and fibrinogen level slightly lowered (189 mg/100 ml). Platelet count, hemoglobin level, blood urea, electrolytes and assays for factors V and VIII were normal. By day 2 nausea, vomiting and muscle tenderness were abated, and by day 3, the patient was alert with decreased drowsiness. Edema in the arm was decreased. On day 4, taste and smell returned, ptosis was incomplete and ocular movement was slight and nystagmic in all directions. At day 5, ptosis was partial, corneal reflex was present but eye movement was still incomplete. By day 6, eye movement and balance were restored. The patient was discharged from the hospital on day 10, despite his pupils remaining fixed and dilated, with no response to light, and the gag reflex still absent. By day 30, pupil activity was normal but the gag reflex was still absent.

A second case history is given by Lloyd (1977) who was bitten on the ventral surface of the left index finger close to the nail by a 32-cm berg adder. Two puncture marks were clearly visible. He reported only slight bleeding which stopped relatively quickly and a burning pain in the finger after 5 minutes. After 1 hour, the finger was swollen with a numb tingling feeling. At 1.5 hours, he was unsteady on his feet and stumbling. Vision was impaired. By 4 hours, eye focusing was impossible, with double vision and heavy eyelids. Speech was slurred and a "drunken euphoric feeling" was reported. He was admitted to the hospital after 6 hours. His speech was normal, but eyelids were heavy and vision was blurred and diplopic with images vertically aligned. Polyvalent antiserum (5 cm^3) was given intergluteally along with an antihistamine. By day 2 diplopic images were side-by-side, balance was impaired and there was general muscle weakness. Swelling was restricted to the index finger which was darkly discolored. Ampicillin and tetanus toxoid were given. On day 3 swelling was reduced but the finger remained numb and tender. By day 8 the finger and vision were almost normal. However, during days 11–16, a rash developed over most of the body along with pain in the legs that made walking unaided almost impossible. This was assumed to be a reaction to antivenin. Antihistamines were given and by the evening of day 16 he was almost fully recovered.

A third case history (Visser and Chapman 1978) involved an individual of unspecified sex and age, bitten on the thumb with one fang by a 15-cm specimen. An incision was made through the fang puncture. At 15 minutes, the thumb was swollen and painful. By 30 minutes, bite site pain eased but the patient was lightheaded with impaired balance and focusing difficulty. At 1 hour 15 minutes, 3 ml of polyvalent antivenin was injected into the thumb and biceps. The patient was now staggering, cross-eyed, with blurred vision and heavy eyelids. Ten minutes later, another 20 ml of polyvalent antivenin was injected intragluteally. At this time the patient was semiconscious with all sense of taste and smell lost. By hour 2 the patient regained consciousness but the pupils remained dilated and unreactive to light. The bitten hand was swollen to the wrist. By day 2 swelling in the hand subsided but the thumb was unchanged. The left eye was closed, and both eyes were still unreactive to light. On day 4 voluntary control of eyelids was possible but focusing was still not normal. The patient was discharged from the hospital on day 7. At this time, focusing improved and the sense of taste and smell were normal. The bite site was still numb but no necrosis occurred. Full recovery occurred.

A fourth case history (Paget and Cock 1979) is atypical. A 31-year-old male was bitten on the right hand between the base of the fingers and the wrist. Although both fang punctures were evident, no bleeding occurred. After 15 minutes, there was slight swelling with mild pain but other signs were normal. Two Tavegyl tablets, and two Panadol tablets were given. An icepack was applied and the limb was elevated, however, the icepack became too uncomfortable and was removed after 1 hour. At 6.5 hours postbite, swelling reached the elbow, with the hand and fingers grossly swollen, hard and painful. No blistering, discoloration or systemic symptoms were noted. The icepack was replaced, two more Panadol were given, and 10 ml of 10% calcium gluconate was administered intravenously. Pain subsided temporarily but by 10 hours it intensified. The patient was alert and cheerful and had a good appetite. One Tavegyl, two Propon capsules and two Orenzyme were given. No ptosis or disturbance in vision, balance or smell was observed even after 13 hours. By day 2 swelling reached the shoulder but no other systemic effects appeared. Medication continued along with another i.v. of calcium gluconate given 24 hours postbite. Although pain subsided, swelling progressed to the armpit causing slight axillary adenitis. By evening of day 2, 36 hours postbite, swelling progressed, becoming tender, gross and hot, with axillary adenitis. In addition to the usual medications, two 5-mg Prednisilone tablets were given to try to control swelling. Swelling remained in check until the afternoon of day 3 when it advanced into the neck and trunk regions. Other signs were normal. In the afternoon of day 4, swelling subsided and the patient could move his fingers slightly. Swelling continued to subside and the patient was released on day 6. Medications were continued until 10 days postbite. Full recovery occurred by day 20. This case showed extensive swelling typical of adder envenomation without any indication of neurological symptoms.

Palmer (1986) described a case of berg adder bite on the outer surface of the right index finger. No immediate pain occurred, but by 10 minutes the finger was swollen and the patient was faint, pale and had difficulty focusing. SAIMR polyvalent antivenin was administered (10 cm^3 i.v. and 10 cm^3 i.m. in the right deltoid) along with 100 mg of Pernilotfan. The finger was cleaned and dressed with Safratulle. After 5 hours, the finger and hand were swollen and painful with serous fluid and blood draining from the wounds. Swelling extended midway up the forearm.

Parenzyme, Orenzyme, Mystechin and Dalorene were given for pain. By 14 hours, vision was blurred, the entire arm was very swollen and tender with axillary adenitis. The patient was unable to sleep. Phenergan (60 mg) was given and cold compresses were applied every hour throughout the night. By 24 hours, vision remained blurred and pupils were fixed and dilated. No other neurological symptoms were observed. The finger and hand remained swollen with blood blisters on the index finger. Alternate warm and cold compresses were applied to the hand, and antibiotic and Orenzyme treatment was maintained. After 2 days, pupils remained dilated and fixed, swelling stabilized, and some blood blisters opened. By day 3 swelling was subsiding and pupils began reacting to light. On day 6, pupils reacted normally, swelling in the arm subsided, and dead skin was surgically removed from the finger leaving some deeper necrotic tissue. After 2 weeks, the wound was healing, swelling was almost gone, but bright light was still uncomfortable and "cigarettes tasted like burnt porridge." After 1 year, the patient continued to experience chronic weakness, lack of stamina and depression.

Treatment and Mortality—FitzSimons (1980) stated that polyvalent antiserum is ineffective. However, Rivers and Koenig (1981) conducted immunodiffusion assays with SAIMR polyvalent antivenin and reported that five components of berg adder venom were neutralized. In general, Spawls and Branch (1995) state that antivenin is not indicated. Some authors believe that treatment should be conservative and include rest, immobilization, cold compresses, elevation, antibiotics to prevent possible cellulitis and analgesics for pain (Broadley and Cock 1975; Ellis 1979). Symptoms usually disappear after 3–4 days (Broadley 1990). Lloyd (1977) cautions medical practitioners that berg adders close to sloughing are easily mistaken for young puff adders (*B. arietans*) and that the treatment for these two species is quite different.

No fatalities have been recorded (Spawls and Branch 1995).

Remarks

These snakes often fail to feed in captivity and therefore do not survive long (Ellis 1979; Broadley 1990). However, Mehrtens (1987) found that they will accept prekilled mice of suitable size, and suggests that housing should contain thin slabs of rock (slate) and/or several clumps of dried grass or similar vegetation arranged to form crevices in lieu of a standard hide box. Temperature should be maintained at 23–26°C with a basking site several degrees higher.

Bitis caudalis
(Smith 1839)

Commonly known as the horned puff adder, or simply horned adder.
Its name is derived from the single hornlike scale over each eye.

Recognition
(Plates 5.3 and 5.4)

Head—Triangular; circumorbital ring 9–20; supraorbital ridge strongly raised anteriorly, forming a soft, 2–3-mm, supraorbital horn (absent in exceptional cases); interoculars 11–17; 2–5 scales between suborbitals and supralabial; supralabials 10–14; sublabials 10–15, the first two in contact with chin shields (Haacke 1975; Broadley 1990).

Body—Dorsal scales strongly keeled except outer row on either side, numbering 20–29 anterior rows, 25–31 midbody rows; ventrals 120–155; subcaudals 18–34, well developed and smooth or slightly keeled at distal edge in males, smaller and distinctly keeled in females; anal plate entire (Haacke 1975; Broadley 1975, 1990; Branch 1992).

Size—Small, robust, with short tail; maximum length over 50 cm, adult males average about 46 cm, females slightly larger around 51 cm; largest recorded specimen 51.5 cm from Botswana (Spawls and Branch 1995); average tail length contained in total length 8–12.4 times for males and 12.5–14.4 times for females (Broadley 1990).

Pattern—Variable, even within littermates; dorsum of head with broad V-shaped to hourglass-shaped dark mark, extending forward to posterior borders of orbits; dark marks or band present posterior to each raised supraorbital; oblique band from behind eye to jaw angle; body coloration sandy gray, buff to pinkish, reddish or dark brown; males may be vivid with blues, reds, grays and yellows; females generally sandy or reddish orange with little or no pattern; dorsally with a vertebral row of slightly elongate quadrangular dark markings with pale edges; ash gray patches present between vertebral markings; dorsolaterally with smaller, rounded to square, pale-centered, or pale-edged marks alternating with vertebral markings; dark brown to grayish brown dorsolateral spots, arranged irregularly forming dark bars in a trelliswork design; venter uniform creamy to yellowish white (Mehrtens 1987; Broadley 1990).

Taxonomy and Distribution

Described from the type species, *Vipera caudalis,* collected from Gabon. A phylogenetic relationship between *B. caudalis, B. peringueyi* and *B. cornuta* has been proposed based on similarity of movement, subsurface concealment and overall sand-dwelling adaptations. However, separation in habitat preferences among these otherwise sympatric species occurs, resulting in their distributions forming an interdigitating mosaic according to biotope with minor areas of overlap along contact zones (Robinson and Hughes 1978).

Horned puff adders are found in South Africa where they range throughout Namibia and southern Angola, southeast across the northern Cape Province, east across the Kalahari into the northern Transvaal, Botswana and southwestern Zimbabwe (Broadley 1990; U.S. Navy 1991; McDiarmid et al. 1999).

Population density appears high. While clearing vegetation for the realignment of a road in the central Kalahari, 189 snakes collected from July to August included 134 *B. caudalis* (71%). In Namibia it is probably the most common viper (Broadley 1975).

Habitat

Horned adders are a desert snake found mostly in sparsely vegetated sandy or stony arid scrub (Spawls and Branch 1995). Gravel plains, loose sandy soils, wooded alluvial soils and rocky outcrops are habitats in central Namibia (Haacke 1975; Hoffmann 1988). In the Bulawayo suburbs of Zimbabwe they occur on red clay soils (Broadley and Cock 1975). Although it appears that *Bitis caudalis* and *B. xeropaga* are sympatric in certain areas, detailed observations show that they are ecologically separated: *Bitis caudalis* occupies sandy flats at the bases of mountains or, at the most, the lower slopes thereof, and *Bitis xeropaga* lives on sparsely vegetated rocky hillsides and mountain slopes (Haacke 1975).

Food and Feeding

Bitis caudalis feeds primarily on lizards, particularly geckos and lacertids (Broadley 1990). Small rodents and amphibians are also taken (Broadley and Cock 1975). Skinks and lacertids are taken by day, mostly by caudal luring. Geckos, amphibians and small rodents are captured when the snakes forage at night (Branch 1977).

Of 134 specimens collected from the central Kalahari, 36.5% had full stomachs, with 54 recognizable prey items in 49 stomachs. Lacertids made up the majority (60%), with *Heliobolus lugubris, Pedioplanis lineoocellata, Ichnotropis squamulosa,* and *Nucras intertexta* being representative species. Geckos (26%) were represented by *Colopus wahlbergii, Ptenopus garrulus,* and *Pachydactylus capensis.* Four species of skink, all *Mabuya* spp., were present. One amphibian (*Breviceps adspersus*) and three *rodents* (*Tatera leucogaster, Mus* spp.) had also been consumed (Broadley 1975).

The lacertid, *Meroles cuneirostris,* is a favored prey (Hoffmann 1988). Envenomation usually causes death within minutes, followed by a leisurely inspection, then headfirst swallowing taking 60–90 seconds. Remains are excreted after 6–9 days. Hoffmann (1988) reported on a lacertid that was struck and held by the tail. Swallowing was attempted tailfirst but the legs jammed the snake's throat causing the lizard to be regurgitated. The lizard ran off and buried itself, showing no signs of envenomation.

Behavior

Under normal undisturbed conditions, these snakes move across sand in rectilinear or caterpillarlike manner; however, they sidewind to escape potential threats.

An individual was clocked sidewinding at a speed estimated to be 1.5 m/sec (approximately 5.5 km/hr) (Hoffmann 1988).

Horned puff adders bask in the early morning sun, but as air and surface temperatures increase they bury themselves in the sand. Burial is by lateral pendulumlike body oscillations from the tail forward and takes 5–15 seconds. During the day, they remain buried with the horns creating small sand mounds leaving the eyes exposed.

Temperature and humidity affect activity. A burst of activity was observed during crepuscular and nocturnal hours at the Namib Desert Research Station, over 3 nights in August when air and surface temperatures were high and relative humidity was low. At cooler temperatures they remained buried (Hoffmann 1988).

Predators are avoided by flattening and remaining quiet. This behavior, along with disruptive coloration, renders them highly cryptic (Branch 1977, Broadley 1990).

If severely molested, *B. caudalis* hisses loudly, forcibly inflating and deflating the body while curling into tight curves (Greene 1988; Hoffmann 1988). The head is drawn back into the coils in preparation for a strike (Mehrtens 1987; Broadley 1990). These snakes are irascible when disturbed and will strike with little provocation (Branch 1977; U.S. Navy 1991). The strike is initiated with a loud expulsion of air and may be so forceful that the entire body is lifted as the snake lunges forward. At the strike midpoint, the snake may be completely suspended 5 or more centimeters above ground. Such jumping is done only by warm, well fed snakes, and is best exhibited at body temperatures between 33 and 35°C. Smaller individuals jump more readily and effectively. After several consecutive jumps, snakes will remain in a coiled defensive position. Similar jumping behavior may occur during normal rectilinear motion, but occurs only during pauses when sidewinding. One to two jumps are generally interspersed in a series of sidewinding movements, but certain individuals engage in a series of jumps, each interspersed within 2–10 sidewinding passages. Sidewinding with jumps rapidly tires these snakes, and resting occurs frequently after movements of 15–20 m (Gans and Mendelssohn 1971).

Reproduction and Development

Bitis caudalis engages in male combat (Akester 1983). The initial stages of combat and mating are similar, differing only in the reaction between males and females. After the courting male reaches the female's anterior third, she may remain passive or try to assist in copulating. Aggressive courting males approach other males in a jerky rapid manner with the anterior third of the body raised to crawl on top. Noncourting males either flee or engage in combat. During combat, males try to force the opponent's head down while simultaneously twisting the body around in a corkscrew fashion. The courting male may strike with closed mouth while lashing its body from side to side. Occasionally, fang penetrations occur and death may result within 30 minutes to 32 hours. Males may chase after several males while courting a female. Combat may last an hour, after which the courting male settles quietly or commences courtship with the female. Once copulating, other males are usually ignored (Hoffman 1988).

Males molt prior to combat. Unlike combat in male gaboon vipers, *B. caudalis* combat is prompted by factors other than sexual rivalry (Akester 1983). Competition for food, sexual domination and territorial defense may prompt aggression. Males guard only their immediate surroundings with movement by other males within 1 m eliciting an immediate combat response. Combat is probably visually initiated, but increased tongue flicking suggests it is maintained via chemical cues similar to those triggering combat in *Vipera berus* (Andrén 1982b).

Courtship is initiated by the male, jerkily approaching the female from behind. After making contact, the male moves up the female's body continuing the same jerky motion while rubbing his chin along her back and flicking his tongue. As he approaches her head he tries to pass his tail underneath hers. If unsuccessful in eliciting a response, he reverses direction and moves back toward her tail. This sequence is repeated until the female becomes responsive, indicated by the raising and moving of her tail from side to side. Tail intertwining is a mutual behavior with both sexes taking an active role. Copulations usually occur midday and last approximately 10 minutes (Akester 1983).

Bitis caudalis mating was observed on 12 September at 1100 hrs (Jacobsen 1986b). Initially the male flicked its tongue over the female while she tried to move away. During copulation, both tail tips lashed about and the female attempted to bury her cloacal region. The male made short movements with his head, placing it across the female's body or on the sand in various directions. At 1130 hrs, she flicked her tail violently, arching it in the air at almost a 90° angle, yet remaining coupled with the male. Soon after, they shifted position with the male on top while she buried her posterior in the sand. At 1144 hrs, the pair uncoupled, and the male moved off, continuing to flick his tail.

Jacobsen (1986b) observed mating activity in a captive pair from 12 to 14 May, with 7 young born after 135 days.

Douglas (1981) introduced a female into a pit containing a male about two-thirds her size. The male immediately moved out from under a rock and circled the female. As the female moved, the male followed, flicking his tongue over her body. Copulation occurred within 10 minutes with the female showing total disinterest.

Horned adders are viviparous, producing varying sized litters during summer from December to February (Broadley 1990). Gestation period is between 98 and 135 days (Jacobsen 1986b). Litter sizes are as follows: 4–7 (Broadley 1975); 16, 27 (with 3 stillborn) (Douglas 1981); 16 (9 females, 7 males), 6 (3 females, 3 males) (Akester 1983); 11 (9 females, 2 males) (Jacobsen 1986b); 7–8 (Haagner 1987). Clutch size variability may be related to food availability (Broadley 1990). *Bitis caudalis* birth coincides with the appearance of young lizards (*Chamaeleo namaquensis, Meroles* sp., and *Rhotropus* sp.) which may be important prey for neonates.

Neonates range in size from 10 to 14 cm (Broadley 1990). Neonates molt within 2 days of birth. After 1 week, Douglas (1981) reported that average offspring length was 14.7 cm (range 14.2–15.2). Horn development begins 3–4 weeks postbirth (Douglas 1981). In captivity, an adult male was observed eating a month-old newborn (Jacobsen 1986b).

Young are difficult to rear, ignoring all food offered (Haagner 1987). Even force-fed pinky mice were regurgitated. Several members of this brood died and the remainder were released. However, Fisher (1982) had luck feeding newborns small toads and later young geckos.

Bite and Venom

Epidemiology—Epidemiological information is not available. It may be assumed that bites are rare.

Yield—Average venom yield for *B. caudalis* is 85 mg (Spawls and Branch 1995).

Toxicity—The venom is reported to be highly toxic (U.S. Navy 1991). However, Branch (1977) stated that the venom is substantially less toxic than that of *B. arietans, B. cornuta, B. atropos,* or *B. gabonica.* LD_{50}s as 20 μg for i.v.-injected 16–18 g mice (Christensen 1967b); 0.15–0.22 mg/kg for i.p.-injected mice (Lee et al. 1983). Spawls and Branch (1995) estimated that 300 mg is required to kill a human adult.

Content—A neurotoxic phospholipase A_2 was isolated by Viljoen et al. (1982). This protein, designated "caudoxin," is a 121-amino acid single-chain protein cross-linked by seven disulfide bridges.

Symptoms and Physiological Effects—Broadley and Cock (1975) reported that envenomation in humans causes swelling, severe pain, nausea, vomiting and shock. Blisters followed by necrotic ulcers may form at the puncture site.

Caudoxin is a presynaptic toxin similar to bungarotoxin but with different binding sites (Lee et al. 1983). It depresses chick nerve-muscle contraction without affecting acetylcholine sensitivity. Its general action produces triphasic changes in transmitter release from motor nerve terminals. This triphasic change consists of an initial inhibitory phase due to caudoxin binding at specific presynaptic sites. The second phase is facilitory due to calcium accumulation and the third phase is attributed to enzymatic hydrolysis of membrane phospholipids. Caudoxin also induces local myonecrosis in mice. However, in comparison with other phospholipase A_2's, its myonecrotic effects are weak, occurring only at high doses (Lee et al. 1983). Fair amounts of sodium and other trace elements such as Ca, Mg, P, Si, Fe, Al, Cu and Ag are present in the whole venom (Christensen 1967b).

Case Histories—Murphy (1974) reported on a 24-year-old male bitten on the right middle finger at 1550 h by a 30-cm female that was being fed at the Dallas Zoo. One fang glanced off the nail while the other imbedded. The victim sat down and within 30 seconds complained of intense pain at the bite site. A tourniquet was applied and mouth suction was attempted for approximately 4 minutes. During transfer to a hospital transient nausea occurred. Upon arrival at 1602 hours there was a dull ache in the finger, hand and forearm with pain extending to the armpit. Initial examination showed slightly elevated blood pressure and pulse (150/90 and 104 respectively). Respiration and temperature were normal. The fang penetration site was mildly swollen and cyanotic with ecchymosis and a tender nail bed. Tenderness and swelling occurred on the dorsum of the hand. An i.v. of 20 cm^3 antivenin was started at 1635 h followed by intravenous doses of penicillin and tetracycline. The patient vomited in the emergency room and twice in the ward. At 55 minutes postbite, white blood count was 5500, comprising 52% segmented neutrophils, 40% lymphocytes, 5% eosinophils, 2% monocytes, and 1% basophils. Several hours later, white blood cell count was 10,500 with 90% segmented neutrophils, 8% lymphocytes, 1% monocytes, and 1% basophils. After 15 hours, these values were 9200 for white blood cells, 87% segmented neutrophils, 9% lymphocytes, 2% monocytes, and 2% eosinophils. Throughout this time hemoglobin, hematocrit, platelet count and serum sodium, potassium, carbon dioxide, chloride, glucose, urea nitrogen and amylase were all normal.

After 5 hours the patient experienced extreme apprehension and pain. Swelling of the hand continued but following elevation stabilized after 12–18 hours. At 15 hours, urine analysis showed moderate bacteria present, red blood cells 0–7, and white blood cells at 10–15. A 2-cm area on the finger remained deeply cyanotic with extensive swelling for 24 hours. After 24 hours, i.v. antibiotics were discontinued and 250 mg tetracycline was given orally. In addition, tetanus toxoid prophylaxis, 250 mg Tigan for nausea and 50 mg i.m. of Demerol 3–4 times daily for pain, were given for the first 36 hours. The next morning vitals were stable. By day 3 swelling decreased. On day 4 a blister formed over the cyanotic fingernail, and by day 5 only minor pain and swelling of the fingertip was reported. Seven days postbite the patient developed a 1.5-cm necrotic area at the blister site. This area was debrided to the muscle, and after being soaked for 1 week it formed a dry black scab. At discharge, white blood cell count was 6800 and all other hematological and urinary parameters were normal. Two weeks following the bite, pain was absent but numbness was noticed around the bite.

In another case history (Blazer 1977), a 20-year-old male was bitten on the left index finger by a 28-cm horned puff adder. One fang penetrated midway between the first two knuckles and the other penetrated the second knuckle (it was later concluded that the second fang hit bone releasing little, if any, venom). The victim immediately reported a sharp pain at the bite site. After 7 minutes, swelling occurred around the bite, bleeding stopped, pain eased and the finger became numb. At 14 minutes, swelling continued through the finger with pain in the joint between the index finger and palm. By 17 minutes, pain spread to the palm. The patient arrived at the hospital 22 minutes postbite and by 37 minutes he had received an antitetanus injection. Swelling continued in the index finger and the finger-palm joint was now extremely painful. At this time the index finger was cleaned. Fifty-two minutes postbite, the hand was noticeably swollen. By 1.5 hours the finger was swollen approximately twice its normal size. At 2 hours, swelling stopped and fingers were painful but movable. The patient began feeling sleepy and slightly dizzy, possibly a reaction to antitetanus injection. By 2 hours, 37 minutes, pain spread to the wrist which was painful to the touch and difficult to move. The arm was elevated in a sling above the head which relieved pain significantly. After 3 hours, painkillers and penicillin were given orally. By 4.5 hours, the midforearm was painful and swollen. After 6.5 hours, the elbow moved freely but remained painful to the touch. Seven hours postbite, a definite pain occurred in the armpit and the patient was given penicillin and a sleeping pill. After 10 hours sleep, pulse and temperature were normal. Pain continued in the armpit but the hand and fingers had reduced pain and increased movement. The hand was removed from the sling for cleansing. This caused renewed pain and throbbing, relieved after placement back in the sling. After 20 hours, swelling in the ring, middle and pinky fingers subsided and movement was possible with minimal pain. After 2 days, the patient was released with swelling in all fingers gone except for the index finger which remained twice the normal size. Armpit pain was completely gone but the hand and forearm were still tender. Penicillin therapy continued and the hand was kept elevated. By day 5, the arm could hang without throbbing and sensitivity in the elbow, forearm, and hand subsided. After day 7, all swelling was gone except for the area immediate to the bite. Muscle spasms occurred between the index and middle fingers for an unspecified time whenever the hand did strenuous work.

Treatment and Mortality—Currently available antivenins are reported to be of limited effectiveness. Postbite healing may take several weeks (Broadley and Cock 1975). The U.S. Navy (1991) reports cases of deaths from the bite of this species.

Remarks

Little is reported regarding care in captivity. Several centimeters of fine granular sand has been recommended as a substrate. Temperatures should be cycled from 21 to 32°C daily. Small lizards, especially geckos, and nestling mice are readily accepted as food (Mehrtens 1987). Akester (1983) kept specimens in outdoor pens. These snakes did not drink from a pool of water but drank from rocks and their own bodies when pens were watered with sprinklers.

Little is reported regarding natural enemies but a horned puff adder was found in the stomach of the cape cobra, *Naja nivea* (Broadley 1975).

Similarities in color pattern and striking posture suggests possible Batesian mimicry between *B. caudalis,* the egg eater, *Dasypeltis scabra,* and the spotted beaked snake, *Dipsina multimaculata* (Greene 1988; Spawls and Branch 1995).

Chapman (1967) reported that tribes of southwestern Africa use the venom on hunting weapons.

Bitis cornuta
(Daudin 1803)
Known as the many-horned adder or western hornsman adder.

Recognition
Plates 5.5 and 5.6

Bitis

Head—Circumorbital ring with 11–19 scales; tufts of 2–7 raised horns measuring up to 4.5 mm in length above each eye; interoculars 12–17; 2–4 scales between suborbital and supralabials, which number 11–15; sublabials 11–16, the first 2–3 in contact with chin shields (Broadley 1990; Branch 1992).

Body—Dorsal body scales 23–27 rows anteriorly, 23–29 rows at midbody; dorsal scales strongly keeled, except outer row which are smooth or weakly keeled; ventrals 120–152; subcaudals 18–37; subcaudals in males smooth proximally, keeled distally, while in females they are more or less entirely keeled; anal plate entire (Broadley 1990).

Size—*Bitis cornuta* is small and stocky, reaching 30–50 cm in length, with a maximum recorded length of 75 cm in a captive specimen; females have shorter tails (tail length contained 11–15 times in total length) than males (tail length contained 8–10 times in total length).

Pattern—Head with symmetrical, dark, pale-edged marking which may form an arrowhead or hourglass; dark oblique streak on each side from behind the eye to angle of the jaw; lower labial margin usually with 3 distinct black patches, the first on either side of mental, the second midway along lower jaw, the third near angle of the jaw; body coloration grayish to brown or reddish brown; with 3–4 rows of brown to black pale-edged spots, of which the upper 2 rows are invariably larger, more angular, and arranged in transverse pairs which sometimes fuse to form large transverse subrectangular markings; between spots there may be contrasting light-colored interspaces; venter white to dirty brownish, occasionally spotted with dark brown (Broadley 1990).

Taxonomy and Distribution

Described as *Vipera cornuta* by Daudin (1803) from specimens collected from the Cape of Good Hope. Currently two subspecies are recognized:
Bitis c. cornuta (Daudin 1803), the western many-horned adder; and
Bitis c. albanica, the eastern many-horned adder.
The latter subspecies has supraorbital horns reduced or lacking, lower ventral scale counts (126–138) and a duller pattern (Branch 1992).
Bitis c. cornuta is restricted to southwestern Namibia from western Cape Province in the south through Namaqualand, north to Moeb Bay on the west coast. *Bitis c. albanica* is restricted to the eastern and southern Cape (Baard 1990; Broadley 1990; Branch 1992).

Habitat

Bitis cornuta prefers a hard substratum of rocky outcrops (Haacke 1975). Favored habitats are rocky deserts in dwarf succulent veld, and mountain slopes in heathland vegetation (Spawls and Branch 1995). Many-horned adders shelter in rock cracks and crevices, rodent burrows in rocky areas and grass clumps (Mehrtens 1987; Broadley 1990; Branch 1992).

Food and Feeding

The diet of *B. cornuta* consists primarily of lacertids, agamids and skinks, with occasional rodents, birds or amphibians (Branch 1992; Spawls and Branch 1995).

Reproduction and Development

In southern Namibia, mating occurs May–June and October–November. Young are born in late summer or early autumn. They number 7–20 and measure 13–16 cm in length (FitzSimons 1980; Branch 1992; Spawls and Branch 1995).

Behavior

Although sometimes found in loose sand, sidewinding rarely occurs. Disagreement exists regarding certain behaviors. Mehrtens (1987) states that *B. cornuta* is diurnal, becoming nocturnal only during warmer months and that it buries itself in loose sand. Broadley (1990) and Branch (1992) believe *B. cornuta* to be most active at dusk and early morning.

Bitis cornuta is "nervous in disposition" (Spawls and Branch 1995). When frightened it hisses loudly and writhes vigorously (Broadley 1990). *Bitis cornuta* is capable of striking with enough force to lift most its body off the ground (Branch 1977).

Bite and Venom

Epidemiology—No data available.

Yield—Spawls and Branch (1995) reported yields of 8–87 mg (mean 45 mg).

Toxicity—The venom is reported to be as toxic as *B. arietans* and 3 times as toxic as *Echis carinatus* venom in mice. A dry venom LD_{50} of 8 μg for i.v.-injected mice weighing 16–18 g is reported by Christensen (1967a, b). Spawls and Branch (1995) reported a wet venom LD_{50} value of 45 mg for i.v.-injected mice.

Content—Whole venom contains 31% solids (Spawls and Branch 1995). Phospholipase A is present in exceptionally high concentrations (Minton 1974).

Symptoms and Physiological Effects—Symptoms in humans are similar to *B. caudalis,* with local swelling, pain and necrosis.

Treatment—No specific antivenin exists and it has been suggested that treatment should be conservative with elevation, analgesics and antibiotics. No fatalities are known (Branch 1977, 1992; Spawls and Branch 1995).

Remarks

Branch (1992) found that these snakes feed well in captivity. Conversely, Broadley (1990) states they seldom survive in captivity and refuse to take food. Mehrtens (1987) suggests that captive care should be as for *B. atropos,* with slightly higher temperatures.

Hornless individuals from southern areas may be confused with berg adders. They can be differentiated by the raised edge over the eye, along with the different color pattern.

Bitis heraldica
(Bocage 1889)
Commonly known as the Angolan adder.

Recognition

Head—Triangular, with noticeably pointed snout; eyes and nostrils laterally positioned; rostral small, wider than high; circumorbital ring with 12–14 scales, supraorbital not hornlike; 11–13 interocular scales; 2–3 scales between suboculars and supralabials, which number 13–14; sublabials 11–12, the first 3–4 in contact with chin shields.

Body—Dorsals 27–31 at midbody, strongly keeled along back, smooth bordering belly; ventrals 124–131; anal single.

Size—To about 40 cm maximum length (Spawls and Branch 1995).

Pattern—Head dorsally marked with dark irregular blotches forming a symmetrical "heraldic" figure, forming a trident anteriorly; a prominent dark bar extends from back of eye to angle of jaw; body colored light brown with prominent blotched pattern; 26–38 round-rhomboid markings along midline, each flanked by a series of fainter, dark, pale-centered spots with adjacent triangular light spots; venter creamy white, heavily marked with gray spots becoming denser on anterior border of scales; tail with 6–8 round blotches dorsally, yellowish white, with or without gray spots ventrally (Spawls and Branch 1995).

Taxonomy and Distribution

Described as *Vipera heraldica* from specimens collected near the Calai River, Caconda, Angola. No subspecies are recognized.

The Angolan adder is known only from the high plateau of central Angola. Within this region it is found in only a few localities (Rio Calae, Caconda, Bela Vista, Bengu, Bihé, and Kalukembe) (Spawls and Branch 1995).

Habitat

Rocky mountain slopes (Spawls and Branch 1995).

Food and Feeding

Nothing is known. Assumed similar to *Bitis xeropaga* (Spawls and Branch 1995).

Behavior

No information is available.

Reproduction and Development

No information is available.

Bites and Venom

Nothing is known about toxicity or composition of the venom. No cases of envenomation are known. SAMIR polyvalent antivenin has been recommended for severe envenomation (Spawls and Branch 1995).

Remarks

Possibly confused with egg-eating snakes, *Dasypeltis* spp., because of the blotched pattern. Enlarged head shields and rounded blunt head on *Dasypeltis* are distinguishing features.

Bitis inornata (Smith 1838)

Commonly known as the plain mountain adder and hornless adder.

Recognition

Head—Flat, triangular; scalation similar to *B. cornuta,* except for its lack of supraorbital horns and prominent ridge over the eye; circumorbital ring with 13–16 scales; supraorbital region conspicuously curved and raised above head level forming a pronounced interorbital groove or depression; occasionally with a few scales on supraorbital ridge elongated, but never forming a hornlike process; interorbital scales 12–17; 2–3 scales between suborbitals and supralabials, which number 13–15; sublabials 11–14, with first 3 contacting chin shields (FitzSimons 1980; Broadley 1990; Branch 1992).

Body—27–29 distinctly keeled dorsal scales at midbody; ventrals 121–140; subcaudals 19–30, usually keeled, at least in females; anal entire (FitzSimons 1980; Broadley 1990; Branch 1992).

Size—Average length of adults is 25–40 cm with maximum length around 45 cm. Females larger than males (FitzSimons 1980; Broadley 1990; Branch 1992; Spawls and Branch 1995).

Pattern—Crown of head with large hourglass-shaped blotch; sides of head black; eyes with a pale vertical bar in front and a diagonal bar behind; chin and throat creamy white; 3–4 irregular black patches on either side of labial margin; body dorsally pale to dark reddish-brown; dorsally covered with a double, longitudinal series of pale-edged, half-moon-shaped, dark markings; laterally marked with similar but smaller series of dark spots with rounded edges pointed outward; space between dorsal and lateral markings ash gray; venter dusky gray with darker infusions concentrated along the anterior scale edges (Haacke 1975; FitzSimons 1980; Broadley 1990; Branch 1992).

Taxonomy and Distribution

Described as *Vipera inornata* from specimens collected from Sneeuberg, near Graaff-Reinet, Cape Province. No subspecies or races are recognized. A supposed relict population from Cedarberg that once was believed to be a valid subspecies (Spawls and Branch 1995), was later described as *B. rubida* (Branch 1997).

Distribution is confined to isolated areas in South Africa. It is found in southern Little Karoo from Matjesfontein in the west to Albany district in the east, extending inland to the Sneeuberg Range near Graaff-Reinet. Inland, it is found on the old escarpment (FitzSimons 1980; Broadley 1990; Branch 1992).

Habitat

Favored habitat is the grassveld of rocky mountain plateaus. These snakes shelter in grass tussocks and beneath rock slabs on mountain slopes (Branch 1992).

Food and Feeding

Feeds on lizards, primarily skinks and lacertids, but it will take rodents of appropriate size (Spawls and Branch 1995).

Behavior

Active during early morning. Hibernates in rodent burrows. *B. inornata* gives an explosive "puff" when disturbed (Spawls and Branch 1995).

Reproduction and Development

From 6 to 8 young, 12–15 cm, are born in late summer (February–March) (Spawls and Branch 1995).

Bite and Venom

Nothing is known about its venom and there have been no reported bites. No specific antivenin exists (Spawls and Branch 1995).

Remarks

Settles down in captivity and does well, but sensitive to high temperatures (Spawls and Branch 1995).

Bitis peringueyi
(Boulenger 1888)

Commonly known as Peringuey's desert adder, sidewinding adder and the Namib dwarf sand adder.

Recognition
(Plate 5.7)

Head—Flat, short, eyes positioned dorsally permitting vision directly upward, an adaptation for hunting; covered with small strongly keeled scales, the smallest located on the forehead; circumorbital ring with 10–13 scales; 6–9 interorbital scales; 2–4 scales between suborbitals and supralabials, which number 10–14; sublabials 10–13, the first 2–4 in contact with chin shields.

Body—Dorsal body scales 21–27 anteriorly, 23–31 at midbody, strongly keeled, except outer row on either side which are larger and smooth; ventrals 117–144, with males having 117–138, and females with 125–144; subcaudals 15–30 (22–30 in males, 15–25 in females) usually keeled, at least toward tip; anal plate entire; a longitudinal ridge or fold present along the ventrolateral surface (Broadley 1990).

Size—A small adder measuring from 20 to 25 cm with a maximum length of 32 cm (Haacke 1975; Broadley 1990; Branch 1992; Spawls and Branch 1995).

Pattern—Head coloration uniform, sometimes with dark spots or trident-shaped mark on crown, followed posteriorly by cross and 2 large markings on the occiput; body coloration from pale buff, chestnut brown to orange-brown, or sandy grayish yellow; 3 longitudinal series of faint, elongate, gray to dark spots; body irregularly stippled with pale and dark spots; venter generally whitish or dirty yellow, speckled with black (Broadley 1990, Branch 1992). Tail tan or black. A 3:1 tan-to-black ratio from 69 collected specimens suggests a simple two-allele dominance-recessive system (Haacke 1975; Robinson and Hughes 1978; Broadley 1990; Branch 1992).

Taxonomy and Distribution

Described as *Vipera peringueyi* by Boulenger (1888) from specimens collected east of Walvis Bay (= Walvisbaai), Namibia. No subspecies are recognized although lower ventral scale counts in northern individuals suggest the possibility of subspecies (Spawls and Branch 1995).

Peringuey's adder is restricted to dry, shifting, sandy areas of the Namib desert. It is found from Rotkuppe and the Great Namaqualand to southern Angola (Haacke 1975; Broadley 1990; Branch 1992).

In certain restricted localities it is common (Brain 1960). During one evening, nine specimens were observed at a watering hole in the Kuiseb River in Namibia. Despite this, these snakes are poorly represented in most museum collections.

Habitat

Found in fine, windblown and leeward sands of the true Namib Desert. During the day they may be collected by digging at the bases of small bushes where they lie concealed (Brain 1960).

Food and Feeding

Bitis peringueyi feeds on the sand-diving lacertid *Aporosaura* (Louw 1972). These lizards contain a high water content, up to 75% of their body weight, and are an important source of water for these desert snakes. Other prey include the sand lizard, *Meroles,* and the barking gecko, *Ptenopus* (Branch 1977; Robinson and Hughes 1978). Prey are seized and held. Individuals with black tails utilize caudal luring to catch prey (Mehrtens 1987).

Adaptations for drinking fog condensation have been experimentally demonstrated by Louw (1972). Captive specimens deprived of water for 2 months became "visibly excited" when misted. Once aware of water droplets, they immediately flattened their bodies to increase the surface area exposed to the spray. The snakes drank by moving their heads back and forth over their body surface, periodically raising their heads to assist swallowing. A 14% weight gain in a 10.5-g individual and 7% gain in a 2.9-g juvenile were measured after these individuals drank pooled water from a crease formed by pressing their rostrum against their lateral body surface (Robinson and Hughes 1978). This behavior of drinking condensing fog was also observed in the field. However, this activity is probably opportunistic, not obligatory, owing to the high moisture content of the lizard diet and the high water conserving efficiency of the kidney (Robinson and Hughes 1978). The ability to do without water is considerable (Louw 1972; Minnich 1982).

Behavior

Sidewinding is the primary mode of locomotion, regardless of surface hardness or texture. Presence of *B. peringueyi* in sandy areas can be inferred from tracks forming a series of parallel grooves with a "hook" at one end made by the head and neck. Track length, including head-hooks, is on average 90% of the total length of the snake. Rectilinear motion is used when climbing (Brain 1960; Bannister 1974; Broadley 1990; Branch 1992).

Bitis peringueyi buries itself in sand leaving only the top of the head, eyes and tip of the tail exposed. Burying is accomplished by initiating lateral pendulumlike oscillations from the tail forward. During this motion, lateral edges of the ventrals, along with a longitudinal fold, form a sharp angle which moves sand. The process takes as little as 20 seconds during which the snake appears to "sink."

Captive specimens adjust their posture when body temperatures reach 38–40°C (Robinson and Hughes 1978). The behavior involves elevating the anterior into a cobralike stance. While in this position, the rostrum elevates and turns to face the sun, shading the dorsal surface of the braincase. Occasionally the mouth opens in the direction of the prevailing breeze, increasing evaporative cooling rate.

These snakes are reported to be primarily nocturnal (Brain 1960). However, Louw (1972) stated they are active both day and night, with diurnal activity mostly on overcast misty days that are common along the coastal Namib. Robinson and Hughes (1978) observed 54 individuals over 18 months and concluded that categorizing *B. peringueyi* as diurnal, nocturnal or crepuscular is inaccurate. Season, daily weather and sand surface temperature affect activity times. For example, during the hottest summer days (December–February) activity is confined to crepuscular and nocturnal hours. During the remainder of the year specimens were commonly observed in the daytime.

Hunting time is synchronous with activity of their lizard prey, the diurnal activity of which is governed by thermal constraints (Robinson and Hughes 1978). Snakes observed diurnally were usually in stationary hunting positions on dune slipfaces or in the bases of dune grass (*Stipagrostis sabulicola*) clumps. Nocturnal temperatures most of the year are too low for effective hunting. However, active snakes and tracks were most often encountered at night, suggesting movement to new hunting sites occurred at this time.

Bitis peringueyi can climb. A specimen was observed lying among branches of a small shrub (Brain 1960).

Reproduction and Development

Robinson and Hughes (1978) provide the only account of reproduction. A gravid female, 24 cm long, weighing 13 g, was captured on 24 February 1976. On 14 March at 1800 hours the first of 4 young was observed attempting to free itself from the egg capsule. The last of the remaining 3 was born by 2000 h. Vigorous lateral head movement broke the external membrane. Young sidewind almost immediately.

Newborn length ranged from 11 to 12 cm (mean 12.1 cm), with weights of 1.6–1.7 g. Three neonates shed within an hour of birth. The fourth shed only the skin covering its head. It never fed or burrowed and died 1 month after birth. Upon shedding, juvenile color and pattern resembled the mother. Four days after birth, one newborn struck and killed a small (7.5 cm) sand-diving lizard, *Aporosaura anchietae,* and promptly swallowed it tail first. The other two newborns followed soon after, regularly killing and eating these lizards.

Three females (285 g, 275 g and 325 g) collected 17–18 March produced offspring 2–3 weeks postcapture (Robinson and Hughes 1978). Brood sizes were 7, 5 and 10 respectively (of which 9 of the latter were stillborn). Lengths of these offspring ranged from 12 to 13 cm. Eight females collected between June and September were not gravid.

Bite and Venom

Very little is known about the venom and its effects. It is considered mild and not dangerous, causing local pain and swelling in humans (Buys and Buys 1980; Branch 1992). Specific antivenins do not exist. Bites should be treated conservatively with elevation, analgesics and antibiotics (Spawls and Branch 1995).

An incident of three specimens biting themselves then dying 3 days later is discussed by Robinson and Hughes (1978). Postmortem inspection showed massive blood clots in the body cavity at the area of the bite.

Remarks

Rarely seen in collections. Captives are presumed to require several inches of fine, granular sand, and temperatures which cycle from 21 to 32°C. They will eat small lizards, especially geckos, and nestling mice (Mehrtens 1987).

Bitis rubida
Branch 1997

Known as the red adder.

Recognition

Head—Subtriangular with a rounded snout; supraorbital region swollen with some reduced elongated scales (horns); neck distinct and narrow. Crown with finely keeled scales.

Body—Stout, cylindrical in cross-section; dorsals keeled, usually 29 at midbody; tail short. Ventrals 126–142, subcaudals 22–35.

Size—It is a small species. Largest recorded female is 41.9 cm total length, largest male is 37.7 cm total length (Branch 1999).

Pattern—On top of the head a rounded black triangle which sometimes is pale-centered can be seen in some specimens. On sides of head there are dark, narrow to broad stripes running from behind the eye to the upper lip in some specimens. The species has a drab, usually reddish, coloration, but coloration is very varied. Specimens are occasionally patternless. One pattern consists of 20–30 paired dorsolateral, pale-centered blotches, that become staggered on the posterior part of the body. Each blotch is darker along the vertebral region and lightens on the flanks. The belly is gray to dirty cream with more dark gray-brown patterns along the sides of the ventrals (Branch 1999).

Taxonomy and Distribution

Bitis rubida is a member of the *Bitis cornuta-Bitis inornata* complex.

The red adder is mainly restricted to western Cape Province, South Africa. It has an altitudinal range from 302 m to 1380 m (Branch 1999).

Habitat

The species is found in mesic forest to mountain boulder habitats (Branch 1999).

Food and Feeding

The red adder feeds on lizards (skinks, agamas, lacertids and geckos) and small rodents (Branch 1999).

Behavior

A terrestrial species that lives in broken ground and that hides under shrubs and boulders, as well as in small mammal burrows. They avoid sandy substrates, and never sidewind.

Reproduction and Development

A large female gave birth to 10 young in early February. The young measured 120–138 mm (Branch 1999).

Bites and Venom

It can be assumed that in nature and severity, envenomation will be similar to that of many other dwarf *Bitis*.

Bitis schneideri
(Boettger 1886)

Commonly referred to as the Namaqua dwarf adder or spotted dwarf adder.

Recognition
(Plate 5.8)

Head—Small, more rounded than other dwarf adders; distinct from the neck; dorsal scales more strongly keeled than *B. caudalis;* eyes relatively small; circumorbital ring with 8–14 scales, supraorbital ridge small, with one scale moderately enlarged and hornlike, less developed in specimens from south of the Orange River in South Africa; interorbitals 11–14; 3 scales between suborbitals and supralabials, which number 9–13 scales; sublabials 9–15 with the first three contacting chin shields (Haacke 1975; Broadley 1990).

Body—Dorsal body scales 21–27 in males, 25–27 in females, strongly keeled except for outer row on each side which are weakly keeled anteriorly and smooth posteriorly; ventrals 104–129, lowest counts of any dwarf adder, showing marked clinal variation with low counts in south, higher counts in north; subcaudals 17–27, distinctly keeled in females, smooth proximately and keeled distally in males.

Size—The smallest species in the genus *Bitis* and possibly the smallest viperid. They measure 18–25 cm with a maximum recorded length of 28 cm (Branch 1977, 1992).

Pattern—Head irregularly speckled with pale and dark markings as in *B. caudalis* but paler and less defined; lower lip clear or weakly spotted; body dorsally gray to brownish gray due to minute specking, making it cryptic among the sand particles of its habitat (Hurrell 1981); 3 longitudinal series (vertebral and 2 dorsolateral) of subequal, squarish to rounded, dark brown to blackish, pale-centered spots. Lateral surfaces are irregularly spotted or marked with black. The venter is grayish to dirty yellowish white with spots or speckles. The tail is occasionally dark (Haacke 1975; Broadley 1990).

Taxonomy and Distribution

Described as *Vipera schneideri* from specimens collected from Angra Pequena (= Lüderitz Bay), Namibia. No races or subspecies are currently recognized (Spawls and Branch 1995).

Namaqua dwarf adders are found in the southern and transitional regions of the Namib Desert, from Lüderitz Bay south to Hondeklip Bay in Little Namaqualand (Haacke 1975; Broadley 1990; Branch 1992). Boycott (1987) collected an adult female in the duneveld, 160 km south of Hondeklip at an altitude of 65 m. This represents a considerably distant distribution record. Other specimens were collected from Oranajemund, Visagiefontein, Holgat, Port Nolloth and Brazil in Cape Province.

Habitat

Bitis schneideri frequents vegetated sands in the southern and transitional Namib Desert south of Lüderitz. It occurs in low dunes stabilized by tussocks of coarse dune grass and small xerophytic shrubs. Mehrtens (1987) stated that *B. schneideri* is restricted to dry powdery sand deserts. It occurs in small pockets where vegetated sand accumulates or bare dunes are scattered in rough broken terrain. Below the Orange River, individuals have been collected from densely vegetated dunes that form a coastal strip with inland extensions (Haacke 1975; Broadley 1990). In some areas its habitat is threatened by alluvial diamond mining (Haacke 1975).

Food and Feeding

Diet consists mainly of lizards, especially nocturnal geckos (*Palmatogecko rangei* in the north, *Pachydactylus austeni* in the south) and diurnal lacertids (*Meroles*) and skinks (*Mabuya capensis*). It uses the black-tipped tail for caudal luring (Branch 1977). The remains of a short-headed frog thought to be *Breviceps macrops* were found in a specimen from Port Lüderitz. A specimen from Daberas contained mammalian hairs, a few body scales of a small adder (presumably *B. schneideri*), and six fangs (Haacke 1975, Broadley 1990).

Behavior

Frequently sidewinds.

Primarily nocturnal but it has been observed during early morning and evening (Haacke 1975; Broadley 1990; Branch 1992; Spawls and Branch 1995).

These snakes flee when confronted (Hurrell 1981). However, if captured or cornered, they hiss, puff and strike.

Reproduction and Development

Very little is known. Broods are reported as 4–7 young (Spawls and Branch 1995). A gravid female from Port Nolloth contained 4 full-term young, 3 female and 1 male. They varied in length from 11 to 13 cm (Haacke 1975).

Bite and Venom

The venom is mildly cytotoxic with symptoms including swelling and pain (Branch 1992).

Hurrell (1981) provided an account of a Namaqua dwarf adder bite he sustained. He was bitten on the left index finger as he was releasing the snake. Within 5 minutes, a painful throbbing was felt, and by 10 minutes swelling began. After 15 minutes, serum was oozing from the puncture site, pain and swelling were pronounced, and a 4-mm area around the puncture was discolored. One hour postbite, there was intense pain and swelling of the distal half of the finger. Subcutaneous bleeding resulted in bluish discoloration around the bite. At 3 hours postbite, the distal two-thirds of the finger was swollen and painful. Allowing the hand to hang loosely resulted in a sensation described as "finger tips exploding." Twenty-four hours postbite, the entire finger was swollen and stiff, with a 5-mm hematoma at the puncture site. Hurrell reported the pain and throbbing similar to "having a finger slammed in a door." The dorsum of the hand and wrist was tender and axillary lymphadenopathy became pronounced. This condition stabilized over the next 3–4 days with pain and swelling subsiding gradually. Complete healing took 2 weeks, with no loss of function or sensation in the bitten finger. Local tissue damage was minimal. At no time were systemic effects noted.

There is no antivenin for this species and Branch (1992) claimed that antivenin is unnecessary. Hurrell (1981) recommended the following treatment: immediate suction; elevation; tetanus toxoid; 5-day course of penicillin; deroofing of the hematoma; twice-daily dressing (initially with Eusol, then Aserbine after day 7); and analgesics. There have been no reported fatalities.

Bitis xeropaga
Haacke 1975

Commonly known as the desert mountain adder.

Recognition
(Plate 5.9)

Head—Subtriangular, with swollen temporal region; fine scalation on neck; rounded to slightly blunt snout; circumorbital ring 14–18 scales; lacks supraorbital horns; interoculars 15 scales; 2 scale rows separate suborbitals from supralabials, which are 13–17; sublabials 13–16 with first 3 pairs in contact with enlarged chin shields (Broadley 1990).

Body—Dorsal body scales elongate with rounded free ends, moderately keeled; anterior and midbody dorsal scale rows equal, numbering 25–27, distinquishing *B. xeropaga* from all other *Bitis;* outer row of lateral scales about as long as wide, bluntly rounded, smooth anteriorly but increasingly keeled distally, with keels tilting slightly downward; ventrals 147–155; subcaudals 22–33, smooth; anal plate entire (Haacke 1975; Broadley 1990; Branch 1992).

Size—Desert mountain adders are typically small, with a relatively slender body. They are usually 40–50 cm with a record size of 61 cm for a female collected near Aus. The largest recorded male was 44 cm (Branch 1992; Spawls and Branch 1995).

Pattern—Head lacks dorsal pattern except for a light patch on the supraorbital ridge and dark patch posterior to it; laterally, head dark brown to charcoal gray; 3 white to pale gray triangular areas in dark lateral area of face; the first triangular area below the nostril, a second below the eye, and a long pointed third triangle passes from the back edge of eye to posterior third of mouth; head dorsum finely peppered with minute dark spots; iris and face similarly marked, camouflaging eye; tongue black with white forks; throat and chin usually unmarked; lower labial markings often similar to facial pattern.

Body coloration is ash to dark gray dorsally, with 16–34 distinct dark rectangular transverse crossbars flanked by whitish dorsolateral spots; occasionally dark vertebral bars subdivided forming out of cycle half-patterns, usually with pale triangular marks adjacent to the ventrals; lateral dark brown to blackish blotches mixed with paler ones between crossbars; tail with dark spots; venter light gray to dusky, often with fine darker gray speckles; edge of ventral scales with irregularly shaped diffused dark spots and speckles; subcaudals irregularly speckled (Haacke 1975; Broadley 1990).

Taxonomy and Distribution

Described by Haacke (1975) from specimens collected from Dreikammberg on the north bank of the Orange River, Lüderitz District, Namibia. No subspecies or races are recognized. Found in South Africa in arid mountains of the lower Orange River basin, from below the Aughrabies Falls into the Richtersveld in the Cape Province and north to Great Namaqualand and Aus in Namibia. They occur on the eastern escarpment, from Aus southward and eastward (Haacke 1975), and their range extends along the lower reaches of the Fish River Canyon in Namibia (Haacke 1975; Broadley 1990; Branch 1992; Spawls and Branch 1995).

Habitat

Occupies sparsely vegetated rocky hillsides and mountain slopes (Haacke 1975).

Food and Feeding

In captivity, mice and skinks are eaten readily. These snakes rarely hold prey after a strike (Haacke 1975). When misted, they drink from stones, sticks, glass sides of terraria and their own bodies or those of other individuals (Haacke 1975).

Behavior

Although a desert inhabitant, *Bitis xeropaga* does not sidewind or bury in sand.

Reproduction and Development

In an exceptionally large female, Haacke (1975) found 5 ova. Branch (1992) reported that 4–5 young are produced in late summer.

Bite and Venom

Haacke (1975) reported that the venom appears "exceptionally effective" on small rodents since they died much more rapidly than from a cobra bite. No specific antivenin exists. There are no records of bites (Spawls and Branch 1995).

Remarks

In captivity they prefer to rest on stones (Haacke 1975).

Subgenus *Macrocerastes* Reuss 1939

The larger taxa, *gabonica, rhinoceros, nasicornis* and *parviocula* are members of this subgenus, which is characterized by having the nasal separated from first supralabial by four scales or more, and from rostral by three to five scales. The members also have triangular heads with one to several pairs of hornlike scales on the snout tip. Lateralmost body scales in oblique rows, with downward-pointing keels (Lenk et al. 1999).

Bitis gabonica (Duméril, Bibron and Duméril 1854)

Commonly known as the gaboon viper, also called the butterfly adder, forest puff adder and swampjack (FitzSimons 1980; Marsh and Whaler 1984; Branch 1992).

Recognition
(Plates 5.10 and 5.11)

Head—Large and triangular; neck greatly narrowed, almost one-third the head diameter (giving appearance of great fragility); sometimes with paired nasal horns; eyes large and movable; 15–21 circumorbital scales; 12–16 interocular scales; 4–5 scale rows between suboculars and supralabials, which number 13–18; sublabials 16–22; fangs large, reaching 5 cm in length, being the longest of any snake in the world.

Body—Dorsal scales 28–46 rows at midbody, strongly keeled except for smooth outer row on either side; lateral scales slightly oblique; ventral scales 124–140, rarely exceeding 132 in males, seldom less than 132 in females; 17–33 paired subcaudals, with males not less than 25 and females with no greater than 23; anal scale entire (Grasset 1946; FitzSimons 1980; Broadley 1990; U.S. Navy 1991; Branch 1992).

Size — Gaboon vipers are the largest and heaviest vipers in the world (Marsh and Whaler 1984; Phelps 1989). Average adult length is 122–152 cm, with a record of 205 cm for a specimen collected from Sierra Leone (Cansdale 1961; Broadley 1990). Adults are very heavy and stout, especially females, which have been recorded with a total length of 174 cm, head width of 12 cm, girth 37 cm, weighing (with empty stomach) 8.5 kg. Sexes are distinguished by tail length, about 12% of total in males, 6% in females (Grasset 1946; Pitman 1974; FitzSimons 1980).

Pattern — Head dorsally pale gray or buff with a dark brown median line; a dark brown triangular region extends from the corner of the eye, widening to the posterior third of the upper jaw; eye silvery; body coloration soft and pastel-like, drab (even newly molted individuals) to vividly colored; varying from brown, beige, yellow, black, blue and purple; dorsum with vertebral series of elongate, quadrangular, yellow or buff patches connected to each other by hourglass-shaped brown markings; laterally with a series of triangular or crescent-shaped dark brown to purple markings, between which are interlaced chains of brown, yellow and purple; venter buff with an irregular pattern of gray blotches. Young are similar to adults in color and pattern (Mehrtens 1987; Broadley 1990).

Taxonomy and Distribution

Described from Gabon by Duméril, Bibron and Duméril in 1854 as *Echidna gabonica,* then placed in the genus *Bitis* by Boulenger in 1896 (Marsh and Whaler 1984).

Currently, two subspecies are recognized:

Bitis gabonica gabonica, the East African gaboon viper, from Tanzania, Uganda, southern Sudan, Zambia, Zaire, Zimbabwe, Mozambique, and South Africa; and *Bitis gabonica rhinocerus,* the West African gaboon viper, from Guinea, Sierra Leone, Liberia, Togo, and Ghana (Broadley 1990). The two subspecies are easily distinguished by large nasal horns in *B. g. rhinoceros.*

In most other morphological parameters the two subspecies are nearly identical (Mehrtens 1987). However, Lenk et al. (1999, 2001) consider them to be separate species as indicated by their phylogenetic reconstructions using mitochondrial DNA.

Although considered scarce, gaboon viper population density might be generally underestimated because of their secretive nature and camouflage coloration (Ionides and Pitman 1965b). These authors collected 698 specimens from a relatively small area in southern Tanzania over 335 days within a 7-year period. Daily captures ranged from one to as many as 13, with 217 collected in 47 days from 11 May to 30 June. Males comprised 48.4% of individuals. Adults comprised 35%, 25% were "fair size" and 23.6% were "half grown." Juveniles comprised 16.4%, with 13% of the total sample being either newly or recently born.

Habitat

Rain forests and adjacent wooded areas are favored habitats (Mehrtens 1987). Although primarily a snake of lowland forests, it sometimes is found above 1500 m. In

Tanzania, *B. gabonica* occurs in secondary thickets, cashew plantations and under bushes and in thickets in agricultural land (Ionides and Pitman 1965b). In Uganda, where environmental conditions differ from Tanzania, these snakes occur in forests and adjacent grassland. Reclaimed forest areas such as cacao plantations in West Africa and coffee plantations in East Africa are favored (Cansdale 1948; Ionides and Pitman 1965b). Broadley (1971) reported finding them in evergreen forests in Zambia. In Zimbabwe, they are found only in high rainfall zones along the forested eastern escarpment. They are locally restricted to coastal forests, dune regions and remnant montane forest (Branch 1992). Gaboon vipers are also common in farmed areas near forested country and on roads at night (Cansdale 1961). *Bitis gabonica* may inhabit swamplands and be found in still and moving waters (Grasset 1946).

Despite reports they are shade loving (Marsh and Whaler 1984), individuals bask in patches of sunlight on the forest floor, often half-buried in leaf litter (Broadley and Cock 1993). Concealment in leaves is achieved by curling and wriggling from side to side (Berry 1963).

Favored habitats averaged 112 cm of rain per year with temperature ranges between 10°C–35°C. During dry months, gaboon vipers withdraw to abandonded mammal burrows, holes, and cavities in ant hills where they lie dormant until the first rains, usually in November (Grasset 1946).

Food and Feeding

Gaboon vipers are generally opportunistic sit-and-wait predators. However, they will actively hunt, usually beginning at dusk (Broadley and Cock 1993). Following a strike, *B. gabonica* is inclined to hold on to its prey until death ensues. Generally, prey are lifted to keep their feet from getting a surface hold. Prey are typically eaten headfirst but the extreme flexibility of the jaws enables smaller prey to be ingested from any angle. Larger, potentially dangerous prey may be released and searched for after 1–2 minutes, allowing time for the venom to act. Prey are recovered by scent, not sight, as evidenced by the pursuit of prey along scent trails around a cage rather than via a "short cut" to prey which may have died centimeters away (Berry 1963).

Gaboon vipers feed on a variety of birds and mammals. Common prey are francolins, doves, giant rats (*Cricetomys*) and cane rats (Cansdale 1961; Berry 1963). Other rodents found in stomachs include *Rattus, Arvicanthus, Mastomys, Thryonomys, Leggada,* and the squirrel *Aethosciurus* (Pitman 1938). Wakeman (1966) reported three captives were primarily fed rodents, with one bird taken. Three robins, one heuglin (*Cossypha heuglini*) and two natals (*C. natalensis*) were found in the gut of a large female (Haagner 1986b).

Small monkeys, hares, mongooses, brush-tailed porcupine (*Atherurus*) and even the small royal antelope (*Neotragus*) are also reported as prey (Grasset 1946; Broadley and Cock 1975, 1993). Instances of predation on full-grown domestic fowl are reported, with several accounts of specimens being killed while foraging inside poultry runs (Mehrtens 1987; Broadley and Cock 1993). These vipers enter burrows and small caves in termite mounds in search of bats, particularly fruit bats, *Epomorphorus* (Berry 1963). Despite reports of predation on toads and frogs,

Wakeman (1966) reported that no attention was paid to toads left in the cage of three captive specimens.

The voracious appetite of gaboon vipers in captivity was evidenced by the simultaneous introduction of two rats into a cage. One rat was immediately caught and killed. While the first rat was half-ingested, the other rat was spotted, resulting in the first rat being swallowed rapidly and the second rat being seized within seconds (Berry 1963). Whaler (1971) maintained specimens for the purpose of milking and found feeding was erratic. Two of his snakes fed readily on dead rats (150–300 g) or guinea pigs (350–800 g) which were introduced into the cage toward evening. Two other specimens did not accept food and were force-fed while anesthetized.

Behavior

The usual mode of progression is by a sluggish "walking" motion of ventral scales. When alarmed, individuals writhe from side to side but this method is employed only for short distances (Berry 1963).

Gaboon vipers tend to be lethargic, often remaining coiled and immobile. Individuals may remain in one place all day and, unless disturbed, they seldom move great distances (Grasset 1946; Sweeney 1961; Ionides and Pitman 1965b). When aroused or hungry, they can move with speed and agility, striking with a force that can lift almost half the body length from the ground (Mehrtens 1987; Broadley 1990). A striking speed of 2.34 m/sec is reported (Broadley and Cock 1975).

Despite their sinister appearance they generally have a placid disposition. Like other adders they hiss loudly as a warning. They are unlikely to strike except under extreme provocation. Sweeney (1961) reported that *B. gabonica* is so docile "it can be usually handled as freely as can a nonvenomous species" (this is in no way recommended). Lane (1963) reported that Ionides believed they could be tread on with bare feet and, if unhurt, they would do little more than hiss and try to get away. Ionides (*in* Lane 1963) described a capture method further attesting to the docility of this snake. He first lightly touched the top of the head with a pair of tongs to test the reaction. Unless anger was displayed, which was rare, tongs were laid aside. The neck was then calmly but firmly grasped with one hand while the body was grasped and lifted with the other. He stated that the snakes seldom struggled. He would also often stroke them before catching them. In some parts of Africa, native children drag live gaboon vipers by the tail to their villages before killing them for food (Coborn 1977).

Gaboon vipers are said to be nocturnal and "difficult to awaken" during the day (Marsh and Whaler 1984; U.S. Navy 1991). However, Berry (1963) reported catching a large specimen on a forest track at 1000 h.

Reproduction and Development

Male combat may occur during periods of peak sexual activity. In captivity, combat is stimulated by mutual contact between two males on a female scent trail. However, it is doubtful that two males will contact each other often at the same time and place on a scent trail in the wild. Thus combat may rarely occur naturally. Physical size and aggressiveness, not age, seem to determine combat challenge acceptance (Akester 1979b).

Combat is initiated by one male rubbing his chin along the back of another. At a point approximately one-third from the head, the bottom male raises its head as high as possible, pushing against the "aggressor" male. Necks are intertwined and, with heads level, they turn toward each other and push. Their bodies intertwine, switching positions as they move together. This continues even if they fall off a surface or into water. Even shy males in combat become oblivious to distractions, human or otherwise, including flash photography. In a pen of captive snakes, no attempts were made by nonparticipants to join the combat. Combat continues until a head is pushed to the ground and the top male raises its own head 20–30 cm. Continuous hissing occurs during combat. Sometimes in violent bouts, males intertwine their bodies along the entire length, squeezing each other tightly causing scales to extend out from the pressure. They may strike each other with mouths closed. Akester (1979b) observed one instance of a male striking another male with an open mouth and extended fangs. However, it appeared the mouth was open due to breathing difficulty from being squeezed. Occasionally while striking, they miss and hit their heads against rocks or other objects with an audible force. When combat is exceptionally aggressive or continues for long periods, both males become visibly fatigued and combat is terminated via "mutual consent." Sometimes males temporarily break off combat, coil together, and rest in the sun. In captivity, bouts may occur 4–5 times per week until courtship and copulation ends (Akester 1979b; Marsh and Whaler 1984; Broadley 1990).

In captivity, males and females should be kept apart for at least 2 months to stimulate courtship and mating behavior in males. Successful pairing is reported in April and May (Akester 1979a, 1984).

During courtship, a male moves over the female, tongue flicking, and rubbing his chin along her back anteriorly in a jerky motion. Upon reaching her head, he turns and moves back down. The male then stretches beside the female. If she is receptive, he passes his tail under hers and inserts his hemipenes. If unreceptive, she may move away or stiffen her body and use her weight to keep her tail pressed to the ground, preventing his tail from passing under hers. Copulation reportedly lasts approximately 5 minutes (Akester 1979a).

Natural hybridization of gaboon vipers with puff adders, *B. arietans,* and *rhinoceros* vipers, *B. nasicornis,* occurs, with coloration and morphology of hybrids being intermediate between the two species (Hughes 1968; Broadley and Parker 1976; Broadley 1990).

Gestation requires approximately 1 year, suggesting bi- and triennial breeding cycles similar to European and North American vipers (Broadley 1990; Branch 1992). Akester (1984) suggested that a 5-year breeding cycle is possible.

A female mated on 3 May 1978 began to feed "reluctantly" in September. During this time, she consistently emerged to bask 2 hours before any other snakes. Her emergences became increasingly earlier so that by December she was emerging 3.5 hours before any penmate. At this time she refused to feed and had not yet shed (as had her penmates). She also became restless, changing basking position every 10–15 minutes, positioning herself at right angles to the sun, presenting first one side, then the other. By January 1979, this female, which was usually even-tempered, became aggressive, standing her ground, hissing loudly and raising her head to face any intruder. However, she never made any attempt to strike. She was becoming thin, taking on a "pear-shaped" appearance, with the anterior thin and the middle and posterior showing considerable bulk. In March, she molted. During April, she drank large quantities of water and increased in weight from 5.0 to 5.4 kg. Two days prior to giving birth she buried herself under leaves. On 9 May 1979, newborn individuals were observed in the pen (Akester 1979a).

Prior to birth the posterior abdomen becomes swollen to the point of obesity, clearly showing the clutch mass near the vent. Muscle contractions occur throughout the body. During birth the tail is displaced by the volume of each baby.

Bitis g. gabonica produces 8–43 live young. Litters of 16, 17, 22, 23, 25 and 26 from wild-caught specimens are reported (Huffman 1974). *Bitis g. rhinoceros* may produce 60 young. However, actual litter size seldom exceeds 24 (Broadley 1990; Branch 1992).

Young are usually born in late summer. A female captured on 1 January gave birth to 25 on 15 March. Time intervals between births averaged 14 minutes, and ranged from 1 minute between the births of snakes 22 and 23, and 78 minutes between the births of snake 24 and 25. Snakes 9 and 10, 13–15, and 18 and 19 were born in clusters. During the process the female stretched out on a level surface, but for the birth of snakes 9 and 10 she climbed to a high point and raised her tail, dropping the babies 4–5 cm. Remaining births occurred with her tail in a basin of water. Snake number 15 was stillborn, and number 20 was an unfertilized egg (Huffman 1974). Akester (1979a) reported that during parturition the female moved around dispersing the young throughout the enclosure. He also reported young were usually born alternating between upright and upside down positions.

Most young emerged from the vent in individual amniotic sacs which were immediately ruptured by jerks from their heads. Snake number 12 was inactive for 3 minutes before it broke through its membrane and moved off. Snake number 13 jerked its head continuously but never broke through the membrane. After 2.25 hours the sac was snipped open and it was found this individual had a deformed head. Young born in an upside down position took considerably longer to free themselves from their amniotic sacs (Akester 1979a).

Molting occurred within minutes of birth but varied in onset time and duration. Many initiated molting by rubbing their snouts along the mother's body scales. "Yawning" also occurred during the first few minutes after birth. Neonates were vigorously active and aggressive, puffing and hissing at each other as they collided (Huffman 1974). The mother remained placid throughout. Ten individuals from this clutch fed 24 hours after birth, each consuming a half-grown mouse (Akester 1979a).

Neonates average 22–37 cm in length and 25–45 g in weight (Branch 1977). Birth sizes from a clutch of 43 offspring were as follows: 33–39 g (mean 36 g), and 24–27 cm (mean 26 cm) for males; and 34–39 g (mean 37 g), and 24–27 cm (mean 26 cm) for females. The female that produced this clutch had pre- and post-parturition weights of 5.4 and 3.2 kg respectively (Akester 1979b).

Growth is relatively slow, with total length reaching 1 m after 2 years and 1.3 m after 3 years, which probably represents adult status (Ionides and Pitman 1965b; Marsh and Whaler 1984). One-year-olds achieved 18–20 times their birth weight (Jacobs and Belcher 1983). Molting in adults takes place 1–2 times per year, though injured specimens may molt more frequently (Berry 1963; Jacobs and Belcher 1983).

Food intake of 2.1 g/kg body weight/day for three captive 3-year-olds resulted in a weight increase from 0.5–1.8 g/kg/day (Marsh and Whaler 1984). Over a 3-year period, mean body weight increased from 3.2 kg to 5.2 kg and length increased from 134 cm to 143 cm. Specimens fed 1.7 g/kg/day lost weight. After 6 years, weight and length remained fairly constant suggesting that *B. gabonica* reaches full size in approximately six years.

Longevity is reported as 10–11 years by Pitman (1974), and Marsh and Whaler (1984) reported three captives that lived 13 years, 10 months 13 years, 2 months and 14 years, 7 months. Akester and Broadley (separate pers. comm. cited in Marsh and Whaler 1984) reported lifespans to be 20–30 years.

Bite and Venom

Epidemiology—There are relatively few recorded cases of human envenomation from this species. No recorded fatalities exist. Only two of seven clinical reports suggest that envenomation was life threatening (Visser and Carpenter 1977). However, the bite may have devastating consequences.

Yield—Glands and fangs are large and and produce the largest quantity of venom of any snake (Marsh and Whaler 1984). Yield is related to body weight, not milking interval. Injected quantities from snakes 125–155 cm long ranged from 200–600 mg (Minton 1974; Broadley 1990). Marsh and Whaler (1984) reported that the largest venom yield was 2.4 g dried weight (9.7 ml wet venom). Anesthetized snakes (133–136 cm long, 23–25 cm in girth and 1.3–3.4 kg in weight) with half-inch "alligator" clip electrodes attached to the angle of the open jaw yielded 1.3–7.6 ml (mean 4.4 ml) of venom. Two or three 5-second electrical bursts were sufficient to empty the gland. Snakes remained healthy, even gaining weight, after being milked 7–11 times over a 12-month period. Venom potency was retained over the course of milkings (Whaler 1971). Extracted venom is milky and viscid, and is canary yellow when dried (Grasset and Zoutendyk 1938).

Toxicity—Gaboon viper venom is of relatively low toxicity with LD_{50} values ranging from 0.55 to 0.71 and 0.86 to 2.76 mg/kg for i.v.-injected mice and rabbits, respectively (Tu et al. 1969; Minton 1976; Marsh and Whaler 1984). Grasset (1946) reported that *B. g. rhinoceros* is "considerably" more toxic than *B. g. gabonica*. However, Hyslop and Marsh (1991) reported *in vivo* studies in rabbits revealed no significant quantitative or qualitative differences between the two venoms. (See Table 1).

Table 1. Lethal doses and time of death for several test animals (Whaler 1971).

Test Animal	Injection Mode	Dose*	Time of Death
Mice (15–30 g)	i.p.	30–36 μg	140–220 min (1.0–2.4 mg/kg)
Rats (50–80 g)	i.p.	150 μg	72 min (1.8–3.0 mg/kg)
Rats (50–80 g)	i.p.	75 μg	124 min (0.9–1.5 mg/kg)
Guinea pig (750 g)	i.p.	900–1800 μg	<22 hours (1.1–2.6 mg/kg)
Guinea pig (750 g)	i.p.	7300 μg	150 min (9.7 mg/kg)
Rabbit (1.3–2.9 kg)	i.v.	600–1200 μg	3–200 min (0.8–4.1 mg/kg)
Vervet monkey (3.0–5.2 kg)	i.v.	600–3000 μg	22–45 min (0.2–0.6 mg/kg)

*Mass of protein, not total venom solids.

Whaler (1971), assuming that monkey sensitivity is equivalent to that of man, estimated 14 mg of venom protein is lethal in humans. This amount is contained in 0.06 ml of venom which is approximately 1/50–1/1000 the amount obtained from a single milking. Branch (1992) reported 90–100 mg whole venom is fatal in humans and Marsh and Whaler (1984) estimated a lethal dose approximating 35 mg for a man weighing 70 kg. This amount is contained in 1/30 of an average venom yield.

Content—Eleven enzymes have been quantified in venom from *B. gabonica* (Marsh and Whaler 1984). Proteinase and peptidase activity hydrolyzed 27 different dipeptides and six tripeptides (Tu and Toom 1968). Other enzymes found are ATPase, ADPase, hyaluronidase, and phosphodiesterase. DNAase and RNAase levels were the highest of ten venoms tested. High levels of 5'-nucleotidase were also found (Christensen 1967b; Sarkar and Devi 1967). Phospholipase A (PLA) activity varies. Botes and Viljoen (1974) and Marsh and Whaler (1984) reported the presence of PLA, while Mebs (1970a, b) found the second highest quantity of PLA of 41 snakes tested (*Agkistrodon piscivorus* had the highest) and gaboon vipers were ranked at 19 of 28 snakes investigated. Adams et al. (1981) failed to detect any activity.

Bitis gabonica venom had the highest carbohydrate content (11.8% dry wt.) of 16 venoms tested (Oshima and Iwanaga 1969). Protein content varied between 13.1% and 28.8%, with most in the 22%–27% range (Whaler 1971). Mohamed et al. (1975) found that *B. gabonica* venom does not contain a cholinesterase inhibitory factor.

Serotonin is an important constituent of many nonsnake venoms. *Bitis gabonica* is the only snake with relatively high (5 mg/gm) concentrations (Minton 1974). A histaminelike substance in the venom was reported by Tilmisany et al. (1986b).

Electrophoretic analysis revealed seven fractions (Chippaux et al. 1982). Using ion-exchange chromatography and gelfiltration, Marsh and Whaler (1974) demonstrated nine different activities. All fractions were identical from left and right fangs but relative quantities varied among 30 individuals from the same litter. Chippaux et al. (1982) speculated that these differences are genetic rather than physiological or ecological.

Symptoms and Physiological Effects—In humans, envenomation causes rapid and marked swelling, intense pain, severe shock and local blistering. Defecation, urination, uncoordinated movements, swelling of the tongue and eyelids, convulsions and unconsciousness may occur (Whaler 1975; Marsh and Whaler 1984).

Tissue damage in mice at the injection site and hemorrhage from serosal and mucosal surfaces are reported (Mohamed et al. 1975). In mammals (monkeys, dogs, rabbits, guinea pigs, mice) petechial hemorrhages occur in the gut walls, testes and connective tissues, and extensive internal hemorrhage was the principle post mortem finding (Whaler 1975; Marsh and Whaler 1984). *Bitis gabonica* venom contained the highest hemorrhagic activity of 25 venoms tested (Oshaka et al. 1966). This activity is dependent on metal content, probably magnesium and zinc. Of three species of *Bitis* tested (*gabonica, arietans, nasicornis*), *B. gabonica* was the only species whose venom retained hemorrhagic activity when heated to 94°C for 5 minutes (Tu et al. 1969).

Although *B. gabonica* venom affects neural activity in guinea pig skeletal muscle (Alloatti et al. 1986), distinct neurotoxic fractions cannot be isolated (Marsh and Whaler 1984; Broadley and Cock 1993). Hemorrhage and intraventricular clotting in the cerebrospinal fluid, and resultant convulsions may contribute to apparent "neurotoxic" effects (Caiger et al. 1978).

Abrupt hypotension develops within seconds in i.v.-injected anesthetized dogs (Grassett and Goldstein 1947; Gatullo et al. 1983). Profound peripheral vasodilation and reduced aortic impedance caused an increase in stroke volume, thereby increasing ventricular discharge. A decline in peripheral resistance was observed during first and second exposures, but repeated low doses failed to show a summation effect, indicating a tachyphylaxic response (Adams et al. 1981).

Tilmisany et al. (1986a) suggested that hypotension in rabbits may be due to histaminelike substances in the venom. Deaths were attributed to progressive reduction in stroke volume and cardiac output over a series of three venom injections (Cevese et al. 1984). Venom dependent changes in cardiac cell permeability were found in monkeys and rabbits (Marsh et al. 1979). Failure to control intracellular calcium movement accounted for reduction of systolic contraction and hindrance of relaxation (Adams et al. 1981).

Reduction of total peripheral resistance in mesenteric and external iliac vascular beds of dogs is reported (Cevese et al. 1983). After 1 minute during the hypotensive period, external iliac vessel resistance decreased to a greater degree than mesenteric vessels. The external iliac vascular bed is typical of many muscle regions suggesting vasodilation in muscle beds is higher than elsewhere. Transient hypotension was suggested due to lessened vasodilatory effects and extravasation of plasma (Cevese et al. 1983). Plasma extravasation due to proteolytic destruction of vessel endothelium increases blood viscosity causing a drop in stroke volume, cardiac output and lowered blood volume. Increases in coronary flow before changes in systemic flow indicate that coronary vessels receive venom before the main arterial vessels (Tu et al. 1969; Mohamed et al. 1975; Abu-Sitta et al. 1978).

Changes in heart rate, arteriovenous (A-V) conduction, excitation transmission, decrease in contractile strength, rise in diastolic tension and general loss of cardiac pumping function occur in envenomated guinea pigs (Alloatti et al. 1984, 1986). Venom concentrations as low as 0.5 μg/ml in perfusion fluid causes failure of diastolic relaxation and ultimately arrests the ventricles (Alloatti et al. 1984). Conversely, atria are relatively resistant, with a concentration of 50 μg/ml required to cause atrial arrest. At this dose, ventricular arrest occurs within 1 minute, with spontaneous contractions restarting after 10–15 minutes. Progressive cardiac failure has also been seen in dogs, rats and rabbits. Two different venom components are responsible for cardiac effects. One is vasodilatory and the other is cardiotoxic. The latter causes reduced stroke volume and possibly damages the sarcoplasmic system (Whaler 1975, Zaki et al. 1976). It is postulated by Marsh and Whaler (1984) that bradykinins and/or prostaglandins may be involved in the overall cardiovascular effect.

Destruction of mammalian lung microvasculature followed by pulmonary edema causes marked hyperventilation, with a progressive decrease in oxygen consumption. This condition may lead to respiratory failure, mimicking a neurotoxic effect (Whaler 1975; Marsh and Whaler 1984).

Contrary to findings by Tu et al. (1969), myotoxic effects at the injection site in mice were revealed by muscle tissue absence, vacuolization, fasciculation, and loss of transverse striations (Mohamed et al. 1975). Reduction in muscle fiber size and subsarcolemmal density also occurred. Sarcolemma effects produce a type of toxic denervation of the nerve supply. This is supported by the presence of degenerated and vacuolated intramuscular nerve trunks and degenerated motor end plates. These histopathological and histochemical changes are apparently irreversible (Mohamed et al. 1975).

At the cellular level, inhibition of the intramitochondrial enzyme succinic dehydrogenase occurs 48 hours after injection in mice (Mohamed et al. 1975).

Intravenous injections of diluted frozen venom protein in rabbits and monkeys cause hypotension and marked increase in pulse pressure. After several minutes, there is a progressive 15%–45% fall in O_2 uptake and CO_2 output. Ventilation rate and tidal and minute volumes increase. These remain elevated in rabbits despite blood pressure recovery. Respiratory failure occurs rapidly with massive pulmonary edema. Ventilation changes in monkeys are less dramatic (Whaler 1972). Recovery is spontaneous except after large doses. Postmortem examination of monkeys and rabbits that died quickly following large venom doses showed no gross changes. However, those that died 30–120 minutes postinjection had extensive hemorrhagic foci in the gut, abdominal walls, diaphragm, ureters, heart and fatty tissue. No hemolysis was observed in plasma or urine. The right side of the heart continued to beat while the left side was powerfully contracted. Based on these results, Whaler (1972) hypothesized that direct cardiac damage and O_2 uptake failure contributed greatly to death.

Hematological Activity—Gaboon viper venom is both coagulant and anticoagulant. In some cases these contrasting effects may result from variations in protein concentration and venom preparation and storage (Marsh and Whaler 1974). For example, Marsh and Whaler (1974) reported that frozen venom and venom dried at 4°C retained coagulant activity while venom dried at 25°C–30°C did not.

A substantial quantity of coagulant material is found in *B. gabonica* venom. Ion-exchange chromatography revealed two thrombinlike enzymes that exhibited coagulant activity *in vitro*, but only one showed this effect *in vivo*. This material, identified as a thrombinlike enzyme named "gabonase," caused incomplete fibrin formation. It releases fibrinopeptides A and B and activates factor XIII. It is not inactivated by heparin. Heparin failed to inhibit *in vivo* effects of this thrombinlike enzyme and even appeared to enhance its effect. No specific fibrin or fibrinogen clotting enzymes have been found in *B. gabonica* venom (Marsh and Whaler 1974; Bajwa et al. 1982; Pirkle 1988).

Addition of venom in a 1/10,000 dilution shortened coagulation time in human hemophiliac blood from 25 to 11 minutes. *Bitis arietans* venom increased hemophiliac blood coagulation time to 1 hour at similar dilutions. Normal human blood, having a coagulation time of 5 minutes, coagulated in 3 minutes (Grasset 1946).

In vitro anticoagulant action results from reduced plasma prothrombin caused by lytic action and effects on factors VII and X (Rosenfeld et al. 1967; Forbes et al. 1969). *Bitis gabonica* venom failed to exhibit clot formation in human plasma and fibrinogen after 15 minutes at venom dilutions from one in 1000 to one in 64,000 (Nahas et al. 1964).

Venom causes rapid defibrination *in vivo,* and antithrombin activity when venom and thrombin are added to fibrinogen (Nahas et al. 1964). Interference with the thromboplastin system is evidenced by defective thrombin generation, defective prothrombin consumption, inhibition of thromboplastin generation and accelerated loss of thromboplastin. The coagulation defect is further complicated by inhibition of the tissue thromboplastin system and platelet aggregation. The venom impairs clot formation by direct proteolytic action on fibrinogen. The resultant fibrinogen degradation products yield defective fibrin polymerization which in turn prolongs thrombin clotting time.

Case Histories—Broadley and Cock (1993) reported on a park attendant bitten on the finger. Within 30 minutes, there was gross swelling and pain in the hand and arm. Respiratory difficulty described as "feeling like a heavy weight was placed upon the victim's chest" followed. Acute kidney and respiratory tract pain ensued, followed by dilated pupils, rapid labored breathing, and an almost imperceptible pulse. The hand, arm and chest were grossly swollen, becoming a black-purple "pulpy mass," with blisters on the hand and arm. The victim was eventually discharged with the finger intact after 3 weeks of treatment. However, marked atrophy of the entire arm persisted for a "long time."

Visser and Carpenter (1977) describe the case of a 16-year-old boy bitten on the thumb with both fangs while placing a water dish into a cage. A burning sensation occurred immediately at the bite. Within 3 minutes, the puncture sites were bleeding freely and the victim felt dizzy. The hand was submerged in warm water. Five minutes postbite, the thumb was swollen and dizziness increased, forcing the victim to sit. Speech became slurred and impaired, breathing was difficult, and swallowing was painful. At 10 minutes, the victim was staggering and his vision was impaired. The victim collapsed approximately 13 minutes following the bite. At this time, both eyelids and surrounding areas were swollen, breathing difficulty increased, and punctures bled profusely. The victim was transported to a hospital, with 10 ml i.m. polyvalent antivenin administered during the journey. The patient soon became unconscious, his pulse was hardly discernible, and he was perspiring profusely. By 20 minutes, his heart had stopped and CPR was started. Within 2 minutes, heartbeat and semiconsciousness were restored, however, vision remained impaired. By 25 minutes, the patient was carried into the hospital without a pulse. Intravenous fluid with 600 ml albumin was provided. Five minutes later, 70 ml i.v. polyvalent antivenin was given. Signs of recovery were observed at 35 minutes postbite and the patient was moved into an intensive care unit where, 10 minutes later, he vomited and showed generalized muscular spasms. At 2100 hours (75 minutes postbite) a sedative was given but severe pain prevented sleep. By 0500 h the following day, sight and speech had returned. The hand and arm were swollen, glands under both arms were tender and enlarged, and the bite site was covered in large blisters. Serous fluid continued to ooze from the fang punctures. Hydrocortisone and ampicillin injections were given at routine dosages for 3 days. On day 4, the patient's arm was still swollen and he was moved to a general ward. Necrotic tissue was surgically removed from the thumb on day 7. This relieved pain and reduced swelling in the limb. The patient was released on day 10. Three weeks after the bite, healing was satisfactory and thumb movement was unimpaired.

A famous case history is provided by Ditmars (1934) who described the bite of Marlin Perkins, curator of the St. Louis Zoo and host of the television series "Wild Kingdom." While treating a large gaboon viper for parasites, Perkins was struck on the left index finger by one fang at 0950 h. Immediate pain in the finger, described as "excruciating," was followed seconds later by pain throughout the hand. A half-inch long incision was made at the fang puncture site, with the victim himself sucking the wound. During a 60-foot walk to his office, Perkins became dizzy and nearly fainted. As he sat in a chair, he felt "extremely giddy." A loose tourniquet was applied. Ten minutes postbite, the zoo veterinarian injected the left hand with 2 cm^3 of American Anti-Crotalic serum (no specific gaboon viper antivenin was available). Other 2 cm^3 injections followed at the wrist and left pectoral muscle. At 15 minutes postbite, the entire arm was in great pain, with shooting pains in the left side of the chest, radiating toward the heart. There was severe headache restricted to the top of the head. Discoloration and swelling began in the finger with swelling extending up the arm. By 20 minutes, Perkins was very pale, and blood was oozing freely from the incision. Perkins was admitted to the hospital 35–40 minutes postbite and was still rational, conversing freely, especially about the pain in his appendage and chest. He began passing into shock and described difficulty breathing, "as though a weight was placed on his chest." A 7.5-grain caffeine sodium benzoate and a 1/30-grain strychnine sulphate injection were given. The arm was placed in a hot saline solution bath which did little to relieve pain. Breathing difficulty increased and the wound bled freely. Ten minutes after admission, his pulse was 70 per minute, temperature was 36.5°C, and respirations were 20 per minute. After 1 hour, he continued struggling to breathe with respirations of 28 per minute and a pulse of 64. These symptoms suggested the presence of a neurotoxin, so an injection of cobra antiserum was administered. By 1055 hours urination was painful and urine was bloody. Pulse rate continued to drop and by 1058 h it was imperceptible. Respiration was rapid, labored, and shallow. Perkins was described as "looking very bad, deathly pale, and bathed in a cold sweat." At 1100 hours he became unconscious with no discernible pulse. His eyes were highly dilated with no reaction to light. It was believed that he was dying. American Anti-Crotalic antivenin was slowly administered in the right median basilic vein, along with 7.5 grains of caffeine sodium benzoate and one-fifteenth grain of strychnine sulphate. Cobra antivenin (10 cm^3) was given in the right pectoralis major muscle. This quieted the respirations but the pulse remained imperceptible. Novocaine infiltration was performed and incisions were made on the swollen hand and arm to relieve pressure and allow blood and lymph to escape. At this time, the upper arm was swollen to nearly twice its normal size and the bitten index finger was nearly black. Dark blue to bluish purple discoloration extended to the back of the hand and up the arm to the armpit. At 1125 hours the original incision on the bitten finger was lengthened, with tissues herniating through the opening. By 1130 hours both eyes were extremely bloodshot and pupils were moderately dilated, the right more so than the left. At 1145 hours he showed signs of consciousness, pupils were normal, and he asked for water. The tourniquet was released for 20 seconds every 10 minutes. From 1202–1230 hours he complained of headache, severe shaking, and chills. The pulse was still weak and respirations were 28. Overall color was cyanotic. At 1245 hours the arm was removed from the saline bath and by 1300 h body temperature was 38°C, pulse was 116, respirations were 28, perspiration was occurring freely, and the urine was very bloody. Most red blood cells had normal contour although many disintegrated. Wounds were oozing and bleeding freely, making compression necessary to control bleeding. At 1500 hours his pulse was 134 and thready and respirations were 28. By 1515 hours hemorrhagic disturbance was so pronounced that a 10-cm^3 polyvalent *Bothrops* antiserum was administered in the right deltoid. Owing to marked anemia, general weakness, and shock, a 500-cm^3 citrated i.v. blood transfusion was given in the right arm at 1645 h. As this occurred, he began to look and feel better, with lips and cheeks regaining color. Pulse rate strengthened. During the next several days, he remained weak, the hand remained swollen, and the entire arm was discolored from extravasated blood. These conditions slowly diminished and he slowly showed overall signs of improvement. Perkins improved and was discharged after 20 days. The final report of this case concluded that death would have occurred had it not been for i.v. administration of Anti-Crotalic sera as well as the anticobra serum given to check respiratory failure.

Treatment and Mortality—In regions where puff adders, *B. arietans,* and gaboon vipers occur sympatrically, treatment of bites has proved difficult due to uncertainty regarding the species of offending snake.

The large amount of venom delivered, complicated by the great injection depth necessitates a large quantity of antivenin for effective treatment (Broadley 1990).

There is negligible cross-neutralization of *B. gabonica* venom by bivalent antivenin from puff adder, *B. arietans*, and Cape cobra, *Naja flava* venoms (Grasset and Zoutendyk 1936, 1937; Grasset 1946). However, rabbits injected with *B. gabonica* antivenin are protected against *B. arietans* venom in doses 10 times lethal (Grasset 1946). Christensen (1967a) reported good specific neutralization of venom from polyvalent antivenin made from *B. gabonica, B. arietans, Naja nivea* and *Haemachatus haemachatus* venoms. New World viper polyvalent antivenin provides protection against 2–4 LD_{50}s of *B. gabonica* venom and has been recommended in emergency situations (Minton 1976). Monovalent *Cerastes vipera*, and bivalent *C. cerastes and C. vipera* antivenins show surprisingly high neutralization against *B. gabonica* venom (Mohamed, Abdel-Baset and Hassan 1980).

Visser and Chapman (1978) summarized treatment in three steps: 1) adequate doses of antivenin; 2) prompt fluid replacement, especially blood; and 3) respiratory and cardiac supportive measures. Broadley and Cock (1993) recommended that antivenin be used only in cases involving extensive painful swelling. Marsh and Whaler (1984) stated that antivenin should be used only if systemic envenomation, indicated by hypotension and dyspnea, occurred.

Remarks

Pitman (1938) reported that nearly all wild-caught specimens harbored numerous pentastomid worms, *Armillifer armillatus*, and cestodes, *Proteocephalus gabonica*, are common. Inspection of fecal material from wild-caught specimens yielded ascarids, strongylids and cestodes (Jacobs and Belcher 1983). Thiabendazole (60 mg/kg) followed by fenbendazole (50 mg/kg) via stomach gavage is effective in treating these parasites.

Bitis gabonica adapts well to captivity, reproducing readily under proper conditions. Whaler (1971) had success keeping four specimens together in a cage 115 cm long, 33 cm wide, and 33 cm deep. He used extracted, dried sugar cane fiber as a substrate, with a small bowl of water. Room temperature was 24°C, with no exposure to sunlight.

For breeding, Akester (1979a) kept three females and seven males together in a 16-m^2 outdoor pen. The pen, planted with thick vegetation, had a pool of water and an underground refuge. During winter (May to September) the pen was covered with plastic sheeting to stabilize the minimum temperature at 2°C. Even at these low temperatures, snakes were observed out all night. However, although active, they did not feed from April to September (Akester 1979a).

Jacobs and Belcher (1983) housed a pregnant female in a fiberglass cage measuring 118 cm long × 92 cm wide × 86 cm high. The substrate consisted of gravel 1–2 cm in diameter. The enclosure contained leaf litter, several artificial plants and two cottonwood log sections fastened to boards that were buried under large rocks. This arrangement provided two substrate levels for more open space during the birthing process, reducing the possibility of crushed babies. A water bowl 20 cm in diameter and 7 cm deep was provided at all times. Ambient temperature in the surrounding room was 25°C–30°C. Lighting was provided by two 40-W "Vitalites" plus two 40-W "daylight 64 tubes" for 4 hours during midday. Diurnal lighting cycles varied from 9.5 hours in June to 16 hours in December.

Although viperids typically swallow shed fangs without incident, Elkin (1979) reported on a 1.4-m gaboon viper that died, apparently of complications arising from swallowing its fangs. The autopsy revealed a moderately inflamed lung, and pancreas with a large central abscess indicating septicemia, the probable cause of death. Further investigation revealed two fangs, each 2.5 cm long, that had pierced the gut wall.

Cooper (1971) described surgical treatment for an adult male gaboon viper that was attacked with a large knife, resulting in deep wounds near the cloaca and the head. Six weeks postsurgery, the snake was eating and growing normally.

Bitis gabonica seems to have few natural enemies. The Benis people of Uganda consider gaboon vipers a delicacy, which they eat in soup (Pitman 1938). Gaboon vipers have a wide range of eye movement, more so than many other snakes. This may be due to the large head, necessitating greater movement for peripheral vision. Eye movement is retained along a horizontal plane even when the head is rotated up or down up to an angle of 45°, after which a compensatory correction is made. Rotating the head 360° results in one eye tilted up and the other down, depending on which side of the head is rotated. In general, eyes flick frequently back and forth in a rapid jerky fashion. In addition, as one eye moves forward, the other moves proportionally backward as though both were connected in a fixed position on an axis between them. Sleeping vipers fail to move their eyes and the pupil is strongly contracted. Upon awakening, the pupils dilate sharply and eye movement occurs (Parry 1975).

Range cattle are frequently bitten. Grasset (1946) described a bull bitten on the ankle by a large specimen. Based on the pattern of trampled ground, the bull struggled considerably, covering at least 10 m from the bite site. When it finally died 2 hours later, blood exited through its anus and edema was prominent throughout the body.

Bitis nasicornis (Shaw 1792)

Commonly known as the rhinoceros viper or river jack.

Recognition
(Plates 5.12 and 5.13)

Head—Narrow, flat and triangular, small relative to body; with a cluster of 2–3 pairs of nasal hornlike scales; eyes small, set forward; pupils vertically elliptical; iris green or gold with black flecks; circumorbital ring with 19–20 scales; interocular scales 14; 4–5 scales between suborbitals and supralabials, which number 15–17.

Body—Body scales strongly keeled (so much so that Akester [1989] described them as resembling "miniature shark fins," and Spawls and Branch [1995] stated they sometimes inflict cuts on handlers when the snake struggles); dorsals 31–43 rows at midbody; ventrals 117–140; subcaudals 16–32. There is sexual dimorphism in subcaudal count, males 25–30, and females 16–19.

Size—Ranges in length from 72 to 107 cm. Females longer than males (Mehrtens 1987; U.S. Navy 1991). Pitman (1938) reported the weight of a 42-cm male as 1.4 kg and a 60.0-cm male as 1.6 kg.

Pattern—Head dorsally blue or green, with a vivid black forward-pointing arrow mark; body ground color varies through assorted shades of blue, purple, pink and green (western individuals more blue, eastern more green); complex dorsal pattern made up of a vertebral series of 15–18 paired yellow-edged blue blotches, with a lateral series of light-edged dark triangles extending up from the belly; many lateral scales white-tipped, giving a velvety appearance; venter is marbled and mottled; newly hatched juveniles vividly marked miniatures of adult. The vivid coloration of newly molted individuals fades rapidly, the result of silt accumulating on the rough keeled scales when these snakes spend time in moist habitats (Mehrtens 1987; Akester 1989; U.S. Navy 1991; Spawls and Branch 1995). Easily distinguished from *B. gabonica rhinoceros* by the large dark arrow-shaped mark on the crown of *B. nasicornis* (Villiers 1950b; U.S. Navy 1991; Spawls and Branch 1995).

Taxonomy and Distribution

Described as *Coluber nasicornis* from "central Africa." No subspecies are recognized. Its range is contiguous rain forests of sub-Saharan central Africa, from southern Sudan, western Kenya, Uganda west through Zaire and Angola to Guinea. Records are available from Liberia, Ghana, Ivory Coast, Cameroon, Gabon, Congo Republic, Rwanda and Burundi (U.S. Navy 1991; Spawls and Branch 1995).

Current information on population density is not available. Loveridge (1933) called it "fairly common," reporting 18 captured by natives in Kenya in 3 days, estimating he could have captured at least 100 in a month.

Habitat

Rhinoceros vipers occur in forested areas, especially evergreen swamp forests (Broadley 1971; Spawls and Branch 1995). Microhabitat is variable, including relatively dry forested areas, and moist habitats such as riverbanks, swamps and shallow pools. In Kenya, it has been collected on valley slopes and in small forest patches in grassy valleys. Villagers in some areas of Kenya tolerate their presence, permitting habitation in roofs of huts. In Ghana, cacao plantations and wooded valley slopes are favored habitats (Pitman 1938; U.S. Navy 1991; Spawls and Branch 1995).

Food and Feeding

Rhinoceros vipers are sit-and-wait predators, catching prey via ambush while lying hidden among leaf litter. Froesch (1967) reported on a captive specimen that would seldom leave its hidebox even when hungry. This specimen once waited 3 days for a live mouse to enter the hidebox before striking.

Small mammals including *Mastomys, Leggada, Lophuromys* and *Crocidura* form the main diet (Pitman 1938). In wetland and riparian habitats, frogs, toads and even fish are eaten. A long-term captive specimen, which fed regularly on freshly killed mice and frogs, always held on to its prey for several minutes following the initial strike. This occurred whether prey was introduced live or dead (Matthews 1968).

Behavior

Rhinoceros vipers are primarily nocturnal. During the day they hide among leaf litter, around fallen trees, in holes, or among tangled roots of forest trees. Despite vivid coloration, they can remain virtually invisible among the leaf litter and spotty lighting on the forest floor. Although primarily terrestrial, they do climb several meters into thickets and trees to hide among clumps of leaves or in tree crevices (Matthews 1968; Spawls and Branch 1995). Slight tail prehensibility enhances their ability to climb (Akester 1989).

Conflicting opinions exist regarding disposition. Pitman (1938) reported them as "peculiarly placid and inoffensive in disposition, and most reluctant to bite." Copley (quoted in Pitman 1938) stated that these snakes are seldom known to strike, and that two prospectors brought him a "huge" specimen that was kept as a pet which "could be handled with ease." However, Cansdale (1961) and Ionides and Pitman (1965b) stated that these snakes are much less placid than gaboon vipers and it is a "snake to be handled with extreme caution at all times." Matthews (1968) and Akester (1989) stated they are placid in temperament but can be aggressive if aroused or molested. They hiss loudly (the loudest of any African snake) and puff if approached, often revealing their presence before they are seen (Spawls and Branch 1995). They can strike quickly, both forward and sideways, without first coiling or giving prior warning. It is not safe to hold them by the prehensile tail as they can throw themselves upward to strike (Akester 1989).

Based on back and forth head movements in response to external stimuli, Froesch (1967) determined that *B. nasicornis* can perceive movement up to 6 m away.

Captive specimens were seldom found in, near or drinking from their water pool (Akester 1989). They did, however, become active during rainfall or sprinkler activity, usually taking the opportunity to drink from the rocks, vegetation and their own bodies. They are powerful swimmers (Akester 1989).

Reproduction and Development

In West Africa live young are born March–April, the beginning of the rainy season. In East Africa the breeding season is indefinite (Spawls and Branch 1995).

The breeding season of captive *B. nasicornis* in Zimbabwe lasts from March until early June. This coincides with the onset of the colder months and the end of the rainy season. However, it is uncertain if this is natural or a result of artificial conditions (Akester 1989).

A detailed account of courtship and mating is given by Akester (1989). A pair kept in an outdoor pen spent most of the time apart, often in one place for long periods. On 21 March, the pair copulated in the afternoon with the male being dragged through the vegetation by the female. The next day at noon, soon after sprinklers were turned on, the male moved over the female with rapid tongue flicks and head jerks. She first moved away but he persisted. They coupled in the afternoon, again with the male being dragged. Six days later they coupled again. No further mating occurred until 11 April, when they coupled. A second lull in mating occurred until 1 June when, at 1715 h, the last coupling for the year occurred.

The mated female fed well until December when she began to refuse food. Her daily behavior consisted of shuttling between sunlight and shade, then coiling tightly in late afternoon, remaining coiled through the night until the sun was once again on the enclosure. During this time she became aggressive.

A poor fragmented molting occurred from late March to mid-April. This fragmented molting has been reported in *B. gabonica* and appears to be an indication that parturition is imminent. Birth occurred on 6 May, making the gestation period somewhere between 11 and 14 months. A total of 42 neonates (27 males, 15 females) were produced and all molted immediately after birth. In addition, there were 4 stillborn, 3 infertile eggs, and 1 very small, live, but unmolted baby that died 2 weeks later.

Litter size varies from 6 to 38, babies measuring 18–25 cm (Spawls and Branch 1995). A 96-cm female contained 22 eggs measuring 36 × 29 mm, each with a well developed embryo (Pitman 1938). Two large females were captured in Kenya in early February, each containing 38 large embryos (Pitman 1938). A female collected in mid-February from the Congo region contained 31 young, arranged in rows of 15 and 16 (Schmidt 1923).

Some neonates feed readily on pinky mice offered a few days after birth, but the majority must be coaxed. Females feed more readily than males. In one case, 7 months after birth most still required hand-feeding, however, a number of problem feeders fed readily when offered small toads. These feeding problems were far more numerous than those from over 150 *B. gabonica* births described by the author of the report (Akester 1989).

Growth rates vary with feeding. In Akester's study, littermates that fed voluntarily (group 1) grew faster than those induced to feed by placing food into their mouths (group 2), which grew faster than littermates that were force-fed (group 3). (See Table 2.)

All littermates were maintained at the same temperature and received approximately the same amount of food. Most molted at 3 months and some molted a third time by 7 months (Akester 1989).

Substantial growth of two males and one female kept for milking is reported by Marsh and Glatston (1974). The snakes were fed one to two 50–60-g mice while

Table 2. Mean mass and length of study groups 6 months after birth.

	Group 1	Group 2	Group 3
Weight	69.9 g	42.8 g	33.2 g
Length	340.6 mm	305.4 mm	296.3 mm

anesthetized following each milking. Milking intervals varied from 2 to 21 days, with feeding occurring at alternate milkings when the milking period was less than 7 days. After 10 months, length increased by 17%–41% and weight by 100%–260%. These increases accounted for 82% of the food intake weight for the female and 75% and 89% for the males. After 7 months, the larger male did not increase in size whereas the other two continued to grow. (See Table 3.)

Natural hybridization between *B. nasicornis* and *B. gabonica rhinoceros* has been reported by Hughes (1968) who collected a subadult from Ghana that displayed characteristics of both species.

Bite and Venom

Epidemiology—The restricted range has resulted in few reported bites. Statistics are not available.

Yield—Maximum wet venom yield is 200 mg (Spawls and Branch 1995). Decreasing the milking interval from 21 to 2 days increases dried venom yield from 14 to 36 mg per snake respectively, with substantial changes occurring between days 7–14. Increasing the milking interval from 2 to 21 days decreases mean dry venom yields from 45 to 27 mg dry weight respectively, or 30%. This relatively low drop in yield occurring after a 10-fold increase in milking frequency suggests that substantial venom is produced in short periods of time. Fangs are not large, rarely exceeding 1.5 cm in length (Pitman 1938).

Toxicity—Intravenous LD_{50}s of 19 µg for 16–18-g mice, and 1.1 mg/kg are reported (Christensen 1967b, Spawls and Branch 1995 respectively). An i.m. LD_{50} of 8.6 mg/kg in mice was the highest of five viperid venoms tested (*B. arietans, B. gabonica, B. nasicornis, Daboia russelii, Vipera aspis*) (Tu et al. 1969). A lethal dose in mice was recorded as 440 µg in 100 minutes. In vivo studies with mice

Table 3. Growth of a captive male and female gaboon viper (Akester 1989).

	Male		Female	
	Weight	Length	Weight	Length
Initial size	658 g	80.0 cm	622 g	75 cm
Five months	1042 g	92.5 cm	1017 g	85 cm
Three years	2500 g	117.0 cm	N/A*	113 cm

*A final weight was not taken because the female was near parturition.

showed no significant difference in lethality of venom obtained from milkings taken after 2- or 21-day intervals. Dried venom doses of 1.7 mg obtained from a 21-day milking cycle killed in 41 minutes and 1.5 mg dried venom taken from a two day milking cycle killed in 54 minutes (Marsh and Glatston 1974).

Bitis nasicornis venom is slightly more toxic to rabbits than *B. gabonica* venom, with 500–1000 μg protein of the former killing in 1–2 minutes whereas 600–1200 μg of the latter killed in 3–200 min. Doses of 750–1000 μg dried venom produced death in rabbits within 1–2 minutes. A larger total amount of venom, up to 3 mg, could be administered without immediate death provided it was given in a series of smaller doses not exceeding 200–250 μg (Marsh and Glatston 1974).

Content—Protein content, 95–130 mg per ml, is relatively low, with yields decreasing and is inversely proportioned to milking frequency. Electrophoresis reveals that females contain an extra unidentified protein band (Marsh and Glatston 1974).

Several enzymes have been detected. Peptidases were revealed by hydrolysis of 25 of 29 dipeptides and six of six tripeptides (Tu and Toom 1968). Proteolytic enzymes have been confirmed by Kirk and Corkill (1946), Mackay et al. (1970) and Marsh and Glatston (1974). Venom obtained after a 21 day milking interval has greater proteolytic effects than that from a 2-day milking interval (Marsh and Glatston 1974). An enzyme with collagenolytic activity is reported (Simpson et al. 1972) and phospholipase and hyaluronidase have been identified (Christensen 1967b).

Crude venom fractionation reveals six fractions. One possesses marked anticoagulant activity via antiplasmin action. The other five shortened plasma recalcification time. Mackay et al. (1970) assumed that crude venom also contained a labile factor which enhanced plasmin activity. However, this factor is apparently lost upon fractionation. All fractions exhibit striking but variable effects on platelets, with the variability being caused by the presence of fractions with opposing actions. Four fractions enhance platelet adenosine diphosphate (ADP) reactivity and produce platelet aggregation in the absence of ADP. One fraction has no platelet action in the absence of ADP but reduces platelet ADP reactivity. The sixth fraction inhibits plasmin and ADP reactivity and is anticoagulant (Mackay et al. 1970).

Symptoms and Physiological Effects—There are few detailed records of human envenomation. Massive swelling, which may lead to necrosis, is reported (Spawls and Branch 1995). Postmortem examination of mice shows hemorrhage in the diaphragm, with trace signs of hemorrhage elsewhere in the abdominal cavity and local punctate hemorrhages in kidneys and lungs (Marsh and Glatston 1974).

Intravenous injection of 3.2 mg/kg in rats produced an immediate atropine-sensitive bradycardia lasting 10–15 seconds, with an associated fall in mean arterial blood pressure. Heart rate recovered rapidly but arterial blood pressure remained depressed at 40% of preinjection levels. Cessation of respiration occurred 9–11 minutes postinjection. Along with bradycardia there was transient hyperventilation. This preceded a brief apnea followed by a large increase in breathing depth and rate for the first minute, which gradually decreased until all respiratory activity ceased. Lower doses of 0.4, 0.8 and 1.6 mg/kg show similar initial cardiovascular and respiratory effects but these are not lethal during a 2–4-hour postinjection period. Extensive hemorrhage and blood pooling occurs in test animals at all doses (Ghalayini and Whaler 1985). In vitro studies by Marsh and Glatston (1974) revealed that venom caused the majority of fibroblast cells to deform resulting in cell death.

Intravenous injection of anesthetized rabbits produced an immediate and profound drop in blood pressure. The degree of hypotension and the extent of recovery was dose-dependent, with recovery usually taking 3–4 minutes. Respiration ceased for 5–10 seconds immediately following injection, then gradually increased over the next 2 minutes to a level above resting rate. Postmortems showed no macroscopic abnormalities (Marsh and Glatston 1974).

Of four venoms tested (*B. arietans, B. nasicornis, Cerastes cerastes,* and *Echis carinatus*), only *B. nasicornis* produced irreversible contraction of isolated rabbit aortic strips. The venom apparently increases calcium ion influx and, hence, concentration in smooth muscle cells resulting in aortic strip contraction (Tilmisany et al. 1986a).

Hematological Effects—There is no *in vitro* effect on human whole blood clotting time (Kirk and Corkill 1946). Studies by Mackay et al. (1970) showed that venom probably has its main effect on the intrinsic thromboplastin mechanism, since much higher concentrations of venom were required to produce an effect on the one-stage prothrombin time. It is speculated that the venom interfered with thromboplastin production and destroyed thromboplastin as it formed. There was no evidence that venom directly interfered with the conversion of fibrinogen to fibrin under the action of thrombin. The venom activated plasminogen and enhanced its activation by streptokinase. Absence of coagulant activity was demonstrated by Marsh and Glatston (1974).

In vitro studies using high concentrations of crude venom significantly interfered with platelet function, delaying aggregation and inhibiting its ADP reactivity. However, microscopic examination of platelets failed to show any morphological changes (Mackay et al. 1970).

Treatment—Polyvalent antivenin from *Naja nivea, Hemachatus haemachatus, B. gabonica* and *B. arietans* are reported to have a neutralization effect for bites in humans (Christensen 1967b).

Remarks

Remains of B. nasicornis have been recovered from stomachs of civet (*Civettictis c. schwarzi*) and mongoose (*Ichneumia albicauda ihbeana*) (Pitman 1938).

Marsh and Glatston (1974) kept two males and one female in a glass and steel enclosure measuring 83 × 40 × 40 cm, at an ambient temperature between 18°C and 23°C. Akester (1989) housed specimens in glass cages at a minimum temperature of 16°C in very high humidity, achieved by daily spraying. Initially, he housed them in outdoor pens, 14 m^2, during warm months only, but eventually they were kept outdoors permanently with a clear plastic sheet covering the pens in winter. Pens had a large pool of water and a sprinkler system was used during dry months. Specimens were housed outdoors even at temperatures as low as 2°C. Mehrtens (1987) recommended keeping specimens at 24°C–27°C, believing that higher temperatures inhibited feeding. A substrate of shredded bark mulch covered with a layer of dried leaves to provide normal contact security is recommended. Mehrtens (1987) reports these snakes breed easily in captivity and may live in excess of 8 years. He recommends feeding prekilled mice at night.

Froesch (1967) kept specimens in boxes containing "airy" turf as the only bottom covering, a heat rock, and light up to 5 hours per day. He stated that long periods of darkness provided a secure feeling and calmed the snakes. Temperatures ranged from 24°C to 30°C in the day and from 20°C to 24°C at night. He sprayed the cages every third day with warm water. On occasion the snakes would face the spray bottle and drink directly from the spray, making a type of chewing motion with their mouths. Terraria kept too humid at cool temperatures may cause respiratory problems (Froesch 1967). Although treatable, these problems can be chronic and contagious. Froesch cautions about amoebic dysentery, evidenced by watery stools and specimens spending long periods of time sitting in water and drinking.

Pentastomid, nematode and linguatulid infestations are common. Loveridge (1933) found 25 large linguatulids, (*Armillifer armillatus*) in the stomach, intestine and in viscera just behind the head of a single snake. Imported specimens invariably contain a large number of parasites (Mehrtens 1987).

Jacobson and Spencer (1983) reported dystocia, an abnormality in formation, location or positioning of developing young, which can be fatal. They provide an excellent detailed account of diagnosis and treatment of this condition in a rhinoceros viper. Furthermore, Jacobson (1985) reported on the use of a polyvalent autogenous bacterin treatment of mixed Gram-negative bacterial osteomyelitis.

Bitis parviocula
Böhme 1977

Commonly known as the Ethiopian mountain adder.

Recognition

Head—Long, flat triangular, distinct from narrow neck; eye large and dark, set well forward, vertically elliptical pupil, brown iris; nostrils large and prominent; crown with small imbricate keeled scales.

Body—Stout, cylindrical in cross-section, with slight vertebral ridge; dorsals keeled, 37–39 at midbody; tail short.

Size—Largest recorded specimen is 75 cm. Maximum length "probably" exceeds 1 m (Spawls and Branch 1995).

Pattern—Head brown, with dark triangle between eyes; behind triangle, extending onto the nape, is a dark hammer-shaped mark; side of head dark with narrow pale stripe running from behind the eye to the upper lips, which are white; chin and throat are white with black speckling; body ground color light to dark brown; dorsally, extending along the midline is a series of black hexagons or diamonds which sometimes have pale crossbars; between black hexagons is a chain of yellow butterfly-shaped markings; along flanks is a series of black triangular to subtriangular spots with white centers; lower flanks with series of green-gray upward pointing triangles, often yellow-edged, especially the tips; between triangles flanks are mottled green; venter green-gray, clear or with dark speckling (Spawls and Branch 1995).

Taxonomy and Distribution

Described from specimens collected at Doki River bridge near Yambo, on the road from Meru to Bedelle, Illubabor Province, southwestern Ethiopia.

The Ethiopian adder is known only from three localities in southern Ethiopia, from altitudes between 2000 and 3000 m. It is assumed to be more widespread in highlands (Spawls and Branch 1995).

Habitat

Of the three specimens known, two were from forested areas in the west and one was from grassland on the east of the Rift Valley. One specimen was captured in an old coffee plantation in a forest clearing and another was hiding in grass on the grounds of a brewery in a forest town. The third specimen was collected near a rocky stream running through high grassland (Spawls and Branch 1995).

Food and Feeding

No information is available.

Behavior

Nothing is known. Assumed to be terrestrial (Spawls and Branch 1995). Thought to be nocturnal. A single captive specimen was calmer than a puff adder, *Bitis arietans,* of comparable size, was hesitant to strike, but hissed when teased and struggled "furiously" when restrained.

Reproduction and Development

Nothing is known. Spawls and Branch (1995) assume it produces live young.

Bite and Venom

Nothing is known about toxicity or composition of the venom. No cases of envenomation are known, but it is assumed to cause some of the reported bites in densely populated regions of southern Ethiopia. Local people consider it highly dangerous (Spawls and Branch 1995).

Subgenus *Keniabitis*
Lenk, Herrmann, Joger, Wink 1999

Bitis worthingtoni is the single species in the subgenus, which is a sister group of the other subgenera. It is characterized by having nasal in direct contact with rostral and by having undivided subcaudals. It has a single horn-like scale above each eye and a different cytochrome b (Lenk et al. 1999). The geographical distribution is restricted to Kenya.

Bitis worthingtoni
Parker 1932

Commonly known as the Kenyan horned viper.

Recognition
(Plates 5.14 and 5.15)

Head—Broad, flat and triangular, distinct from narrow neck; eyes small, pupil vertically elliptical and set far forward; crown with small, imbricate, strongly keeled scales; supraorbital is raised and bears a small horn; sides of head from eye with black triangles, separated by off-white regions.

Body—Stout, especially females; dorsals 27–31 at midbody, rough and strongly keeled; subcaudals undivided; tail thin and short, terminating in a spike.

Size—Relatively small, 20–40 cm; maximum 50 cm (Isemonger 1962; Spawls and Branch 1995).

Pattern—Head ash gray dorsally, with a blackish brown, narrowly white-edged, forward-pointing triangular blotch; body dorsally brownish gray to olive gray; with 30–35 alternating triangular to semicircular black blotches, may be white to yellow along lower edge, fusing in some specimens; an additional series of butterfly-shaped dark blotches occur along flanks which these may be directly below or alternating with dorsal triangular markings; venter off-white heavily mottled with black; tail ventrally uniformly yellow (Isemonger 1962; Spawls and Branch 1995).

Taxonomy and Distribution

Described from specimens collected near Lake Naivasha, Kenya (Golay et al. 1993). No subspecies are recognized.

Restricted to high altitudes (over 1500 m) of Kenya's high central Rift Valley. The southernmost record is from the northwestern Kedong valley. Ranges north along the floor and eastern wall of the Rift Valley through Naivasha and Elmenteita to Njoro, where it extends up the western wall and out of the Rift Valley to Kipkabus and Eldoret (most northerly record). It also occurs on the Kinangop around Kijabe, on the hills west of Lake Naivasha, and possibly on the eastern Mau escarpment (Spawls and Branch 1995).

Habitat

High grassland and scrub are preferred habitats. It is also found on the valley floor and at the edges of acacia woodland (Spawls and Branch 1995).

Food and Feeding

Captive specimens readily feed on rodents and lizards. It is a sit-and-wait predator, lying coiled in cover capturing prey from ambush (Spawls and Branch 1995).

Behavior

Nocturnal, crepuscular and periodically diurnal, especially during periods of low nighttime temperatures (Spawls and Branch 1995).

Reportedly bad-tempered, hissing and puffing when disturbed. Struggles "wildly" when restrained (Spawls and Branch 1995).

Reproduction and Development

From 7 to 12 live young are produced in March or April, the start of the rainy season. Hatchlings are from 10 to 14 cm long (Spawls and Branch 1995).

Bite and Venom

A single case history has been reported. The victim was an amateur herpetologist bitten on the hand. Moderate pain and mild swelling occurred at the bite site. Treatment involved only analgesics and i.v. fluids. Symptoms were resolved without serious complications.

Chapter 6
Cerastes

The genus *Cerastes* contains snakes commonly known as horned vipers. However, horned and hornless individuals occur within the same population and indeed within the same litter (Werner et al. 1991; Sterer 1992). Horns are usually short or absent in individuals from stony deserts (Schnurrenburger 1959). When present, the horn is located over each eye and consists of a single long, ribbed, spikelike scale. The horns permit a buildup of sand above the eye, without impairing vision. These supraocular horns may be folded back in response to direct stimulation, effectively streamlining the head facilitating passage through burrows. The medial attachment of the scale serves as a supple hinge, stretching with pressure and scale movement, and an indentation in the postocular scale houses the folded horn (Cohen and Myres 1970).

In general, members of this genus are not particularly bad-tempered or dangerous. Although bites are painful, little venom is injected and it is not highly toxic. They will stand their ground if threatened. Fatalities are rare.

The following sections discuss the ecology of the species separately. Following the species-specific accounts of ecology, the venom and bite of these species are discussed together.

Recognition

Head—Broad, flat, triangular; distinct from neck; dorsally covered with small, irregular, tubercularly keeled scales, which usually number 15 or more across; snout short and broad; eye small to moderate with vertically elliptical pupil; supraorbital horns present or absent.

Body—Cylindrically depressed, short and stout, with a short tail that tapers abruptly behind the vent; with oblique lateral row of serrated keeled scales (U.S. Navy 1991).

Taxonomy and Distribution

The genus was established by Laurent in 1768 from the type species *Coluber cerastes*. Historically, records date back 5,000 years (Werner et al. 1991). Three species are widely recognized: *Cerastes cerastes, C. gasperetti* and *Cerastes vipera,* (Werner et al. 1991; Leviton et al. 1992).

Horned vipers are restricted to desert regions of North Africa and southwestern Asia. *Cerastes cerastes* and *C. gasperetti* are ecologically similar but mainly with an allopatric distribution. Although *C. vipera* and *C. cerastes* have a general sympatric distribution, microhabitat differences are a major factor in their geographic partitioning. The stony Negev desert functions as a "filter zone" between *C. cerastes, C. vipera,* and *C. gasperetti* populations (Werner and Sivan 1992; Werner 1995). Several subspecies have been described (Werner et al. 1999).

Cerastes cerastes
(Linnaeus 1758)

Commonly known as the Saharan horned viper or desert horned viper.

Recognition
(Plates 6.1 and 6.2)

Head—Supraorbital horn present, reduced or absent; individuals without horns with prominent brow ridge; suborbitals separated from supralabials by 3–4 small scale rows; with pair of enlarged tubercles near midline of head (lacking in *C. gasperetti*); nostrils directed upward and outward; interoculars 14; eye situated on side of head; significant sexual dimorphism in head size and eye diameter (males > females). Statistically significant differences occur in several characteristics of African and Mideastern individuals. Relative head sizes of *Cerastes cerastes* individuals are larger and there is a greater frequency of horned individuals (48% compared to 13%). Individuals of *Cerastes gasperetti* have flatter and broader heads with larger nasal shields and darker cheek blotches (Werner et al. 1991).

Body—Dorsal scales heavily keeled, with apical pits, 23–35 rows at midbody; lateral scales smaller, serrated and oblique; ventrals laterally keeled, 102–165 (*C. cerastes* males <147, females <154); significant sexual dimorphism between male and female ventral scale counts (females > males); subcaudals posteriorly keeled, paired, 18–42; anal scale entire, rarely divided or partially divided.

Size—Most specimens 30–60 cm. Maximum reported length 85 cm with females larger than males (Joger 1984).

Pattern—Ground color from yellowish, pale gray, pinkish or pale brown, and nearly always matching the color of the substrate from inhabiting areas; dark semirectangular dorsal blotches extend posteriorly and may fuse into crossbars; ventrally whitish. Tail tip may be black (Khalaf 1959; Mehrtens 1987; Phelps 1989; U.S. Navy 1991; Werner et al. 1991; Spawls and Branch 1995).

Taxonomy and Distribution

Described as *Coluber cerastes*. Two subspecies are described,
 Cerastes c. cerastes and
 Cerastes c. hoofieni (Werner et al. 1999).

Two other subspecies have been proposed: *Cerastes c. mutila* from southwestern Algeria, and *C. c. karlharti* from southeastern Egypt (Mehrtens 1987). These last two subspecies are mostly considered to be synonymous with *Cerastes c. cerastes* (Werner and Sivan 1992).

Cerastes cerastes is found in the Sahara Desert, from Egypt (the Sinai) to Morocco, south to Mauritania, northern Mali, Niger, Chad, Sudan, Libya, and Israel. It also occurs in the southwestern corner of the Arabian peninsula. In Egypt, the species occurs along the desert adjacent to the Nile River and the shores of the Red Sea from Suez to the Sudan (Kramer and Schnurrenburger 1963; Mohamed and Khaled 1966; Mehrtens 1987; Werner et al. 1991; Spawls and Branch 1995). The isolated Arabian populations belong to the subspecies *Cerastes cerastes hoofieni* Werner and Sivan (in Werner et al. 1999).

Habitat

Prefers dry sandy areas with scattered rocky outcroppings (Warburg 1964). Avoids coarse sand (Kramer and Schnurrenburger 1958). Inhabitants of arid areas, they occasionally occur around oases. Found to 1500 m of altitude (Joger 1984; Spawls and Branch 1995).

Microclimate is important in regulating distribution. In northern Israel, they are absent from sand dunes which appear suitable, but maximum and mean temperatures are too low and humidity is too high. *Cerastes cerastes* favors cooler temperatures with annual averages down to 20°C or lower (Warburg 1964).

Food and Feeding

They feed primarily on rodents, birds, geckos, agamas, chameleons, and other lizards, with the exception of skinks. In captivity, they readily feed on the lacertid, *Eremias olivieri,* and the gecko, *Stenodactylus petrii* (Kramer and Schnurrenburger 1958, 1963; Schnurrenburger 1959). A full-grown specimen in captivity for 14 years in the Giza Zoological Gardens in Israel was fed only weaver finches, *Passer domesticus niloticus* (Flower 1933). When struck, the birds died in 27–90 seconds. Swallowing took place only "after much deliberation" by the snake. During warm months, it ate 1 bird per month, requiring only 6–8 meals per year. It refused food during winter months. *Cerastes cerastes* also eats jerboas, *Jaculus jaculus,* and the stomachs of 7 specimens contained yellow wagtails, *Motacilla flava thunbergi,* and chiffchaffs, *Phylloscopus collybita* (Kramer and Schnurrenburger 1958, 1963; Schnurrenburger 1959).

Behavior

The primary method of locomotion is sidewinding (Schnurrenburger 1959). During sidewinding, their weight presses into sand (rather than sliding on it sideways) leaving whole-body impressions from which ventral scales can often be counted (Kramer and Schnurrenburger 1958, 1963; Schnurrenburger 1959; Werner and Sivan 1992).

In winter, activity is limited to several hours each day. Hibernation occurs from January to February and is frequently interrupted on warm days. Increased activity occurs in late March (Kramer and Schnurrenburger 1958; Mermod 1970). In late fall and early spring, they forage mainly at night, basking during the day. As a rule they expose themselves only during high humidity, which, just prior to sunrise, may reach 90% in Libyan deserts. High humidity minimizes water loss in these snakes, their only source of water being their prey (Kramer and Schnurrenburger 1958; Spawls and Branch 1995). Days are spent mostly buried in sand, in rodent holes, under rocks, or at the base of shrubs (Schnurrenburger 1957; Kramer and Schnurrenburger 1958; Spawls and Branch 1995).

Cerastes can "sink" into sand quickly using its keeled, angled, and serrated lateral scales in a rocking motion. The process begins posteriorly, extending anteriorly until the entire head is buried with just the eyes and nostrils exposed. Burrowing can occur whether the body is outstretched or coiled.

Cerastes cerastes is fairly placid. When disturbed it either retreats to a hiding place or remains still. Sometimes when angered, individuals assume a C-shaped posture and rub the inflated body loops together, producing a rasping noise similar to saw-scaled vipers, *Echis*. They can strike quickly (Greene 1988; Spawls and Branch 1995).

Reproduction and Development

In captivity, copulation was observed during the first week of April and always occurred while buried under sand (Schnurrenburger 1959). One pair remained in copulation 4 days.

Cerastes cerastes is oviparous, depositing eggs under rocks or in abandoned mammal or reptile burrows. From 8 to 23 eggs are layed, which hatch after 50–80 days incubation (Spawls and Branch 1995). Sterer (1992) reported an incubation period of 53 days at 24°C–28°C. Eggs incubated between 23°C to 25°C at night and 28°C and 30°C by day hatched in 48 days. Hatching requires incubation at 70% humidity (Sterer 1992). Oxygen consumption during embryonic development exhibited an exponential pattern of increase (Packard and Packard 1988). Hatchlings are 15–18 cm and feed on small lizards and nestling rodents (Mehrtens 1987).

Saint Girons (1986) provided a detailed summary of the reproductive cycle from individuals from the Algerian Sahara: vitellogenesis occurred from May to June; mating from mid-May to mid-June; ovulation during early July; egg laying during late July to early August; and hatching during September. Flower (1933) reported a *C. cerastes* laying a single egg on 13 August and 12 eggs on 16 August.

Cerastes gasperettii
Leviton and Anderson 1967

Known as the Arabian horned viper or desert horned viper.

Recognition

Head—Supraorbital horn present, reduced, or absent; individuals without horns with prominent brow ridge; suborbitals separated from supralabials by 3–4 small scale rows; lacking pair of enlarged tubercles near midline of head (present in *Cerastes cerastes*); nostrils directed upward and outward; interoculars 14; eyes situated on sides of head; significant sexual dimorphism in head size and eye diameter (males > females).

Statistically significant differences occur in several characteristics of *Cerastes gasperetti* and *C. cerastes*. Relative head sizes of *Cerastes gasperetti* are smaller, and there is a smaller frequency of horned individuals (13%). In addition they have flatter heads, with a larger nasal shields and darker cheek blotches (Werner et al. 1991). The subspecies *Cerastes g. mendelssohni* totally lack horns.

Body—Dorsal scales heavily keeled, with apical pits, 23–35 rows at midbody; lateral scales smaller, serrated, and oblique; ventrals laterally keeled, 102–165 (males >147, females >154); significant sexual dimorphism between male and female ventral scale counts (females > males); subcaudals posteriorly keeled, paired, 18–42; anal scale entire, rarely divided, or partially divided.

Size—Most specimens 30–60 cm. Maximum reported length 85 cm with females larger than males (Joger 1984).

Pattern—Ground color from yellowish, pale gray, pinkish, or pale brown, and like in *Cerastes cerastes,* often matching color of substrate from inhabiting areas; dark, semi-rectangular dorsal blotches extend posteriorly and may fuse into crossbars; ventrally whitish. Tail tip may be black (Mehrtens 1987; Phelps 1989; U.S. Navy 1991; Werner et al. 1991; Spawls and Branch 1995).

Taxonomy and Distribution

Described as *Cerastes cerastes gasperetti* by Leviton and Anderson (1967) and raised to species level by Werner (1987) and Werner et al (1991). Two subspecies are described,
Cerastes gasperetti gasperetti and
Cerastes gasperetti mendelssohni (Werner et al. 1999).

Cerastes gasperetti is found in west Asian deserts including most of the Arabian peninsula. The subspecies *Cerastes g. mendelssohni* is restricted to the sand areas within the Árava Valley of Israel and Jordan (within the Great Rift Valley, between Sedom in the north and Elat in the south) (Werner et al. 1999) while *Cerastes g. gasperetti* is distributed along the eastern edge of the Sinai Peninsula, south and east across the Arabian Peninsula to Iraq, Kuwait and southwestern Iran (Mohamed and Khaled 1966; Mehrtens 1987; Latifi 1991; Werner et al. 1991; Leviton et al. 1992; Werner et al. 1999).

Habitat

Cerastes gasperetti usually avoids soft dunes. Inhabitants of arid areas, they occasionally occur around oases. Found to 1500 m altitude (Joger 1984; Spawls and Branch 1995).

Microclimate is important in regulating distribution. *Cerastes gasperettii mendelssohni* from Israel and Jordan occurs only in the Arava Valley, which is the warmest part of these two countries, where annual average temperature is 24°C.

Food and Feeding

Actively sidewinds to track prey or utilizes sit-and-wait behavior while buried. If necessary, they will travel long distances in search of prey, sometimes as much as a kilometer or more in a single night (Gasperetti 1988).

They feed primarily on rodents, birds, geckos, agamas and other lizards, with the exception of skinks.

Behavior

No information is available.

Reproduction and Development

No information is available.

Cerastes vipera
(Linnaeus 1758)

Commonly known as the Sahara sand viper or Avicenna viper.

Recognition
(Plates 6.3 and 6.4)

Head — Supraorbital horns lacking; 9–13 interocular scales; eye small, set far forward on the junction of the side and top of the head.

Body — Dorsal scales rough and heavily keeled, in 23–27 rows at midbody; ventral scales less than 130, with slightly obtuse lateral keels.

Size — Stocky, flattened, smaller than *C. cerastes;* 20–35 cm, maximum length 50 cm; females larger than males (Joger 1984; Mehrtens 1987; Spawls and Branch 1995).

Pattern—Coloration generally more faded with fewer spots than *C. cerastes;* light brown, yellowish, or cream; dorsally marked with pale brown blotches and spots. Sexual dimorphism occurs in *C. vipera,* with female tails darker than males (Marx 1958; Kramer and Schnurrenburger 1963; U.S. Navy 1991; Spawls and Branch 1995).

Taxonomy and Distribution

Described as *Coluber vipera* from specimens collected from Egypt. No subspecies are recognized.

Restricted to north African sandy deserts. It is found from Israel, the Sinai, west to Morocco, Mauritania, Mali, and Niger (Kramer and Schnurrenburger 1958; Mohamed and Khaled 1966; Mehrtens 1987; Werner et al. 1991; Spawls and Branch 1995).

Habitat

Exclusively a desert dune dweller. Fine silty sand is an essential requirement. *Cerastes vipera* favors relatively cool temperatures, with annual averages 20°C or lower. Maximum altitude appears to be 1500 m (Kramer and Schnurrenburger 1958; Joger 1984; Gasperetti 1988; Spawls and Branch 1995).

Food and Feeding

Diet is similar to *Cerastes cerastes*. *Cerastes vipera* utilizes caudal luring to catch prey and is unusual in that the dark tail and luring behavior persist through adulthood. Small lizards, especially the abundant *Acanthodactylus pardalis,* are susceptible to luring and are heavily preyed upon (Heatwole and Davison 1976). In captivity they accept prekilled mice (Mehrtens 1987).

Behavior

Locomotion is primarily by sidewinding with the tracks of *C. vipera* more angular than those of *Cerastes cerastes* (Schnurrenburger 1959).

Active mainly at night, spending most of the day buried in sand, often at the base of a shrub or below a dune slipface (Schnurrenburger 1957; Kramer and Schnurrenber 1958; Spawls and Branch 1995). During late fall, winter and early spring, they may be seen during the day, exposing themselves only during high humidity.

During a 2-year study, Schnurrenburger (1959) observed that at temperatures between 18°C and 21°C, *Cerastes vipera* left their hiding places at about 2000 h to wander. They alternately crawled and buried themselves until 0100 hours when they stayed buried under about one-half inch of sand. At approximately 0400 hours many started crawling again with activity continuing until around 0600 h. Thereafter, all stayed in their hiding places.

Spends the winter months of January and February hibernating in dark retreats with frequent interruptions to bask on warm days (Mermod 1970).

Home range is correlated with the quality of the physical environment. An individual *C. vipera* remained for a 3-month period in an area less than 2 hectares in size which was heavily inhabited by small shrubs and prey items (Mermod 1970).

Upon being disturbed, *Cerastes vipera* immediately burrows into the sand. Sometimes when angered, individuals rub the inflated loops of their body, producing a rasping noise similar to that of the saw-scaled vipers, *Echis* (Greene 1988).

Reproduction and Development

Very little is known. It is reported to be both oviparous and viviparous (Shine 1986; Sterer 1992). One captive female layed 8 yellow-brown eggs. Most eggs hatched a few hours after deposition, a phenomenon not previously known in African snakes (Spawls and Branch 1995).

Bite and Venom

Epidemiology—Horned vipers are not particularly bad tempered or dangerous although bites do occur, especially in the vicinity of the Suez Canal. No bite statistic data are available (Kramer and Schnurrenburger 1958; Mohamed and Khaled 1966; Mohamed, Abdel-Baset and Hassan 1980).

Yield—Yields for *Cerastes cerastes* are reported as follows: 19–27 mg dried venom (Calmette 1907); 70 mg (Spawls and Branch 1995); 40–70 mg (Minton 1974, 1990). Labib et al. (1979) collected 100 mg total from 5 individuals. A range of 9.9–20.9 mg (mean 16.8 mg) collected over 1 year from 199 individuals is reported by El-Hawary and Hassan (1974a).

Cerastes vipera yields are smaller, with a range of 0.5–5.9 mg (mean 3.5 mg) collected over 1 year from 743 individuals (El-Hawary and Hassan 1974a). An average yield of 6.7 mg from 15 individuals is reported by Labib et al. (1979).

Cold weather affects venom yield from *C. vipera* more profoundly than *Cerastes cerastes*. During cold months, *C. cerastes* produced 64.1% of its summer yield, while *C. vipera* produced only 8.3% (El-Hawary and Hassan 1974a).

Toxicity—*Cerastes cerastes* venom is not very toxic (Mohamed, Kamel and Ayobe 1969a). Intravenous LD_{50}s are reported as follows: 0.50 mg/kg (Minton 1974, 1990; Spawls and Branch 1995); 5.0–6.0 µg/mouse (Mohamed et al. 1973); 0.3–0.4 mg/100-g rat (Mohamed et al. 1966); 9.6 mg per 20-g mouse (Christensen 1955); 0.48 mg/kg (Mebs 1978). LD_{50}s for subcutaneous and intraperitoneal injections are 15.0 mg/kg (Minton 1974) and 0.2–0.3 mg/kg (Mebs 1978) respectively. An estimated lethal dose for humans is 40–50 mg (Spawls and Branch 1995). LD_{50} values for *C. vipera* are not known (Spawls and Branch 1995).

Times of death for 18–20-g white mice i.p.-injected with 0.025 ml of a 1%

venom solution was 55 minutes for *C. cerastes* and 60 minutes for *C. vipera* (Hassan and El-Hawary 1977).

Content—Specific gravity of *C. cerastes* and *C. vipera* venom is slightly above 1.0 (1.095, 1.150 respectively), suggesting a high protein content (El-Hawary and Hassan 1974a). Protein comprised 85.6% and 80.4% dry weight and 4.2% and 4.02% of whole venom respectively. The venoms are more soluble in saline than distilled water, suggesting the major protein portions are globulins. Free amino acids are similar between the two species, with 18 detected from *C. cerastes* and 20 from *C. vipera*. *Cerastes vipera* venom contained a high concentration of histidine, while *C. cerastes* had high alanine and cystine levels (El-Hawary and Hassan 1974b; Hassan and El-Hawary 1977).

Gel filtration revealed eight venom fractions from both *Cerastes* species. The first fraction was the most toxic, having hemorrhagic activity. It contained approximately one-third the total protein and accounted for about two-thirds the "recovered lethality" after fractionation. This fraction had a molecular weight of 75,000, which is much higher than most isolated viper toxins (Labib et al. 1979, 1981a, b). Further studies involving venom fractions revealed 60% of the recovered fractions were lethal, suggesting venom lethality as a whole is due to synergistic action of two or more components. Two fractions were anticoagulant and three were procoagulant for both species. Daoud et al. (1988) isolated an anticoagulant proteinase from *C. cerastes* venom that hydrolyzed both fibrin and fibrinogen.

Cerastes cerastes venom exhibits proteinase and transaminase activity (Mohamed, El-Serougi and Khaled 1969b) and Labib et al. (1981a, b) reported that the proteases found belong to the metal chelator group. A powerful enzyme, ophio-L-amino acid oxidase, capable of oxidizing L-amino acids is present. A homogeneous single polypeptide, anticoagulant proteinase possessing both fibrinogenolytic and fibrinolytic activities was purified by El-Asmar et al. (1985). In addition, the venom is rich in phospholipase A_2 (Spawls and Branch 1995). Phospholipase A activity is higher in *C. cerastes* venom than in those of seven other venomous Egyptian snakes (Mohamed, El-Serougi and Hannah 1969).

Symptoms and Physiological Effects—Despite the relatively wide range and common occurrence of *Cerastes,* little information is available on bite symptoms. In humans, *C. cerastes* venom is reported similar in action to saw-scaled vipers, *Echis* (Mohamed, Kamel and Ayobe 1969a). Envenomation usually causes swelling, hemorrhage, necrosis, nausea, vomiting and hematuria. The high phospholipase A_2 content may cause cardiotoxicity and myotoxicity (Spawls and Branch 1995).

An isolated hemorrhagic toxin caused extensive damage to mouse capillary endothelium, which became thin and highly degenerated causing gaps for erythrocyte leakage (Rahmy et al. 1991).

Effects of *Cerastes* venom in guinea pigs are multisystemic, affecting various organs and tissues at cellular and molecular levels. Phasic contractile inhibition of pulmonary arterial rings occurred, but cardiac output was minimally affected, with a brief reduction in heart rate and contractility occurring only after high venom doses. This vasodilatory effect in other blood vessels may cause hypotension (Abdalla et al. 1992).

Dog arterial blood pressure dropped from 160/140 mm Hg to 50/40 mm Hg following i.v. injection of 2 mg of venom from *C. cerastes* (Mohamed and Khaled 1969). It is not clear whether hypotension is a direct venom effect or the result of liberated hypotensive agents from tissues or blood. Atropine failed to prevent hypotension, suggesting that the hypotensive mechanism is not mediated through activation of the cholinergic autonomic-parasympathetic nerve endings' inhibitory effect on myocardium. Possible mechanisms include the following: the presence of a free, active peptide; venom phosphodiesterase; or released histamines, which occur in varying amounts, increasing with dose and contact time. A positive pressure response following adrenalin administration, even after large injected doses, shows that arterioles and capillaries are not irreversibly damaged (Mohamed, El-Serougi and Hanna 1969). A definite neurological effect occurs. In rats, venom from *C. cerastes* blocked muscle contraction by inactivating fine nerve branches. Postsynaptic membranes remain sensitive to transmitter substances following addition of acetylcholine (Mohamed and Khaled 1966), suggesting that motor response depression may be due to inhibition of presynaptic transmitter release or to a neuronal blocking effect (Tilmisany and Najjar 1982). Cholinesterase activity was found in venom from both *C. cerastes* and *C. vipera,* but this activity was relatively low in comparison to venom from the Egyptian cobra, *Naja haje* (El-Hawary and Hassan 1976).

Venom from *C. cerastes* also showed a myonecrotic effect (Rhamy et al. 1991). Mice injected with crude venom displayed variable degrees of muscle structure alteration, with myofibrils becoming fragmented and separated. The characteristic banding pattern of skeletal muscle was lost, sarcomeres became undifferentiated, and the sarcoplasmic reticulum disintergrated. In addition, skeletal muscle lactate levels increased significantly (Abu-Sinna et al. 1993). Hemorrhage occurred within the endomysium, which contained extravasated red blood cells and remnants of broken capillaries (Rhamy et al. 1991). An isolated hemorrhagic toxin that damages capillaries left myofilaments normal, suggesting myonecrosis is due to a component separate from the hemorrhagic toxin (Mohamed, Fouad, Abbas et al. 1980).

At the cellular level, venom from *C. cerastes* produced a strong uncoupling and reverse acceptor control response in rat liver mitochondria (Augustyn and Elliott 1967). Mitochondria became dense to the point of obliterating the cristae (Rhamy et al. 1991). The venom caused statistically significant inhibition of oxygen uptake in brain tissues (Mohamed, Kamel and Ayobe 1969b).

Sublethal doses (0.2 mg/kg) increased sodium levels and decreased potassium levels, indicating stimulation of the adrenal cortex. Conversely, lethal doses (0.5 mg/kg) caused decreased sodium and increased potassium levels, an indication of adrenal cortex inhibition (Mohamed, Fouad, Abdel-Aal et al. 1980).

Venom from *Cerastes cerastes* targets plasma proteins, especially high and low density lipoproteins (HDLs and LDLs respectively). However, despite extensive lipoprotein degradation, LDLs remained intact (El-Asmar and Swaney 1988).

Venom from *C. cerastes* inhibited phasic contractions of the ileum, causing significant relaxation of its tone, possibly explaining diarrhea that often occurs following envenomation (Abdalla et al. 1992).

In rats, sublethal and lethal i.p. injections of venom from *C. cerastes* simultaneously depleted liver and muscle glycogen and inhibited glucose utilization (Mohamed, Fouad, Abbas et al. 1980). Sublethal i.p. doses caused marked hypoglycemia in only 15 minutes (Abu-Sinna et al. 1993). This condition continued for at least 24 hours. During this time, hypoglycemia was accompanied by significant increases in plasma, liver, and muscle glycogen.

Hematological Effects—The major hematological disturbance of *C. cerastes* envenomation is thrombinlike activity causing hypercoagulability followed by incoagulability due to fibrinogen degradation (Minton 1974). At low concentrations venom acts as a procoagulant (Rao et al. 1959; Boquet 1967a), and at high concentrations as an anticoagulant (Labib et al. 1981b). Procoagulant activity is due to action at multiple sites in the blood coagulation cascade. A component that coagulates human plasma that is deficient in several clotting factors was isolated. This compound was otherwise nontoxic and had no effect on platelet aggregation (El-Asmar et al. 1986).

In vivo coagulation was reported 15 minutes after lethal injections in rabbits and dogs. However, the venom could not coagulate a pure fibrinogen solution, indicating a lack of thrombinlike compounds. Rather, it acts as an extrinsic thromboplastin, containing a compound capable of activating prothrombin (Mohamed, Kamel and Ayobe 1969a).

Anticoagulant activity is due mainly to direct and/or indirect fibrinogenolysis (Labib et al. 1981a). Venom from both *Cerastes* species caused hypofibrinogenemia followed by fibrinolytic activity, which appears due to plasma activation of fibrinolysins. An *in vitro* fibrinogenolytic effect mediated by venom proteases is reported (Mohamed and El-Damarawy 1974) with the accumulated fibrinogenolytic fragments blocking fibrin polymerization (Fletcher et al. 1962). An enzyme, "cerastase F-4," exhibits both fibrinogenolytic and fibrinolytic activities *in vitro* in an interesting pattern. It degraded the "A" alpha-chain rapidly, followed by the "B" beta-chain. This is in contrast to other purified fibrinogen proteases, which degrade only the "A" alpha-chain or the "B" beta-chain (Daoud et al. 1986,1988). Differences in fibrin and fibrinogen degradation products indicate that cerastase F-4 differs from plasmin. In addition, cerastase F-4 degraded prothrombin, but in a manner too slow to make it a significant contributor to overall anticoagulation. Dogs heparinized prior to venom injection showed no sign of hypofibrinogenemia (Mohamed, Kamel and Ayobe 1969a). Daoud et al. (1986) suggested that the low toxicity and hemorrhagic activity of cerastase F-4 could make it useful as a thrombolytic agent.

Venom from *C. vipera* i.p. injected into rabbits produced an initial rise in erythrocytes after 15 minutes, followed by a progressive drop (Mohamed, Saleh, Ahmed and El-Maghraby 1977). This drop was associated with the occurrence of transient spherocytosis.

The proteolytic activity of venom from *Cerastes* on hemoglobin is higher than from any other venomous snakes found in Egypt, and causes marked hemolysis *in vitro* (Mohamed, Saleh, Ahmed and El-Maghraby 1977; Labib et al. 1981a).

Significant leucopenia and thrombocytopenia occurs in rabbits following injection of lethal doses of venom from *C. vipera*. Sublethal doses resulted in similar effects, however, they were only transient, reattaining preinjection levels after 24 hours (Mohamed, Saleh, Ahmed and El-Maghraby 1977).

The effects of venom from *C. vipera* on rabbit bone marrow revealed erythroid hyperplasia 5–6 hours after injection (Mohamed, Saleh, Ahmed and El-Maghraby 1977).

Case Histories—Harrison (1992) described an account of a 31-year-old 82-kg man bitten by an 84-cm *C. cerastes* on the left middle finger by one fang at 1400 hrs. The victim had been bitten twice previously, once by a copperhead, *Agkistrodon contortrix,* and once by a cottonmouth, *Agkistrodon piscivorus*. Initial examination revealed an extremely deep laceration that exposed the bone. Thinned blood and serum drained from the wound and the victim complained of burning pain. No first aid was given.

After 1 hour the patient arrived at the hospital. Antivenin was withheld due to the mild systemic involvement. By this time, the finger was discolored and swelling reached the middle of the hand. A saline i.v. was started. Blood was first drawn after 80 minutes and revealed clot formation breakdown, a result of decreased serum fibrinogen. However, the coagulapathy resolved itself within 2 hours, by which time swelling extended 13 cm up the arm.

After nearly 3.5 hours, the entire hand showed ecchymosis, and swelling progressed 30 cm from the bite. Leucocyte count was slightly elevated, but clotting time, blood pressure, heartbeat, and EKG were all normal. By approximately 5.5 hours, the victim's vision was slightly blurred, and edema and ecchymosis involved the entire hand and wrist, with bleb formation at the bite site and on the hand. Swelling progressed 46 cm from the bite. The arm was swollen to 53 cm after 7.5 hours. Antibiotics (i.v.) were given 10 hours postbite and the patient was released 17 hours after admission. Swelling disappeared within 3 days but blebs remained for 2 more weeks. Some necrosis appeared at the distal portion of the finger, which after 4 weeks became gangrenous, necessitating amputation at the first joint. Although the patient had been bitten in mid-June, a rash developed in November in the inguinal and axial lymph nodes which remained swollen and sore (such a rash also occurred following the previous two bites).

Treatment and Mortality—*Cerastes* antivenin has high specific neutralization against its homologous venom but weak neutralization power against heterologous venoms (Mohamed et al. 1974b). *Cerastes cerastes* antivenin was 92% effective on venom from *C. cerastes* but only 53% effective on *C. vipera* venom, with similarly poor cross-neutralization effects found on a variety of venoms from other snakes. A bivalent antivenin against venom from *C. cerastes* and *C. vipera* proved very effective against bites of these species and moderately effective against other viperid venoms. This antivenin was 3 times more effective when concentrated and refined (Hassan and El-Hawary 1975). However, Mohamed, Fawzia et al. (1977) reported bivalent antivenin increased mouse LD_{50}s neutralized for

C. vipera (from 140 to 150) but decreased them for *C. cerastes* (from 141 to 96). Polyvalent antivenin from *Naja haje, N. nigricollis, N. naja, Cerastes vipera, C. cerastes, Echis coloratus, E. carinatus, Walterinnesia aegyptia, Bitis gabonica, B. arietans,* and *Trimeresurus flavoviridis* was equally effective in neutralizing venom from *Cerastes* when produced from either horses or goats (Mohamed et al. 1973; Mohamed, Fawzia et al. 1977). New World polyvalent antivenin (from three species of *Crotalus* and one *Bothrops*) provided a high degree of protection against venom from *C. cerastes* (Minton 1976). Concentrated *Echis* antivenin neutralizes venoms of both *Cerastes cerastes* and *C. vipera* (Schwick and Dichgeisser 1963).

Remarks

When *Cerastes cerastes* and *Cerastes vipera* occur sympatrically in simple uniform habitats like sand dunes, they usually differ significantly in size, with *Cerastes cerastes* being the larger. This occurs in the Moroccan and Algerian Sahara, the northern Sinai, and western Negev deserts. In northern Libya the two species are sympatric, but their habitats are usually partitioned such that *C. cerastes* lives in stony desert while *C. vipera* is found almost exclusively in sandy habitats with few rocks. When sympatric, *C. vipera* outnumbers *C. cerastes* 15:1. In the northern hills of Morocco and Algeria, *C. cerastes* coexists with *Macrovipera mauritanica*. In the Arava Valley in Israel, the sandy habitat of *C. gasperettii* is also occupied by the smaller saw-scaled viper, *Echis coloratus,* and the somewhat larger false horned viper, *Pseudocerastes persicus fieldi* (Werner and Sivan 1992).

Cerastes cerastes has only a few natural enemies. These include the honey badger, feral cats, wildcats, and the monitor lizard, *Varanus griseus* (Gasperetti 1988).

Resting metabolic rates of *C. cerastes* were below predicted values (Al-Sadoon 1991). The Q_{10}, while low in the lower part of the metabolism-temperature curve, increased sharply between 20°C and 25°C, followed by a sharp decrease. This suggests this species may be less sensitive to temperature change, since it is nocturnal and therefore active over a wide range of temperatures. The jump in Q_{10} during the 20°C–25°C temperature range is correlated with its voluntary preferred body temperature, which averages 30°C during the day and 27.6°C at night.

Hematological and biochemical comparisons of active and hibernating horned vipers showed interspecific differences in a few blood parameters (Al-Badry and Nuzhy 1983). Hematocrit values dropped for *C. cerastes* from 32.99% during summer active periods, to 21.79% during winter hibernation. Similarly, *C. vipera* exhibited a drop from 28.70% in summer to 18.66% in winter. A corresponding fall in hemoglobin concentrations from summer to winter were also recorded with *C. cerastes* dropping 23.84% and *C. vipera* dropping 20.0%. These decreases in hematocrit and hemoglobin percentages were statistically significant and are probably due to reduced metabolic rates and lower oxygen demands, bringing about a reduction in erythropoiesis. Leucocyte counts showed significant differences both interspecifically and seasonally, with a significantly higher overall count in both species during the winter months. Lymphocyte counts, the most common leucocyte in both species, dropped significantly in both species during hibernation.

Blood water levels remained nearly constant during active and hibernating times. Water content, 86.26% for *C. cerastes* and 88.12% for *C. vipera,* is relatively high and is thought to function as a temporary water depot. Electrolyte analysis showed little interspecific difference, but during hibernation, blood chloride, sodium, and bicarbonate concentrations dropped significantly, whereas potassium, calcium, magnesium, and inorganic phosphorous increased. Bicarbonate level decrease is characteristic of hibernating reptiles, which use it as a buffer against lactic acid buildup while active. Other electrolyte "drifts" result from reduced hibernacular metabolism reducing Na-K pump activity resulting in potassium leaving cells, while sodium enters and the release of magnesium from inactive muscles (Dessauer 1970). Blood sugar levels dropped appreciably during hibernation, suggesting that hypothermia may impair glycogenolysis during winter. Hibernation also resulted in a significant decline in lipid reserves which is probably due to increased lipolysis functioning as an important energy source during hibernation (Al-Badry and Nuzhy 1983).

During hibernation, there is a significant rise in total nonprotein nitrogen and uric acid concentration, with a concomitant drop in creatinine and total free amino acids. These fluctuations result from several factors including hypothermia, decreased glomerular filtration, and increased protein catabolism, which occurs in hibernating vipers. In addition, albumin and globulin decreases during hibernation in both vipers resulted in an overall decline in blood protein (Al-Badry and Nuzhy 1983).

Horned vipers do well in captivity, with some individuals living more than 17 years. Captive specimens should be maintained on several inches of fine nonabrasive sand with stacked rocks to permit hiding in the crevices. Water should be offered weekly but not left in the cage. Prekilled mice are readily accepted (Mehrtens 1987). Flower (1933) reported that during his studies of Egyptian reptiles he observed large numbers of *Cerastes* for sale at tourist resorts.

Native snake charmers, using *C. cerastes* in their acts because of their horns, will sometimes use *C. vipera* when the former species is in low supply. Porcupine or hedgehog quills are forced through the head to provide suitable but eventually fatal cranial adornments (Mehrtens 1987).

Chapter 7
Daboia

Daboia was first used by Gray in 1842 and remained in the literature until the late 1800s. The genus was reviewed by Obst (1983) and the extension of *Daboia* to include *lebetina, palaestinae, xanthina,* and allied taxa was proposed. However, Herrmann et al. (1992) retain *Macrovipera* for *lebetina* and *Vipera* for *xanthina,* and employ *Daboia* only for *russelii.* In this book, we follow Herrmann et al. and place *russelii* in the genus *Daboia.* In addition we also include *palaestinae* in this genus (see below).

Characters on which *Daboia* is based include body size, scalation, and biochemical relationships (Herrmann et al. 1992; Lenk et al. 2001). The present range for the genus Daboia is Asian, but during Pliocene it had a European distribution as shown by the discovery of the Spanish *Vipera maxima,* a species very similar to *Daboia russelii* (Szyndlar 1988).

Recognition

Head—Flat, triangular; snout obtuse and rounded: head distinct from neck; nostrils large, characteristic nasal scalation and shape; head dorsally covered with small, keeled scales, large supraoculars bordering eye, separated by 6–9 scales across; eye moderate with vertically elliptical pupil; distinct head color pattern (Groombridge 1986; Brodmann 1987).

Body—Stout and massive. Reduction or loss of peritoneal pigment.

Daboia palaestinae
Werner 1938

This is the Palestine viper. The generic position for *Vipera palaestinae* is under evaluation. Groombridge (1980, 1986) united *russelii* and *palaestinae* as a clade based on shared apomorphies (reduction or absence of black peritoneal pigment, nasal plate arrangement, snout shape and head color pattern). In a recent study, Lenk et al. (2001) found support for that idea based on molecular studies, suggesting the unification of *palaestinae* with *russelii* (together with *Macrovipera mauritanica* and *M. deserti*) in the genus *Daboia.*

Such a scenario would mean a mixture of a tropical ovoviviparous viper (*russelii*) with temperate adapted oviparous vipers (*palaestinae, mauritanica* and *deserti*). However, a mixture of ovoviviparity and oviparity occurs also in another Viperid genus, *Echis.*

An alternative possibility could be to retain *russelii* alone in *Daboia,* and place the clade *M. mauritanica, M. deserti,* and *V. palaestinae* in a separate genus.

In recent times, phylogenetic reconstructions have currently given new and different interpretations of relationships, and until a final phylogenetic pattern is verified by additional research we retain the present view for the genus *Macrovipera* (see below).

Vipera palaestinae show considerable phyletic distance from the small Eurasian vipers (subgenera *Pelias* and *Vipera*), as well as from the *Vipera xanthina* complex (subgenus *Montivipera*) (Herrmann and Joger 1997). In addition with its phylogenetic position close to *Daboia russelli* (Groombridge (1980, 1986; Lenk et al. (2001), we include *palaestinae* in *Daboia* as the most suitable position for that taxon.

Daboia is morphologically distinguished from *Macrovipera* by having large supraoculars.

Recognition
(Plate 7.1)

Head—Overall elongate, triangular, distinct from neck, with many small keeled scales dorsally, supraocular intact and bordering eye, 2 scale rows between eye and upper labials.

Body Scales—24–25 dorsal rows at midbody, 162–172 ventrals, 36–44 subcaudals.

Size—Average length 70–90 cm, maximum 130 cm, but occasionally bigger (180 cm—captive specimen in Bern Zoo—Nilson, pers.obs.), tail length 8.5–11.1 cm, largest adults tend to be males but there is considerable overlap in size, and sexual dimorphism is not clearly established.

Pattern—Head with a relatively constant pattern consisting of a thick, dark, inverted V dorsally, often with a dark blotch apex of the V, sometimes connected directly to the dorsal band, with a dark blotch below the eye and a dark line extending from the eye to corner of mouth.

Body dorsal ground color gray, reddish brown, beige, olive, or yellowish, usually with a wide central zigzag band of dark gray, blackish, brown, or reddish, sometimes edged with dark black, band may be broken into oval spots, saddles, or diamonds, edging may sometimes be pale, sides usually with spots or bands, alternating with the saddles or spots of the back, usually some yellow visible in the pattern of wild-caught snakes regardless of general coloration, yellow may not develop in captive-reared individuals unless a carotene-rich diet is fed, there is generally little or no discernible sexual dimorphism in dorsal pattern. Young are consistently a different color from adults, with the pattern exhibiting little or no yellow, all browns lacking reddish tones, and all having a gray dorsal ground, blotches or zigzags of the central band may tend to fuse and smooth yielding a single dorsal stripe, particularly on the anterior of some individuals (Mendelssohn 1963; Joger 1984; Groombridge 1986; Mehrtens 1987).

Taxonomy and Distribution

Over the past century, this species has been the subject of great taxonomic controversy and has been incorporated into several taxa. It was formerly included in *Vipera xanthina*. *Vipera xanthina*, s. l., was synonymized with *lebetina* by Boulenger in 1896 and the Palestinian populations remained in that group until they were given specific status as *Vipera palaestinae* by Werner in 1938. Since then, Mertens (1952) returned it to *V. xanthina* as a subspecies. Based on certain chemical features of venom, Nilson and Sundberg (1981) characterized *xanthina, palaestinae, raddei, and latifii* as distinct species but did not suggest what allocation among genera might be appropriate. Also, Obst (1983) suggested that it rated its own species status and should be moved to *Daboia* with *russelii* and *lebetina*. The controversy of names has got the result that many studies on venoms and envenomation, based on *D. palaestinae*, have been published under the name Vipera *xanthina*.

This species has a relatively restricted range. It is found from the Mediterranean coastal plains to inland hills of Lebanon and Israel and in adjoining regions of Syria and Jordan (Disi 1983; U.S. Navy 1991). Its distribution in the southern desert is poorly known but it apparently reaches the border of Egypt in the northern Sinai. It is absent from the Dead Sea region and the Negev Desert (Gitter and de Vries 1967).

Habitat

It is speculated that this species was historically a resident of canopied openfloored oak forests which covered large areas of the eastern Mediterranean landscape. Today, such forests are restricted to remnant patches, and density of *D. palaestinae* in such forest remnants is higher than in other regional ecosystems (Mendelssohn 1963).

Daboia palaestinae is found on rocky slopes, in pastures, in fallow and planted fields, and in wetlands. Soil type is not correlated with distribution. Variation in precipitation over the range is great, from under 300 mm/yr to over 1000 mm/yr. Although not an animal of dry sandy deserts, it occurs in human settlements in such landscapes. It occurs from below sea level in the Jordan Valley to over 1000 m in Galilee (Mendelssohn 1963).

Favored habitats are associated with humans and water. *Daboia palaestinae* is common in agricultural settlements throughout the region and is found in barns, stables, animal pens, along canals and irrigation ditches, and on the banks of streams (Mehrtens 1987). It is apparently transported effectively on agricultural machinery and in straw, hay, and fodder. Of specimens in the Tel Aviv University collection through the early 1960s, 80% were from settlements despite active collecting in remote areas (Mendelssohn 1963).

While population data are lacking, there is clear variation among different areas.

Daboia palaestinae is rare in the hills west and south of Jerusalem and common on coastal plains west to the Mediterranean. It is sparse in the Jordan Valley desert and elsewhere south of Israel. Relatively high population levels are reached in rodent-rich farm and village areas, with up to 20 adults being captured in a single season in less than a hectare of ground in such habitats. *Daboia palaestinae* is basically a lowland species which does not occupy mountain or high ridge areas. It should be noted, however, that large areas of the range are remote, and present distributional understanding may reflect poor collecting efforts as much as true habitat preferences and range (Mendelssohn 1963).

Food and Feeding

Warm-blooded prey are preferred in the wild, and birds and mammals are generally taken in relation to abundance and/or accessibility. Lizards are eaten, but less readily, especially by adults. Observations on captive animals suggest a preference hierarchy for food items: birds over mammals, gerbils over white mice, mice over rats, hamsters, or voles, and all of these preferred to spiny mice (*Acomys*) (Mendelssohn 1963).

Stomachs from field-caught specimens are usually empty (this may be due to behavioral patterns in which recently fed individuals are inactive and in hiding, and so are seldom caught). Stomachs yielding food items most commonly include the house mice (*Mus*) and nestling weaver finches (*Passer*). Because most snakes in collections were taken around human settlements, the dominance of these prey items is not surprising (Mendelssohn 1963).

Other recorded prey include rodents (*Microtus guentheri, Merione tristrami,* and *Rattus rattus*), birds (the goldfinch, *Carduelis carduelis*), shrews (*Crocidura russula*) and lizards (*Laudakia stellio* and *Acanthodactylus shreiberi,* the latter in young *palaestinae*) (Mendelssohn 1963).

Young captive and wild-caught individuals fall into one of two clearly differentiable feeding categories. Some eat and thrive on baby mice even though lizards are preferred when offered. Another group eats only lizards during the first year, starving to death if offered only mice. Lizard-feeders begin to accept rodents only in their second year of life. These habits are not correlated with obvious physical parameters such as body size, sex, or color pattern and seem to be distributed randomly among individuals within broods. In much of the habitat, lizards are abundant during the late summer hatching period of *D. palaestinae* and form the trophic basis for growth. Around human settlements, rodent-feeders may have a distinct growth advantage due to the presence of large numbers of mice and relative scarcity of lizards. It is not known, however, whether differential feeding has any genetic or evolutionary significance.

The initial feeding strike is usually repeated several times in rapid succession. Birds, characteristically are held until venom takes effect. The same is true of mammals if the snake has been starved or is otherwise exceptionally hungry. In general, however, mammals are struck, released, then followed via the scent trail for consumption. Swallowing capacity has been documented: a 450-g snake was incapable of swallowing past the shoulders of a 206-g rat but was able to swallow a 170-g rat without complication (Mendelssohn 1963).

It drinks readily when water is offered in a dish. However, it also obtains water from dew or droplets sprinkled on vegetation. Drops are retrieved from leaf surfaces while the animal exhibits normal drinking movements of the head and mouth (Mendelssohn 1963).

Behavior

Generally nocturnal, due largely to extremely high daytime temperatures throughout its range. When conditions are suitable, it is found abroad during daylight hours, usually basking in the morning in early spring. Adults prefer ambient temperatures around 30°C, and will feed to temperatures as low as 23°C degrees. Young prefer slightly lower temperatures between 27–29°C (Mendelssohn 1963).

A large individual sometimes basks with part of its body inside a retreat hole. This may be both a defensive and thermoregulatory response. Basking often occurs in branches of shrubs or trees, and Mendelssohn (1963) reported an individual encountered 4 m from the ground in a tree, which retreated into a deep hole in the trunk when disturbed.

Daboia palaestinae is not usually aggressive, preferring to flee even when provoked. It exhibits a characteristic coiling and hissing threat display, ceasing to hiss when uncoiling to escape. If disturbance continues, the display increases until the forward part of the body is held off the ground in the prestrike coil and flattened by rib spreading. The animal may continue elevating and hissing when retreating from this posture. There are reports of aggressive individuals pursuing people who disturbed them during the mating season. (Mendelssohn 1963).

Reproduction and Development

They are oviparous with eggs are usually being laid in August, rarely in July. Morning is preferred for oviposition although afternoon egg-laying sessions have been observed. In wild-caught specimens fertile egg number ranges from 7 to 22. Some individuals possess more than 22 eggs in the oviduct. A well-fed captive-reared individual had 35 well-developed ova. Sperm storage apparently does not occur and females not mated in a particular year lay only infertile eggs. Eggs are elongate, from 38 to 50 mm in length and 22–28 mm in width. Average length of 20 eggs in 4 clutches was 43.5 mm, average width 25.2 mm. Eggs weigh between 12 and 16 g, averaging 13.6 g, increasing in weight as they incubate. There is no relationship between the egg number per clutch and the size or weight of individual eggs. Larger females generally produce larger clutches although individual condition causes this relationship to vary greatly (Mendelssohn 1963).

Hatching takes approximately 6–9 weeks. Hatchlings weigh between 8 and 12 g. Total hatchling length ranges from 19 to 23 cm (mean 22 cm). Female hatchlings are generally larger with shorter tails. However, this sexual dimorphic tendency is variable and cannot be relied upon (Mendelssohn 1963).

Hatchling growth is relatively rapid through autumn. There is little or no growth from January through March, although the animals are capable of feeding and do sometimes feed at relatively low temperatures. At about 1 year old, young reach more than 40 cm in total length. Those hatched late in the previous year are smaller, about 35 cm (Mendelssohn 1963).

Initial molts take place 8–12 days after hatching. Those hatched early in the wild may feed heavily and molt a second time before winter, usually in November, and sometimes a third time before January. There are usually 3–4 molts in year 2, 2–3 in year 3. Adults molt once or twice per year. In captivity, if maintained without a cool period in winter, individuals will molt throughout the year (Mendelssohn 1963).

Bite and Venom

Epidemiology—Over a 7-year period in Israel (1969–1975), 615 snakebite cases were reported. The offending animal was identified in 179 cases, and of these, 165 (92%), were *D. palaestinae*. Clearly, this species is medically important within its range although bite incidence has decreased substantially in recent years (from 165 to 37) from 1969 to 1975 (Efrati 1979).

In a detailed report of 61 cases of viper bite in Israel, men were found to be at greater risk than women. Men represented 44 of the cases (72%), women 17 (28%). By far the greatest proportion of bites occurred during June, July, and August (Efrati and Reif 1953), which are the peak months of activity in the wild (Mendelssohn et al. 1971). In Jordan from 1970 to 1972 and from 1975 to 1980, 112 venomous snakebites were reported, not including those treated at military facilities or by folk medicine practitioners. The majority of the 31 bites from the northern district (from Amman to the Syrian border) were attributed to *D. palaestinae*. Similarly, 25 cases from the Salt Mountains east of Amman were likely *D. palaestinae*, because this is by far the most abundant venomous snake in this area (Amr and Amr 1983).

Yield—Average venom yield per first strike is approximately 32 mg. This represents less than 11% of the total content of the venom gland at the time of the strike (Kochva and Gans 1966). Individuals 90–120 cm produced yields 90–140 mg (Minton 1974).

Individuals ranging in weight from 428 to 678 g (autumn) and in length from 99–110 cm yielded up to 217 mg dried venom at a single milking. Maximal dry matter concentration of whole venom was 23.9%. Venom yield per milking was only partially dependent on individual snake size. Temperature played a major role in determining extracted venom quantity. Other determining factors included milking sequence and number of bites per extraction (Kochwa et al. 1960).

Toxicity—Venom is extremely toxic. Intravenous mouse LD_{50} is approximately 0.3 mg/kg (Ovadia and Kochva 1977), comparable to that for *D. russelii* and *V. ammodytes* and substantially less than most reports available for *lebetina* and *aspis* (Khole 1991). Minton (1974) reported LD_{50}s for i.v.- and s.c.-injected mice as

0.18 mg and 9.40 mg respectively. Intraperitoneal mouse toxicity levels for 15-g mice are 0.55 mg at the LD_{100} level and 0.038 mg for LD_{50} concentration. Injection of a single LD_{100} dose causes death in 15-g mice within 3 to 7 hours (Gitter and de Vries 1967).

Minimum lethal doses for human beings have not been specifically determined. Based on relationships between venom quantities commonly extracted and reported rates of fatality, a minimum lethal dose for humans has been estimated as 75 mg fresh weight (Kochva and Gans 1964). This is well within the injectable quantity for a single bite. Relatively light symptoms were associated with bites believed to inject 40 mg fresh weight or less.

Investigations employing radiolabeled snake venom indicate about 50 mg fresh venom weight is injected per bite by adults. This is about 8% of the venom usually present in the glands. Venom quantity injected is not clearly related to prey size although first bites usually inject a single lethal dose, at least for smaller prey items. Some individual snakes adjusted venom output to prey size but most individuals tested did not (Allon and Kochva 1974).

Content—The venom gland has a prominent accessory gland at the anterior end (Bdolah 1979). The function is currently not well understood. Its contents seem to have little or no effect on whole venom action.

Lytic enzymes of *D. palaestinae* venom are created and secreted directly from cells lining the venom gland walls. There is apparently not a necessary biochemical step involving the lumen of the gland (Bdolah 1979), and the open areas within the gland itself function primarily as a storage and preinjection "staging" area (Gennaro et al. 1963, Bdolah 1979).

Venom has anticoagulant, procoagulant, protease, lecithinase, and L-amino acid oxidase fractions in addition to major neurotoxic and hemotoxic components. A proteolytic factor has been isolated and is clearly separable from hemorrhagic venom components. Zinc and copper contents are very high (Gitter et al. 1963; Friedrich and Tu 1971).

Daboia palaestinae venom has clear protein-degrading activity measured by its ability to break down gelatins and caseins (Iwanaga and Suzuki 1979). Hemorrhagic and proteolytic actions may be due to the same venom components. However, the mechanisms by which these fractions are neutralized by blood sera apparently differ, indicating that the mechanisms of action, if not the molecules themselves, likely differ (Huang and Perez 1980).

Phospholipases A and B are present and phospholipase A can also act as phospholipase B, indicating a broad range of lipid substrates can be hydrolyzed. Phospholipase A activity in *D. palaestinae* is less dependent on calcium ions than those of other venoms and it may hydrolyze lipids in the absence of calcium ion. However, cell membrane substrates used to demonstrate this property were severely ruptured, and it is unclear if this finding has significance for bite symptomology or pathology (Barzilay et al. 1978).

Daboia palaestinae venom has two distinct components which act as presynaptic neurotoxins, affecting neurotransmitter release from motor nerve terminals.

In general, this kind of toxin may increase or decrease neurotransmitter release. In *D. palaestinae* it appears to inhibit release (Hawgood and Bon 1991).

A powerful neurotoxic fraction, "viperotoxin," has been purified. It is a large molecule (molecular weight approximately 11,600) consisting of a single crosslinked polypeptide chain with 108 amino acid residues (Moroz et al. 1966a). When tested individually the neurotoxic fraction induced paralysis and respiratory failure (Kochva and Gans 1964).

In venoms fractionated by ion exchange chromatography, neurotoxins represent approximately 60% and hemorrhagic factors about 35% of the residue. There are two distinct classes of neurotoxins in this residue, one causing paralysis and one causing generalized tremors. Anticoagulant fractions have been separated from the neurotoxins, indicating that at least some anticoagulant properties of the venom are due to direct effects rather than to effects mediated by the nervous system (Kochwa et al. 1960).

Hemotoxic and procoagulant fractions have been isolated (Grotto et al. 1967). The hemotoxic fraction is both hemorrhagic and proteolytic. The procoagulant fraction reduced whole blood clotting time and prothrombin time (Kochwa et al. 1960). Hemorrhagic factors may be divided into three components: one basic, one weakly acidic, and one strongly acidic. Two possess both gelatinase and caseinase activity, one has no quantifiable proteinase activity. In purified form these venom constituents are an order of magnitude more active than whole venom (Ovadia 1978b).

Symptoms and Physiological Effects—Following a bite there is intense pain and local swelling at the site within 15–20 minutes, followed by vomiting, discoloration, shock, fever, anemia, abdominal pain and cramping, bloody and serous blisters, edema, lymphangitis, weakness, and cold perspiration. Diarrhea occurs within hours and stools are bloody or watery. Hypotension is characteristic and may last several days. The bitten extremity may swell enormously, and become sensitive to touch and motion. After several days, blisters filled with blood or serum appear. Dehydration, remote swelling (usually involving the face), and fever may also occur.

Hypotension is a characteristic symptom and in *D. palaestinae* bites it is induced primarily by neurotoxic effects on vasoregulatory centers (Bicher et al. 1963, 1964; Vick et al. 1964; Moroz et al. 1966a). Neurotoxic fractions in *D. palaestinae* venom reduce blood pressure by depressing control center function in the center medulla of the brain (Bicher et al. 1966). Hypotension occurs secondarily as a result of extravasation due to vessel wall disruption (Chugh and Sakhuja 1991).

Shock is characteristic and may occur with a delay of up to 1 week following apparent recovery. In this "delayed shock" syndrome, the patient becomes restless with rapid respiration and develops fever and profuse sweating. Blood pressure falls and blisters and edema reappear. The delayed shock reaction may be fatal (Efrati and Reif 1953; U.S. Navy 1991).

Prognosis depends on intensity of circulatory failure, primarily the degree of extravasation and resultant hypotension. If internal bleeding ceases quickly, patients generally recover within 2–3 weeks. If bleeding is severe, death may follow within days (Efrati and Reif 1953, Efrati 1979, U.S. Navy 1991).

Other symptoms, occurring in fewer than 10% of cases, include urine retention, mucous membrane and skin hemorrhage, restlessness, bradycardia, and difficulty in swallowing (Efrati and Reif 1953; Efrati 1979; U.S. Navy 1991).

Hematological Effects—Hemorrhagic action is due to direct effects of venom on cells of capillary walls. The venom lyses capillary endothelium, leaving gaps through which plasma and blood cells escape into the surrounding tissues. However, it is presently unclear whether lysis of capillary endothelium or disruption of intercellular material by the venom permit erythrocytes passage between the cells. Perhaps both mechanisms operate, or one or the other predominates (McKay et al. 1970, Oshaka et al. 1975).

Daboia palaestinae venom is also a powerful anticoagulant (de Vries and Gitter 1957). Purified hemorrhagin from *D. palaestinae* venom impairs thrombin formation, fibrinogen clotability, platelet clot activity, and platelet release. These actions combine to yield a clinical picture of massive hemorrhage with normal platelet count when hemorrhagin is administered to guinea pigs (Grotto et al. 1969).

Case Histories—In one case (Efrati and Reif 1953), a 24-year-old man was admitted to the hospital 30 minutes after being bitten on the leg. On admission, his tongue and lips were swollen, he was vomiting, and exhibiting severe diarrhea. Pulse was impalpable and blood pressure was very low. First aid prior to admission included enlarging fang marks, application of potassium permanganate crystals to the bite area, and injection of a polyvalent antiserum. About 12 hours postbite, diarrhea ceased and swelling of the tongue and lips was still present as was vomiting. By 24 hours postbite, blood pressure recovered and swelling of the tongue was considerably reduced. Local swelling continued and spread to the upper leg. Two days postbite, leg tissue was swollen and hemorrhaging, vomiting began again, and large blisters filled with serum appeared. At 4 days, severe anemia accompanied by distinct pallor appeared. The patient was discharged in relatively good health 18 days after the bite.

In a second case, a 17-year-old woman was admitted 12 hours after being bitten on the tip of the finger of the right hand. Her tongue swelled immediately and she had trouble swallowing. Polyvalent antiserum was administered in the field. Within 30 minutes she began to vomit and have diarrhea. On admission, she had severe abdominal pain. Pulse was not palpable and blood pressure was too low to measure. After about 28 hours, hemorrhage of the affected arm began. She did not pass urine until 48 hours postbite, and she continued to experience bloody stools or diarrhea for more than 2 weeks postbite. She recovered and was discharged after 3 weeks (Efrati and Reif 1953).

Treatment—A number of folk medicine procedures are employed by local people who are bitten by this species. Bedouins feed victims cooked skins of desert monitor lizards (*Varanus griseus*) boiled with goat's or camel's milk. An alternative Bedouin treatment is burial of the victim in sand. Non-Bedouin villagers apply scrapings from gazelle horns to the bite area and olive oil and milk are sometimes prescribed for consumption. An even more unusual treatment is the consumption of milk containing scrapings from the nasal projections of *Pseudocerastes persicus*. This is called "Karn Al-Halteet," or "horn of the snake" (Amr and Amr 1983).

Antivenin for *D. palaestinae* is available through Ministry of Health, Department of Laboratories, Jerusalem, Israel (Chippaux and Goyffon 1991).

Mortality—Specific antivenin for *D. palaestinae* was not available until the 1950s, and until that time, fatalities occurred in about 6% of bite cases. Since antivenin availability, fatalities are rare (Hadar and Gitter 1959; Chippaux and Goyffon 1991). Children seem to tolerate the venom better than adults (Efrati 1979).

Remarks

The Palestine viper is relatively easy to maintain in captivity. It requires low humidity, good ventilation, hiding boxes, and structure to aid in shedding and for basking. It readily feeds on killed mice and it may tend to become unhealthily obese (Mehrtens 1987).

The blood serum of *D. palaestinae* contains molecules that deactivate hemorrhagic components of the venom. These factors are likely responsible for observed immunity of the snakes to their own bites (Ovadia 1978a).

Daboia russelii (Shaw and Nodder 1797)

Commonly called Russell's viper.

Recognition
(Plate 7.2)

Head—Flat, triangular with short snout; snout obtuse, raised, and rounded; neck constricted, head clearly distinct; nostrils large, located centrally in large undivided nasal, meeting nasorostral at lower edge; supranasal strongly crescent-shaped, separating nasal from nasorostral anteriorly; rostral as broad as high; supraocular narrow and undivided; head covered with small, irregular scales; supraoculars separated by 6–9 scales across the head; 10–15 circumorbital; supralabials 10–12, fourth and fifth notably enlarged, separated from eye by 3–4 rows of suboculars; 2 pairs of chin shields, anterior pair much enlarged; eye large, flecked with yellow or gold color, with vertical pupil; each maxilla generally bearing at least 2 and up to 5 or 6 sets of fangs, one active and others replacement; 9–10 pterygoid teeth, 12–13 mandibular teeth, and 3 palatine teeth.

Body—Stout and massive, rounded to cylindrical in cross-section, belly distinctly rounded; scales strongly keeled except lower row which is smooth, 27–33 midbody scale rows, 25–29 anteriorly and 21–23 posteriorly, ventrals 153–180, anal single. Tail short (approximately one-seventh of total length), divided subcaudals 41–68.

Size—Reaches 166 cm total length, averages about 120 cm throughout the mainland range, island subpopulations are smaller.

Pattern—Ground color deep yellow, tan, or brown of various shades with 3 series of large ovate spots, 1 vertebral, 2 costal. Large spots usually 23–30 in number. Spots brown in center, margined successively by black and white or buff. Dorsal spots may coalesce and side spots may fragment. Head with distinct dark patch or patches and pinkish, salmon, or brownish mark in V or X form with apex on snout. Dark streak, margined with white, pink, or buff behind eye. Dark stripe from eye to lip. Conspicuous white, buff, or pink line from gape converges to V above snout. Lips whitish, pinkish. Belly white, whitish, pinkish, or yellowish, often with black spots or a few half-moon marks on the margins of the anterior ventrals. Populations from Indonesia, Taiwan, China, and Thailand are more grayish or olive, with small spots between the large spot rows, belly suffused with gray posteriorly (de Silva 1980; Daniel 1983; Mahendra 1984, 1990; U.S. Navy 1991).

Taxonomy and Distribution

Described as *Coluber russelii* (Shaw and Nodder, 1797). Described independently as *Vipera daboia* in 1803 (Daudin 1803). First identified as *Vipera russelii* by Gray (1831) and e.g. Strauch (1869), and under which name it was generally known until the 1990s. Recently *Daboia* has been resurrected (Obst, 1983; Herrmann et al. 1992).

Based on statistical analysis of a number of morphometric characters, Wüster et al. (1992) and Wüster (1998) concluded that there are two distinguishable forms, which might be subspecies or even distinct species:

Daboia russelii russelii occurs across the Indian subcontinent through Pakistan and Bangladesh to Sri Lanka; and

Daboia russelii siamensis is found from Myanmar through Thailand, Cambodia, Indonesia, and southern China.

Eastern forms are distinguished from western by an additional row of small spots on the dorsum and substantial additional irregular spots along the sides. Populations on the Sunda Islands in Indonesia are distinguishable from the remainder of the eastern populations and may represent a definable form. Other subspecies have been described, including:

D. r. pulchella, restricted to Sri Lanka;

D. r. nordicus, endemic in northern India;

D. r. formosensis on Taiwan; and

D. r. limitis in Indonesia.

Habitat

Daboia russelii is not restricted to any one habitat but it avoids densely vegetated areas of closed canopy forest (Gharpurey 1962). It is largely found in open, grassy, or brushy habitats (Mahendra 1984; U.S. Navy 1991), but may also be found in "scrub jungles" (second growth forest), and on forested plantations, and in planted fields (de Silva 1990). It is most abundant and widespread in plains and coastal lowlands, and on hills of suitable habitat. It is generally not found in extreme highlands but it has been reported at altitudes of 2300–3000 m. Marshes, swamps, or permanent wetlands are unsuitable habitats, the high humidity of true rain forests also excludes it (Tweedie 1983). It occurs in most physiographic zones recognized for the Indian subcontinent, although not with equal abundance (Das 1996).

Russell's viper is often found associated with highly urbanized areas and countryside settlements. Consequently, landscape, farm, and plantation workers are most at risk of bites. However, *D. russelii* is not as close an associate of humans as certain cobras and kraits (Whitaker 1978). It is often found crossing hardtop roads and footpaths at night, possibly following rodent runs.

Food and Feeding

Preferred food throughout the range is small mammals, with murid rodents most commonly taken (Daniel 1983; Mahendra 1984). However, Russell's viper has an extraordinary range of food items. Reports include rats, mice, shrews, squirrels, domestic cats, land crabs, scorpions and other arthropods. Juveniles forage actively during evening and morning hours feeding primarily on lizards (Vit 1977).

Adult Russell's vipers specialize on mammalian prey, primarily rodents. The abiding preference for rodents is a primary reason for its association with human habitation (Daniel 1983). In southern India it restricts itself largely to gerbils (*Tatera* sp.).

Behavior

Active primarily as a nocturnal forager. During daylight it basks quietly, often in grassy vegetation or at scrub or woodland edges. At higher altitudes and latitudes, it may alter its behavior and be more active diurnally during cool weather.

Russell's viper is reported as persistently sluggish and slow to respond until stimulated past a behavioral threshold. When sufficiently irritated, it becomes fierce and aggressive, even "malicious" in nature (Gharpurey 1962). When disturbed, it throws its body into a series of S-loops, raises the forward third of its length, arches its neck, and hisses. Defensive responses include rapid and forceful breathing resulting in a deep-toned defensive hiss, reported to be louder than any other snake. Its defensive strike is forceful, often bringing most of the body of even large individuals off the substrate (Daniel 1983; Mahendra 1984).

In general, adult behavior is quiet and peaceful until pressed, when it responds in a violent manner. However, Kuntz (1963) interviewed hunters and villagers in Taiwan who stated that in the wild, Russell's viper may strike aggressively without warning. Young, in contrast, are quite active and inclined to aggressive biting with minimal provocation.

Reproduction and Development

Mating usually occurs in the early part of the year but gravid females are present during all months. Young are born from May to November, with a peak of birth activity in June and July. Gestation exceeds 6 months (de Silva 1990).

Russell's viper is prolific, with broods of 20–40 common and a reported maximum of 65 (Wall 1906; de Silva 1990).

Near-term embryos measure about 20 × 43 mm. At birth, young are 215–260 mm in total length. The smallest gravid females recorded are approximately 100 cm in total length and maturity is apparently achieved at 3 (possibly 2) years of age (de Silva 1990). Delivery of 11 babies in another case took nearly 4.5 hours (Naulleau and van den Brule 1980).

Bite and Venom

Epidemiology—*D. russelii* is by far the most dangerous viperid snake in Southeast Asia and is of great medical importance throughout its range. It is a major cause of snakebite injury and mortality throughout the region, particularly in densely populated areas of the Indian subcontinent (Wuster 1992, Jena and Sarangi 1993). Approximately 80% of bite victims reported circumstances involving direct disturbance of the animal, especially with actively foraging snakes during evening hours. About 5% reported unprovoked bites during sleep. Most victims were aged 10–40 years and males were victims 4 times as often as females. Bites are most common during late summer rainy seasons in India and Myanmar. Bites are rare during winter months (Jena and Serangi 1993).

In western Bengal, the "snake catcher's society" reported that in a 6-year survey (1982–1988) 17% of all snakebites of known origin were by *D. russelii* (Saha 1989). In one district of India, 34% of reported bites were attributed to *D. russelii* vs. 23% for *Naja* (cobras) and 15% for *Bungarus* (kraits). Of viper bites, approximately 30% were fatal while 35% of cobra bites and 34% of krait bites resulted in death (de Silva 1980). This contrasts with data from Wuzhou, China, where *D. russelii* was responsible for only 1 of 640 snakebite cases reported from 1973–1984 (Yu, Tang, Liang et al. 1989). Throughout most of subtropical China, *D. russelii* accounted for only 46 of 10,118 snakebites over the same period (Yu, Tang, Yue et al. 1989).

In Burma and Sri Lanka, *D. russelii* is responsible for the majority of snakebite deaths. Approximately 2000 bites and 900 deaths per year respectively are attributed to this species (Warrell 1985).

Yield—Venom yield per individual can be large. Snakes 100–125 cm in length produced yields of 130–250 mg (Minton 1974). Thwin et al. (1985, 1988) reported the average yield per snake was just over 0.2 g dry venom per animal. Total yield on a per-bite basis averaged 72 mg and commonly reached 150–250 mg. Dried venom yields for 25 adult *D. russelii* averaging 111 cm in total length ranged from 21–268 mg (mean 127 mg). For 13 juveniles averaging 79 cm, total dried venom yield was 8–79 mg (mean 45 mg).

In parts of India, 52,000 snakes were captured for venom extraction, of which 2% (948) were Russell's vipers, which yielded 212 total g of dried venom. *Echis carinatus*, representing 85% of snakes collected (44,200), yielded 255 g. This clearly demonstrates the enormous venom load from a single *D. russelii* bite (Dravadamani 1989).

Adults generally injected more than 45% of venom gland contents in the first bite, equating to an average of 63 mg dried venom. For juveniles, first bite dried venom mass averaged 41 mg (with a range of 3–138 mg). There is a correlation between dried venom yield per milking and individual length. Individuals 70 cm or less in total length yielded 60 mg or less of venom per milking, while individuals greater than 110 cm yielded up to 270 mg venom (Pe and Cho 1986).

Toxicity—Intravenous LD_{50}s for 20-g mice range from 0.08–0.31 µg/g. Intraperitoneal LD_{50} is approximately 0.40 µg/kg, and s.c. is approximately 4.75 mg/kg. Doses of 40–70 mg are lethal in most humans, making *D. russelii* venom more toxic than most other species including rattlesnakes, moccasins, coral snakes, *Bothrops*, cobras, kraits, *Echis*, and *Dendroaspis*.

In general, the toxicity of *D. russelii* venom depends on complex synergy among chemical constituents. When whole venom is separated into five easily obtained fractions, the toxicity of each fraction is lower than a mixture of all, but when reassembled, lethality is restored (Master and Rao 1961; Dimitrov and Kankokar 1968a). Toxicity varies within populations, across years, and with different milkings within each year (Than et al. 1985).

In a series of monitored bite tests from older literature, the following times-to-death were reported (Ewart 1878): "fowls" (presumably domestic poultry)—35 seconds to several minutes, dogs—7 minutes to several hours, cat—57 minutes, horse—11.5 hours. Death was not as rapid as in comparable tests with cobras but the fatality rate was apparently equivalent.

Content—*D. russelii* venom composition varies geographically (Woodhams et al. 1990). Generally, the venom contains up to 32% solids. Of that, 57%–70% by dry mass is protein (Brown 1973). Symptoms attributable to protein constituents include coagulation, hemorrhage, and fibrinogenolysis. Major active enzyme constituents are coagulases, L-amino acid oxidases, and phospholipases (Teng et al. 1984, 1989). Certain proteins from *D. russelii* venom are powerful esterases and proteases (Teng et al. 1989). Proteinases are present at high concentrations, however, their activity is relatively low compared to those of *Bothrops, Crotalus, Echis,* and *Vipera berus*. Fewer protein substrates are hydrolized by *D. russelii* venom than any other reported viper, although the activity is substantially greater than that of elapid venoms (Meier et al. 1985).

Other active proteins of *D. russelii* venom include powerful proteinase inhibitors. Polypeptides with molecular weights between 6000 and 7000 have been isolated and purified (Takahashi et al. 1974a, b). These proteins, when challenged with bovine plasma, inhibit the activities of kallikrein, trypsin, plasmin, and chymotrypsin. The activities of these inhibitors are very high in *D. russelii* venom, far higher than in elapids. Biogenic amines are present and are largely responsible for severe pain accompanying viper bites. Serotonin is generally present at only low concentrations, although it is consistently detected by both bioassay and spectrofluorometric techniques in *D. russelii* (Tu 1991).

Whole venom contains powerful coagulants and anticoagulants. Among the most important proteins symptomatically are those causing coagulation (Furukawa et al. 1976). Coagulants rapidly activate clotting mechanisms and break down refractory molecules such as casein (Teng et al. 1984). One key protein functions as an enzyme, activating Factor X in the presence of calcium ions. This protein is large (molecular weight 100,000–130,000) and consists of four polypeptide chains. A hydrolytic process activates this substrate (Esnouf and Williams 1962).

The strong blood clotting action of *D. russelii* venom at high doses is maintained by redundant biochemical mechanisms. These clotting mechanisms are associated with two separate protein-degrading fractions of the venom. One, which is most effective under relatively acidic conditions (pH 3.6), is inhibited by heparin. The other, with a pH optimum at a basic 9.0, is only partially heparin-dependent. The pH sensitivity of these reactions suggests that venom fills the role of an enzyme acting on blood proteins in inducing coagulation (Ghosh and Chaudhuri 1968).

Coagulant proteins are present as complexes, which may be separated into individual enzymes (Jackson et al. 1971). As a diagnostic tool, radioimmunoassay methods have been developed to detect individual coagulant proteins in the blood of victims (Pukrittayakamee et al. 1987).

Anticoagulants present in the venom are active enzymes, requiring other substances to enhance membrane penetration for maximal activity. Phospholipase anticoagulants are also present and their activities are highly complex. A distinct platelet function inhibitor has been purified from *D. russelii* venom (Li et al. 1985). This enzyme is large (molecular weight 13,800) and its activity is dose-dependent. At relatively high levels blood clotting is severely inhibited.

A phospholipase A_2 (PLA_2) enzyme specifically responsible for platelet aggregation failure has been identified and characterized (Li et al. 1985). Intravenous LD_{50} of this enzyme in mice is 0.5 mg/kg. The enzymatic activity of this component is not associated with anticoagulant properties. It inhibits platelet aggregation by several physiological mechanisms, including those associated with ADP, adrenaline, thrombin, and collagen.

An isolated phospholipase with a molecular weight of 11,800 is the major phospholipase A constituent, comprising 24% of the whole venom mass. This phospholipase is responsible for 45% of the PLA_2 activity of the venom, although this is largely attributable to its high concentration. On a per-volume basis it is the least toxic of the phospholipases purified. The intraperitoneal LD_{50} of this fraction is 5.3 mg/kg body weight. At concentrations just sufficient to induce symptoms it is neurotoxic. At higher concentrations, it directly damages lung, liver, and kidney tissue. At sublethal concentrations it induces muscle necrosis and general tissue edema.

Phospholipase enzymes causing substantive neurotoxic effects have been isolated and characterized. Of several phospholipase A_2 isozymes present, acidic forms are the most potent at blocking neuromuscular communication (Ghosh and Chaudhuri

1968; Huang and Lee 1984). Neurotoxic phospholipase activity can be inhibited with or without affecting the catalytic activity of the same molecule, indicating that there are multiple active sites of pharmacological importance on the phospholipase molecule (Bhat et al. 1991). The neurotoxic and catalytic activities of phospholipases can be dissociated chemically, although it is unclear what relationship the altered molecular structures have to the original venom (Jayanthi et al. 1989).

Another PLA_2 enzyme, called VRV PL-V, has been isolated from *D. russelii* venom and studied in detail (Jayanthi et al. 1989). Individual molecules of this enzyme are somewhat smaller, with a molecular weight of about 10,000. This component is relatively toxic compared to whole venom, with an LD_{50} of 1.8 mg/kg body weight, compared to the toxicity of whole venom. At acidic pH, this enzyme forms enzymatically active polymers but they completely lose their neurotoxic activity.

An enzyme system has been isolated which appears responsible for edema (Warrell 1986; Vishwanath et al. 1988). In test animals at doses as low as 0.85 µg, substantial edema was induced with maximal activity at 45 minutes and an undiminished effect after 4 hours.

In addition to proteins, a complex array of carbohydrates are present but mucopolysaccharides are generally absent (Oshima and Iwanaga 1969; Tu 1991). Nondialyzed crude venom contains metals such as sodium, zinc, magnesium, phosphorus, copper, iron, and calcium (Gitter et al. 1963; Devi 1968; Friedrich and Tu 1971).

At times, the whole venom may be colorless (white), although it is ordinarily distinctly yellow. Yellow venom contains proteolytic enzymes which the white lacks, and it has relatively much higher necrotizing activity, indicating the dependence of these symptoms on venom enzyme proteins (Dimitrov and Kankokar 1968b).

Symptoms and Physiological Effects—*D. russelii* venom has neurotoxic, hemotoxic, systemic, and local effects. Victims become symptomatic rapidly and symptoms are often prolonged. In humans, symptoms begin with pain at the site immediately followed by swelling of the affected extremity (Reid 1968a; Jena and Sarangi 1993).

Bleeding, especially from the gums, is a common symptom (Jena and Sarangi 1993). Blood pressure drops, heart rate falls, and blistering occurs at the bite location. In severe cases blistering develops along the bitten limb. Blisters may be filled with serum (clear) or blood. Necrosis is generally superficial and confined to muscles near the bite, but it may also be severe in extreme cases. Vomiting and facial swelling occur in about one-third of bites.

An excellent summary of symptoms is available (U.S. Navy 1991). Severe pain may last 2–4 weeks. Localized pain may persist depending on the level of tissue destruction, swelling, and edema. In many cases, local swelling peaks between 48–72 hours postbite and involves both the affected limb and the trunk. If swelling to the trunk is rapid (within 1–2 hours), massive envenomation may be presumed. Edema is due to effects of PLA_2 enzymes on cell membranes, including destruction of membrane phospholipids and the subsequent action of membrane breakdown products. Discoloration caused by leakage of red blood cells and plasma into muscle tissue may occur throughout the swollen area. Pulse may become rapid, variable, and weak, breathing rapid and irregular, and there is muscle weakness, nausea, and vomiting. Pupils dilate and are sensitive to light. Skin becomes cold and beaded with sweat. Unconsciousness follows rapidly at sufficiently high doses. Other symptoms include intense burning or stinging, abdominal pain, diarrhea, thirst, chills or fever, albuminaria, proteinuria, hypotension, shock, constant oozing of thin bloody serum from the punctures, hemorrhage from intestines and urinary tract, anemia, ecchymosis, lymphadenopathy, and decreased platelet count.

Initial hemmorhage reactions in humans include bloody sputum, which begins within 20 minutes, is characteristic of systemic involvement and generally precedes other symptomatic bleeding (Sitprija et. al. 1974). In many cases, coughing must be forced to support the diagnosis. However, in some cases hemoptysis is profuse. Bleeding in the early stages of a bite may be severe (into the brain, peritoneum, or other vital organs) or into the bitten limb which (when intensely swollen) can accommodate up to half the circulatory fluid. Hence, peripheral circulatory failure may occur from general loss of blood. Bleeding into the brain and subarachnoid cavity may be associated with convulsions in pediatric cases involving relatively high venom doses. Later bleeding is confined to the gums, injection sites and other sores. Internal bleeding is common due to major and "striking" arterial lesions, accompanied by necrotizing arteritis of the interlobular arteries (Sitprija et al. 1974).

Circulatory system effects result in substantial systemic damage to internal organs, including the kidneys. Kidney damage (in moderate cases) and failure (in severe cases) are common (Jena and Sarangi 1993). Acute renal failure is often characteristic of bites which are not rapidly fatal (Chugh et al. 1975). In 10% of cases, immediate renal failure is associated with direct necrosis of the kidney cortex. Less severe kidney involvement is associated with acute tubular necrosis and secondary effects of blood loss. Affected kidneys may cause substantial interstitial edema in other tissues, especially muscles (Sitprija et al. 1974). Severe hypotension may be associated with renal failure, although hypotension has also been observed in the absence of kidney failure. Tubular and cortical necrosis within the kidneys may be due to vascular coagulation and reduction in renal blood flow (Clarkson et al. 1970). While hemolysis may be responsible for acute kidney failure, there is a powerful direct toxic effect of venom on renal cells (Sitprija et al. 1974). Also, immunoglobulin M deposits in large granules in the glomeruli (Chugh et al. 1975). In general, venom depresses both glomerular and tubular functions even when the kidney cells' ability to regulate blood flow is not inhibited (Sanguanrungsirikul et al. 1989).

Of five autopsied fatalities, four resulted directly from renal failure (Sant 1978). Most obvious kidney effects were proliferative, with clear fibrin thrombi. In animals receiving chronic doses of *D. russelii* toxin, hemorrhagic infarcts of the kidney are also observed. Additional structural changes include meningitis, systemic edema, and necrosis with granulomatous reactions in brain tissue.

In distribution studies with radiolabeled *D. russelii* toxin, Thwin et al. (1985) found most venom proteins in the stomach, intestine and liver. Only moderate concentrations are found in the kidneys. Lowest concentrations are in the brain and testes.

Major routes of excretion are urine and feces. Excreted venom is dissimilar to the administered form, indicating that little whole venom passes the glomeruli of the kidneys (Thwin et al. 1988).

Hematological Effects—After a bite, red corpuscles are destroyed, clotting is reduced, blood vessel linings degrade, and there is internal hemorrhage and external bleeding from body openings (Reid 1968a). At moderate doses venom is an effective anticoagulant. At high doses it is a powerful coagulant and death may occur from massive internal agglutination. However, death by clotting is rare in humans, and when it does occur it occurs very rapidly. The high-dose blood clotting effect is poorly understood. Current theory suggests that clotting is induced by a reaction between venom components and the thrombin system (Rosenfeld et al. 1967).

Case Histories—A 50-year-old individual was bitten on the right ankle. Swelling reached the upper leg within 2 hours, accompanied by extreme pain. Bleeding from the gums was observed after 9 hours. Urination ceased after about 10 hours. Upon cessation of urination, the individual was admitted to the local hospital where a diagnosis based on hemolysis and incoagulability was made. Antivenin administration and conservative treatment including diuretics and fluids were ineffective in reversing kidney failure, and dialysis was begun. The patient eventually recovered (Jena and Sarangi 1993).

In a second case history, a 49-year-old man was admitted to the hospital 2 days postbite. The patient was bleeding from the gums, presented bloody urine and considerably constrained urine volume (50 ml in 24 hours). The left leg (site of the bite) was swollen. Blood pressure, body temperature, pulse, and respiration rate were within normal limits. Upon admission, antivenin, hydrocortisone, and whole blood were administered, resulting in rapid and effective bleeding control. Kidney function returned to normal 17 days after admission. Postbite renal biopsy revealed considerable tubular necrosis (Sitprija et al. 1971).

Treatment—In the absence of effective medical treatment, recovery from severe *D. russelii* bite can be protracted. Following antivenin administration, systemic symptoms subside relatively quickly. Local symptoms (necrosis) can require months to years, and permanent scars and tissue damage may remain (Reid 1968b).

Heparin has a variable effect on *D. russelii* venom (Ahuja et al. 1946). Heparin reduces clotting in test tubes and protected dogs and rabbits from clotting effects. However, guinea pigs and pigeons were not similarly protected. Apparently, a clotting component of the venom, which acts through a separate, calcium-independent mechanism, was not affected by heparin. This factor can coagulate blood by a pathway quite distinct from that mediated by thromboplastin and is partially controlled by epinephrine and coumarin. Thus heparin therapy may be ineffective or even counterproductive, especially in comparison with the immediate and proven value of antivenin in controlling symptoms (Rosenfeld et al. 1967; Ghosh and Chaudhuri 1968).

Injection of large quantities of venom and the rapid onset of symptoms makes treatment difficult if antivenin is not immediately available. Development of some immunization prophylaxis would be very valuable for humans at risk of bites. Even a relatively low level of initial immunity might help delay the onset of symptoms, allow more effective postbite treatment, and provide better and less traumatic prognosis.

Rabbits have been immunized against *D. russelii* venom by scheduled administration of a toxoid produced by irradiation of whole venom (Hati et al. 1993). Similarly, venom detoxified by photo-oxidation generated substantive immunity in rabbits (Kocholaty and Ashley 1966). Venom is also detoxified by dihydrothioctic acid treatment (Sawai et al. 1967b).

In many bites, secondary infections are common due to disruption and damage of various tissues and possibly to the presence of microbes in the bite itself. In these cases, antibiotic therapy is useful (Reid 1968b).

Mortality—The immediate cause of death in bites fatal to humans depends on venom dose relative to victim body mass. When death is not associated with internal bleeding or massive clotting, it may result from phospholipase-induced hypotension leading to circulatory failure. Death from cardiac or respiratory failure or septicemia occurs usually in 1–14 days but may come even later (Huang 1984).

In parts of India, mortality rates associated with *D. russelii* bites reach 30% (de Silva 1980). Of those cases that resulted in death, 19 of 34 occurred within 24 hours of the bite. Death can occur very rapidly and has been recorded to occur as quickly as 15 minutes postbite (Wall 1921). Jena and Sarangi (1993) reported mortality estimates for India at 2300 of 10,000.

Remarks

Russell's viper may have a mimic in the Russell's sand boa (*Eryx conicus*) which is similarly patterned and colored. On close examination, however, the harmless species has narrow ventrals and a very short tail, and so may be readily distinguished.

Russell's viper thrives in captivity. It needs only a simple hide box and water dish. Babies feed readily on pink mice. Adults eat rats, mice, and birds. They breed readily and easily but are quite dangerous captives (Cox 1991).

Despite their abundance in some localities and tolerance of human presence, there are conservation problems with some populations of *D. russelii* (Seigel et al. 1987). Primary threats are collection for leather production or use as food. Urbanization may impact very localized populations but is not currently a major threat.

Chapter 8
Echis

Echis (Merrem 1820) contains snakes commonly known as saw-scaled or carpet vipers. "Saw-scaled" is derived from the obliquely keeled and serrated lateral body scales. When threatened, the body coils into parallel loops that are rubbed together to produce a "sizzling" warning noise which becomes stronger and quicker in proportion to the degree of threat or excitement. This warning is possibly an adaptation for minimizing evaporated water loss from the respiratory system which might occur if open-mouthed hissing were the main threat response.

Saw-scaled vipers are relatively small, but they are very aggressive and quick to strike, and produce highly virulent hemotoxic venom. These traits, coupled with their extensive geographic distribution (especially in heavily populated regions) make them some of the most dangerous snakes in the world.

Recognition

The head is short, broad, somewhat spade-shaped, widening slightly posterior to eyes, distinct from the narrow neck. The snout is short and rounded and the eyes are relatively large with vertically elliptical pupils. Dorsally, the head is covered with small, irregular, imbricate scales that may be either keeled or smooth. The body is cylindrical and moderately slender. Lateral scales are distinctive, being small, strongly oblique, and keeled with minute serrations. Subcaudals are undivided and the tail is short (Marx and Rabb 1965; Haas 1973; Bellairs and Kamal 1981).

Taxonomy

The taxonomy of *Echis* is unsettled. Traditionally, only two species were recognized: *E. coloratus* Gunther 1878 and *E. carinatus*. In recent times, *E. carinatus* has been divided into several species and subspecies. However, body-scale number and pattern may be purely meristic adaptations to climatic ecoclines. Cherlin (1983) and Auffenberg and Rehman (1991) found a significant negative correlation between ventral scale counts and annual rainfall. However, differences in venom composition, as evidenced by the failure of antivenins from Asian *Echis* to neutralize venoms from African *Echis* (and probably vice-versa) supports formation of new *Echis* species. Currently ten species are recognized. Very little information is available for four of the species: *E. hughesi, E. jogeri, E. khosatzkii,* and *E. megalocephalus,* and these are just briefly discussed in this text.

Echis carinatus (Schneider 1801)

Commonly known as the saw-scaled viper.

Recognition
(Plate 8.1)

Head—Circumorbital ring with 14–21 scales; supraocular variably developed; interoculars 9–14; subocular with 1–3 vertical scales between eye and supralabials; supralabials 10–12; sublabials 10–13 (Latifi 1991; Auffenberg and Rehman 1991).

Body—Dorsal scales keeled, with apical pits, numbering 25–39 rows; lateral scales numbering 3–8 rows; ventral scales rounded, extending full width of belly, numbering 143–189; subcaudals 21–52 and single.

Size—Between 38–80 cm, typically up to 60 cm (Auffenberg and Rehman 1991; Jena and Sarangi 1993).

Pattern—Dorsal head markings form a light cross, arrowhead, or trident; faint stripe from eye to the angle of the jaw; body coloration pale, buff, tan, olive, and chestnut, even reddish or gray; belly color white to pinkish, with stippling ranging from faint to deeply pigmented; laterally with small, dark brown spots usually in 1–3 rows; usually with dorsolateral series of white bows; with middorsal series of dark variably colored blotches, most being white with dark edges separated by lighter interblotch patches (Joger 1984; Auffenberg and Rehman 1991). Certain color phases are restricted to specific geographic regions (Auffenberg and Rehman 1991), and Greene (1988) contended that color patterns in *Echis* are concurrently cryptic, mimetic, and/or aposematic.

Taxonomy and Distribution

Described as *Pseudoboa carinata* by Schneider (1801) from specimens collected at Arni, Madras, India. Current taxonomy defines *E. carinatus* as strictly an Asian species with four recognized subspecies:

E. c. carinatus, from peninsular India;

E. c. astolae (Mertens 1970), from Astolae Island, Pakistan;

E. c. sinhaleyus (Deriyagala 1951), from Sri Lanka; and

E. c. sochureki Stemmler 1969, from southern Afghanistan, Pakistan, northern India, south and central Iran, Oman, and U.A.E.

Population densities can be very high. Over 2000 specimens were collected in five days in the Ratnagiri district near Bombay. Rewards were paid on an average of 255,721 vipers per year over a 6-year period, with no apparent reduction in local numbers. In 1962, 115,921 rewards were paid in just 8 days (Whitaker 1973b).

Habitat

Found on varied substances such as sand, rock, soft soil, and in scrublands (Latifi 1991). They are often found under rocks. Whitaker (1973b) reported that practicing with native "Mahrs," who collect *Echis* regularly for antivenin production, enabled him to identify rocks that consistently yielded specimens. These rocks were not embedded or surrounded by soil or grass. In the Kavir desert in central Iran a habitat of flat clay *Artemisia*-steppe was used, and specimens surfaced from rodent burrows in the evening (Nilson pers. obs.).

Minton (1966) found individuals in northern Baluchistan at 1982 m altitude.

Food and Feeding

Feeds on small rodents, frogs, lizards, and a variety of arthropods. High population densities of the snake may be aided by its generalist diet, which can be varied according to prey species availability (Bogdanovich 1970). Typical prey are orthopterans (*Schistocerca gregaria, Poekilocerus pictus*), arachnids (scorpions, spiders, solifugids), centipedes (*Scolopendra* spp.), amphibians (*Rana cyanophlyctis*), and reptiles (*Hemidactylus brooki, Typhlops braminus*) (Sharma and Vazirani 1977).

Captives may feed on dead lizards, specifically geckos and agamids (Stemmler 1971b). However, some will not eat, and Murthy and Venkateswarlu (1974) reported that a captive eventually died after refusing food for 5 months.

Behavior

Echis carinatus moves by sidewinding, at which they are very proficient and alarmingly quick. Although they can travel in other ways, sidewinding is the most efficient method of crawling over the smooth sandy surfaces which are prevalent in their habitats. Sidewinding may also help prevent overheating by raising body loops above the heated substrate, leaving only two points of contact with the hot surface (Cloudsley-Thompson 1988a).

These snakes are primarily crepuscular and nocturnal (Whitaker 1973b, 1978), although Gharpurey (1962) reported instances of diurnal activity.

Echis carinatus may climb as high as 2 m into trees and shrubs. During rain, as many as 80% of the adult vipers can be found above ground in bushes and trees and as many as 20 snakes have been observed massed on top of a single cactus or small shrub (Duff-Mackay 1965, Whitaker 1973a).

Reproduction and Development

In India, mating occurs during the winter months and captives have been observed pairing in February and March (Stemmler and Sochurek 1969b).

A seasonal recurrence of testicular development, manifested as changes in weight and gonadal histological texture, occurs early spring through early summer. However, captivity degrades testicular structure, as evidenced by obliteration of spermic channels by an overgrowth of connective tissue. This may result in only small numbers of captive snakes maintaining their ability to produce sperm (Brushko 1968).

Populations from India are viviparous (Stemmler 1971a), with young born from April through August in India and mostly in July in Pakistan (Minton 1966). Litters contained up to 23 individuals (Latifi 1991).

Echis coloratus Günther 1878

Commonly known as the painted saw-scaled viper.

Recognition
(Plate 8.2)

Head—Bell-shaped dorsally, with distinct widening from the eyes posteriorly; circumorbital scales 17–22; no enlarged supraocular scale; interocular scales 13–15; 3–4 scale rows between eye and supralabials, which number 12–15; (Joger 1984, Gasperetti 1988).

Body—Dorsals in 31–35 rows; ventrals 174–214; subcaudals 40–57.

Size—Maximum length 75 cm, with an 8-cm tail; ratio of total length to tail length is sexually dimorphic, with males having longer tails. Few individuals, even wellfed ones, weigh more than 200 g; no significant differences in weight between males and females (Mendelssohn 1965; Joger 1984; Gasperetti 1988).

Pattern—Dorsum of head with a light straight, branched or bifurcate longitudinal line; body yellowish gray to light brownish gray; 40–50 dorsal rows of rhomboid blotches or crossbands; blotches rounded anteriorly becoming irregular distally; blotch coloration grayish white, generally darkly edged with color surrounded by a darker ring; lighter-colored space separating blotches; pattern may be reversed, with blotches being dark and ground color light; blotches sometimes elongated, forming transverse bands with dark edges, or obliquely oriented and interconnected by a dark stripe; venter light, sometimes speckled with minute spots.

Taxonomy and Distribution

Described by Günther (1878) from specimens captured at Sharr Mountain, Midian, Saudi Arabia. *Echis coloratus* is restricted to the Middle East and Egypt, being found from Israel, Jordan, southward through the Arabian Peninsula to Yemen and Oman, and in southeastern Egypt. Although sympatric with *E. pyramidium* in some areas of western Saudi Arabia and Yemen, the two species occupy different biotopes, with *E. coloratus* inhabiting uplands and *E. pyramidium* occurring in sandy lowlands (Flower 1933; Haas 1951; Theodor 1955; Corkill and Cochrane 1965; Klemmer 1973; Harding and Welch 1980; Gasperetti 1988).

It is sympatric with *E. carinatus* in Oman (Gasperetti 1988).

Habitat

Favors hard rocky soil and slopes with scattered boulders and large stones that provide spacious hiding places. It is not found on sandy soil. In spring, it spends the day under large rocks that are not flush with the ground and which permit air circulation. In summer, daytime is mostly spent in holes or crevices to avoid heat. *Echis coloratus* is fairly resistant to desiccation and prefers dry desertlike microclimates; however, increased irrigation and agricultural expansion in Israel has produced increases in *E. coloratus* populations in these areas due to corresponding increases in frog and rodent populations (Mendelssohn 1965; Joger 1984).

Echis coloratus prefers areas with less than 150 mm of rain. However, it has been observed in areas receiving an average of 400 mm rainfall at temperatures as low as 12°C. It has been collected from Oman at altitudes as high as 2600 m. In captivity these snakes prefer dry environs and temperatures of 32°C–33°C (Joger 1984).

Food and Feeding

In the wild, rodents such as gerbils (*Gerbillus*) and spiny mice (*Acomys*) are natural prey (Gasperetti 1988). Feeding on frogs, *Rana ridibunda,* in newly irrigated agricultural areas has resulted in increased populations of *E. coloratus* with individuals having increased body weight and reproductive potential. Less than 50% of females collected outside agricultural regions by Mendelssohn et al. (1971) were gravid, while greater than 50% of females collected inside such areas were gravid.

Other prey items include arthropods (solifugids, scorpions, and centipedes), lizards, birds, and mice (Ishunin 1964).

Captives prefer amphibians, especially *Bufo viridis* and *Rana ridibunda* (Mendelssohn et al. 1971). However, individuals raised in captivity exclusively on mice survive and reproduce normally.

Echis coloratus normally feed only sporadically and always appear thin. Even wild-caught snakes, especially juveniles, are thin, suggesting that food is scarce. This condition may contribute to high juvenile mortality which limits population size. In captivity, neonates feed readily under hot dry conditions, and have been maintained without water for 665 days with moisture provided only through food (1–2 mice during a 7–21-day interval (Goode 1979a). Mallow (pers. obs.) maintained one individual for over 10 years that periodically refused food, once for 8 months (December–August), without any deterioration in health or appearance.

Echis coloratus has not been observed engaging in cannibalism, even in poorly fed, crowded populations of different sized individuals (Duff-Mackay 1965).

Behavior

On rocky slopes movement is serpentine or rectilinear, but on hard flat surfaces the snakes sidewind. When sidewinding, *E. coloratus* minimally lifts its coils and pushes them forward, unlike sidewinding by *E. carinatus* which lifts its coils high off the ground.

Reproduction and Development

Oviparous and reproduces well in captivity. Goode (1979b) reported 11 clutches were produced from 1971 to 1978 at the Columbus [Ohio] Zoo. Mendelssohn (1965) observed several captive matings in May and June. Despite lack of diurnal activity in the wild, captives are frequently observed courting, copulating, or in male combat during the day (Goode 1979b). Pairing lasts approximately 3 hours and occurs day or night. Copulation to oviposition interval is approximately 75 days, and most clutches are laid in August but some have been recorded as early as July and as late as September. Shortly after laying, the white eggs are smooth and pliable, and translucent enough to show blood vessels. On occasion, unshelled masses accompany clutches, and on four occasions a single egg was laid several days prior to the main clutch deposition. Such early eggs never produced hatchlings (Goode 1979b).

Clutch size varies from 6 to 10 (mean 7.6). Clutch weights vary but are relatively large compared to female weight and clutch size. In two instances, clutch total weight was greater than 75% of the female's weight. Proportionally large clutch weight, attributed to large eggs, serves two purposes: 1) large yolks are needed to furnish juveniles with sufficient nutrients in arid scarce food environments (yolk reserves may be up to one-third juvenile body weight); and 2) larger neonates can take larger more readily available prey. During incubation eggs absorb water, gaining up to 36% of their original oviposition weight. Larger eggs, along with scant food resources, result in females not producing eggs every year. There is evidence that females reabsorb unfertilized eggs during times of food shortages (Mendelssohn 1965).

Eggs are adhesive, enabling females to stick them to many substrates including the undersurface of rocks and crevice walls. This enhances exposure to condensation, providing necessary contact humidity. When lacking proper oviposition sites, females use their heads as scoops to dig an oviposition hole, which is subsequently covered after eggs are laid. For captives, Goode (1979b) recommends slightly dampened sphagnum moss as an incubation substrate.

Incubation takes from 43 to 82 days at temperatures of 31.5°C and 28°C, respectively. At a constant temperature of 30°C, incubation within a clutch varied from 44 to 55 days. It is thought this variability may be due to eggs laid at different stages of development (Goode 1979b).

Upon hatching juveniles weigh between 5.5 and 7.6 g (mean 6.74 g), and are 17.6–20.0 cm long (mean 19.1 cm).

Juveniles first molt after 8–20 days and body coil rubbing has been reported as early as 2 days posthatching. In the wild juveniles typically molt 2–3 times per year, while adults typically molt once per year. Captive adults in constantly heated environments molt 2–3 times per year at intervals as short as 2 months (Goode 1979b).

Growth is relatively slow. Average weight gains by captives at 1, 2, and 14 months were 5.6 g, 6.6 g, and 9.3 g respectively. Average length increased 20.8 cm, 21.0 cm, and 25.0 cm respectively. Hatchlings that fed regularly reached 26.0–28.5 cm their first year and 35.0–38.0 cm their second year. Weight loss during the first 2 months was presumably due to depletion of yolk reserves and low food availability. Sexual maturity was reached between 4 and 5 years and longevity is presumed to be around 20 years (Goode 1979b).

Echis hughesi
Cherlin 1990

Hughes' carpet viper, or Somali carpet viper.

Recognition

Head—Scalation similar to *E. pyramidum*.

Body—Mid-dorsal scales 24–25; ventrals 144–149; subcaudals 28–30 (Cherlin 1990).

Size—Around 21–32 cm (Cherlin 1990).

Pattern—Variable in color and pattern. Usually a dorsal series of oblique, pale blotches, with somewhat darker shades in between.

Taxonomy and Distribution

Described from nothern Somalia by Cherlin (1990). The range is given as Somalia: northern Migiurtinia near Meledin (McDiarmid et al. 1999).

Habitat

No information is available.

Food and Feeding

No information is available.

Behavior

No information is available.

Reproduction and Development

No information is available.

Echis jogeri
Cherlin 1990

Joger's carpet viper or Mali carpet viper.

Recognition

Head—Head scalation similar to *E. leucogaster*. Eyes pale yellowish, set near front of the head.

Body—Fairly stout. Cylindrical or subtriangular in cross-section. Dorsal scales rough and heavily keeled, in 27 rows at midbody; ventrals 123–136 (Cherlin 1990).

Pattern—Variable in color and pattern, from brown, gray, reddish, or shades in between. Usually with a dorsal series of oblique pale crossbars or saddles, with dark spaces between. Sides paler, usually with a row of triangular, subtriangular, or circular dark markings with a pale or white edging. Venter pale cream, white or ivory without markings or spots.

Size—Small size, on average around 30 cm (Cherlin 1990).

Taxonomy and Distribution

Described by Cherlin (1990) from specimens caught by Joger at Mali, 3 km from Timbuktu. Its range is given as western and central Mali.

Habitat

No information is available.

Food and Feeding

No information is available.

Behavior

Little is known. Presumed to be similar to *E. leucogaster*.

Reproduction and Development

Little is known. Presumably oviparous (Spawls and Branch 1995).

Echis khosatzkii Cherlin 1990

Known as the Dhofar carpet viper.

Recognition

Head—Scalation similar to *E. pyramidum*.

Body—Mid-dorsal scales 27–31; ventrals 165–189; subcaudals 36–47 (Gasperetti 1988 (as *E.* cf. *pyramidum*); Cherlin 1990).

Size—Average 40 cm.

Pattern—Variable in color and pattern. Ground color brown, gray, or shades in between. Without pattern or with a dorsal series of weak, pale, blotches.

Taxonomy and Distribution

Distribution is given as Dhofar, Oman. Type locality is given as Hadhramaut, Arabia (=Yemen). It has for some time been considered as a synonym of *E. pyramidum*, but recently Lenk et al. (2001) has demonstrated genetical differences from related forms. Also Schätti (2001) consider *E. khosatzkii* to be a valid species from eastern Yemen and Dhofar on the basis of morphological and molecular evidence.

Habitat

Avoids true deserts and other areas of soft sand. Specimens of *Echis* collected from the Arabian Peninsula are found in gravel, rocky, and sandy areas with sparse xerophytic vegetation. Individuals of *E. khozatskii* have been collected in agrarian habitat (Gasperetti 1988).

Food and Feeding

Echis from Saudi Arabia eat centipedes, small mammals such as gerbils, birds, lizards, amphibians, other snakes, beetles, crickets, solifugids, scorpions, worms, and slugs (Corkill and Cochrane 1965; Gasperetti 1988).

Behavior

Primarily crepuscular and nocturnal and seldom found during the day. One specimen was collected from a frankincense bush, 0.5 m from the ground, in a shallow stone-bottomed stream bed (Leviton et al. 1992).

Reproduction and Development

No information is available.

Echis leucogaster
Roman 1972

Commonly known as the white-bellied carpet viper
based on its light unmarked underbelly.

Recognition

Head—Head scalation similar to *E. carinatus*. Eyes pale yellowish, set near front of the head.

Body—Fairly stout. Cylindrical or subtriangular in cross-section. Dorsal scales rough and heavily keeled, in 27–33 rows at midbody; ventrals 165–180 (Hughes 1976).

Pattern—Variable in color and pattern, from brown, gray, reddish, or shades in between. Usually with a dorsal series of oblique pale crossbars or saddles, with dark spaces between. Sides paler, usually with a row of triangular, subtriangular, or circular dark markings with a pale or white edging. Venter pale cream, white, or ivory without markings or spots.

Size—Average size is 30–70 cm with maximum recorded at 87 cm (Spawls and Branch 1995).

Taxonomy and Distribution

Described by Roman (1972) from specimens caught in Boubon, Niger. Its range includes Mauritania, southern Algeria, Upper Volta, Niger, and Nigeria (Golay et al. 1993).

Habitat

Found in arid savanna, semiarid deserts, and wadis. Not found in true deserts but occurs in oases, in elevated vegetated areas within deserts, and along desert edges (Spawls and Branch 1995).

Food and Feeding

Feeds on a variety of vertebrates and invertebrates including small mammals and reptiles, and scorpions and centipedes (Spawls and Branch 1995).

Behavior

Little is known. Presumed to be similar to *E. ocellatus*.

Reproduction and Development

Little is known. Presumably oviparous (Spawls and Branch 1995).

Echis megalocephalus Cherlin 1990

Known as the big-headed carpet viper.

Recognition

Head—Head big; scalation similar to *E. pyramidum*.

Body—Middorsal scales 31; ventrals 186–202; subcaudals 33–37 (Cherlin 1990).

Size—Maximum size around 54–62 cm

Pattern—Variable in color and pattern. Usually a dorsal series of oblique pale blotches, with darker colors in between.

Taxonomy and Distribution

The type locality is Nokra Island (Dakhlak Archipelago, the Red Sea, Ethiopia) (Borkin and Cherlin, 1995). The species is only known for the type locality.

It seems to be sympatric with *E. pyramidum* on Nokra Island (Schätti 2001).

Habitat

No information is available.

Food and Feeding

No information is available.

Behavior

No information is available.

Reproduction and Development

No information is available.

Echis multisquamatus Cherlin 1981

Recognition
(Plate 8.3)

Head—Circumorbital ring with 14–21 scales; supraocular variably developed but not elongated; interoculars 9–14; subocular with 1–3 vertical scales between eye and supralabials; supralabials 10–12; sublabials 10–13 (Nikolski 1964; Joger 1984; Latifi 1991; Auffenberg and Rehman 1991).

Body—Dorsal scales keeled, with apical pits, numbering 34–40 rows; lateral scales numbering 3–8 rows; ventral scales rounded, extending full width of belly, numbering 169–199; subcaudals 28–44 and single.

Size—Maximum length is 86 cm.

Pattern—Dorsal head markings form a light cross; faint stripe from eye to the angle of the jaw; body coloration from pale, buff, tan, olive, and chestnut, even reddish or gray; belly color white to pinkish, with stippling ranging from faint to deeply pigmented; laterally with small dark brown spots usually in 1–3 rows; usually with a continous, dorsolateral white line, and with mid-dorsal series of white narrow transverse stripes on the back (Joger 1984).

Taxonomy and Distribution

This taxon was part of *E. carinatus,* but Cherlin (1981) found two taxa with overlapping ranges in Pakistan, Afghanistan, and Iran, verified by Joger (1984) (Senckenberg Museum specimens). In Pakistan, intergradation between *E. carinatus* and *E. multisquamatus* (Auffenberg and Rehman, 1991) may change taxonomic position. *E. multisquamatus* is distributed from Uzbekistan to Iran in the south, and east to western Pakistan.

Habitat

The species prefers hot lowland habitats, often on clay ground or more stony substrates. The habitat can locally be rich of vegetation, especially grass tussocks. In southern Turkmenistan it is sympatric and syntopic with *Macrovipera lebetina cernovi* and *Naia oxiana*. The same subterranean burrows are used during hot daytime hours (Nilson pers. obs.).

Food and Feeding

Feeds on small rodents, shrews, lizards, small birds, and a variety of arthropods. In some areas, feeding may occur relatively late in the year. In southern former Soviet union, 73% of specimens collected in the fall had food in their stomachs, *Mus musculus* being the predominant prey (Bogdanovich 1970).

Behavior

Echis multisquamatus moves by sidewinding, at which it is very proficient and alarmingly quick. Although it can travel in other ways, sidewinding is the most efficient method of crawling over the smooth, sandy surfaces, which are prevalent in its habitat. Sidewinding may also help prevent overheating, by raising body loops above the heated substrate, leaving only two points of contact with the hot surface (Cloudsley-Thompson 1988a).

Reproduction and Development

Litters contained up to 23 individuals among Iranian *Echis* (Latifi 1991), but it is not clear if the species is *E. multisquamatus* or *E. carinatus*.

Turkmenian specimens can bear between 3 and 15 young (Terentev and Chernov 1965).

Echis ocellatus
Stemmler 1970

Commonly known as the West African carpet viper.

Recognition

Head—Head scalation similar to *E. carinatus*. Eyes brown or yellowish, set near front of the head (Spawls and Branch 1995).

Body—Cylindrical, with short tail; dorsal scales heavily keeled, 27–34 rows at midbody; ventrals 145–154 (Hughes 1976; Spawls and Branch 1995).

Pattern—Ground color brown, gray, or shades between; most with a diagnostic lateral line of small white "eye-spots." Dorsal patterns of two types: one, a series of dark, irregular crossbars on a lighter background; the other, a series of pale saddles with darker interspaces, lighter on the sides. Venter pale, usually with brown or reddish spots (Spawls and Branch 1995).

Size — Average 30–50 cm, maximum 65 cm (Spawls and Branch 1995).

Taxonomy and Distribution

Described from Garango, Upper Volta, by Stemmler (1970b). Its range includes Mauritania, Senegal, Nigeria, southwestern Chad, Mali, Ivory Coast, Guinea, Benin, Burkina Faso, Ghana, southern Niger, Togo, Dahomey, and northern Cameroon (McDiarmid 1999). Rarely found north of latitude 15 N (Spawls and Branch 1995).

This species is fairly common in parts of its range. Of 37 species collected in northwest Ghana it was the most common, comprising 16% of the total catch (Spawls and Branch 1995).

Habitat

Found in open country and rocky hillsides, and in well-wooded and rain forest areas. Sometimes observed in open jungle and occasionally on lakeshores (Warrell and Arnett 1976). In Nigeria it is widely distributed in hilly savanna and it is widespread in the Guinea and Sudan savannas of West Africa (Spawls and Branch 1995). Occasionally enters houses and has been reported from 1800 m (Warrell and Arnett 1976).

Food and Feeding

Feeds on a variety of vertebrate and invertebrate prey including scorpions, centipedes, small mammals, birds, lizards, amphibians, and other snakes (Spawls and Branch 1995).

Behavior

Although primarily terrestrial, it occasionally climbs into low shrubs to avoid hot or wet surfaces (Spawls and Branch 1995).

Crepuscular and nocturnal, it becomes active at twilight, with peak activity during the first 6 hours of the night. Daylight is spent hiding in holes or under groundcover (Pitman 1973; Spawls and Branch 1995).

This species is quite aggressive and will strike continuously and vigorously, even moving toward an aggressor (Spawls and Branch 1995).

Reproduction and Development

Irritability increases during mating (Warrell and Arnett 1976). Lays from 6 to 20 eggs, usually at the end of the dry season (February–March). Hatchlings are 10–12 cm (Spawls and Branch 1995).

Echis pyramidum
(Geoffroy and Saint-Hilaire 1827)

Known as the Northeast African carpet viper, or Egyptian saw-scaled viper.

Recognition

Head—Scalation similar to *E. carinatus*.

Body—Middorsal scales 25–31; ventrals 155–182; subcaudals 27–43 (Parker 1949; Hughes 1976; Spawls et al. 2002).

Size—Average 30–60 cm (Spawls and Branch 1995). After extensive collecting in Kenya, Ionides and Pitman (1965b) reported the largest female collected at nearly 60 cm long, with a girth of 5.4 cm, weighing 90 g.

Pattern—Variable in color and pattern. Ground color yellowish, brown, gray, reddish, or shades in between. Usually a dorsal series of oblique pale crossbars, with dark spaces in between. Laterally the same as *E. leucogaster*. Venter pale, usually with brown or reddish spots.

Taxonomy and Distribution

Described from Egypt by Geoffroy and Saint-Hillaire (1827). It was later placed into *E. carinatus* and has recently been returned to *E. pyramidum*. Three subspecies are recognized:

Echis p. pyramidum (Geoffroy and Saint-Hillaire 1827) from southern Arabia, Somalia, Ethiopia, Sudan, Egypt, Libya, and Tunisia;

Echis p. aliaborri Drewes and Sacherer (1974), from northern Kenya; and

Echis p. leakeyi Stemmler and Sochurek (1969b) from northwestern Kenya.

Population densities vary. In the 6500-km^2 Moille hill area of northern Kenya, nearly 7000 individuals were collected in a 4-month period. Around Cairo, only two specimens were collected in 25 years (Whitaker 1973b; Spawls and Branch 1995). Collections from northern Kenya yielded 41 and 52 individuals over 3-day and 4-day periods respectively, and 218 individuals in a fortnight (Ionides and Pitman 1965a).

Habitat

Occurs in a variety of habitats ranging from open, arid, sandy localities to semi-desert, thorny scrub, light woodland, savanna, cotton soil, grassland, cultivated areas, oases, dry savannas, and rocky hillsides and lava fields (Pitman 1973; Spawls and Branch 1995). Avoids true deserts and other areas of soft sand. In sandy-grassy areas of low coastal plains it rests in small holes beneath the surface during the day.

Specimens collected from the Arabian Peninsula are found in gravel and rocky and sandy areas with sparse xerophytic vegetation. Individuals were also collected around buildings and in gardens (Corkill and Cochrane 1965).

In Kenya, areas of scattered bush with sparse grass covering sandy soil yielded the most individuals (Duff-Mackay 1965; Ionides and Pitman 1965a). The few acacia trees in these areas provided logs favored by these snakes. Heavy well-rotted logs eaten by termites on the underside were preferred. These logs also harbored skinks, geckos, scorpions, solifugids, and centipedes, which are all potential prey. However, apparently suitable logs within thick grass cover invariably failed to yield specimens. Trails to and from rodent holes indicated that most adults spend the daytime in these holes.

Generally not found above 900 m of altitude (Pitman 1973).

Food and Feeding

Feeds on a variety of vertebrate and invertebrate prey including beetles, crickets, solifugids, scorpions, centipedes, small mammals, birds, lizards, amphibians, and other snakes (Duff-Mackay 1965; Stemmler 1971b; Spawls and Branch 1995). Stomach contents of specimens collected from northern Kenya contained the southern long-tailed lizard, *Latasia longicaudata revoili,* the variable skink, *Mabuya varia varia,* and other unidentified lizards (Ionides and Pitman 1965a). *Echis* from Saudi Arabia eat, in addition to the above items, gerbils, worms, and slugs (Corkill and Cochrane 1965; Gasperetti 1988).

Captives may feed on dead lizards, specifically geckos and agamids (Stemmler 1971b) and have rarely consumed raw pieces of goat meat (Duff-Mackay 1965). Captives did well on a diet of one adult mouse per week (Pitman 1973). These snakes readily feed from early June to the end of September. During the rainy season from October to February all food is refused. Cooling snakes during these periods minimizes losses (Stemmler 1971b).

Attempts at cannibalism have been reported. Conspecifics usually bit each other on the head as a prelude to swallowing. However, few deaths occurred following envenomation and attempts at swallowing were mostly aborted (Pitman 1973). In one instance, a large female "powerfully" bit a smaller female around midbody resulting in the death of the smaller snake the next day, possibly due to puncturing of a vital organ (Stemmler 1971b).

Saw-scaled vipers can withstand intense heat with little moisture. However, Stemmler (1971b) reported that *E. pyramidum* often drink water, and Pitman (1973) observed captives drinking freely from troughs.

Behavior

Primarily crepuscular and nocturnal (Warrell and Arnett 1976) and seldom found during the day (Duff-Mackay 1965). In captivity they become active toward sunset and remain so until sunrise. When not active they often conceal themselves beneath the substrate, with just the top of the head showing (Pitman 1973). At night, they are alert and will immediately stridulate when surprised by light, with their head alertly following every movement (Stemmler 1971b).

During rainfall, juveniles are inclined to remain hidden under logs while ground conditions are wet (Duff-Mackay 1965).

Strikes with a great deal of speed and force. Strikes begin with the body coiling in a wide curve and the neck and the head raised and doubled back. During the strike the snake springs forward from the looped position, drawing the posterior into another stridulating loop. They may strike with such vigor that they are propelled upward and forward as much as 30.5 cm, appearing to jump. A strike may extend over 50% of the body length. Propulsion is enhanced by sudden inflation and deflation of the anterior body. The initial strike position is resumed almost immediately. Instances of two quick concurrent strikes, each inoculating a fatal dose, have been reported (Pitman 1973).

Reproduction and Development

Sexually active periods occur in May and November–December, correlating with the rainy season. *Echis pyramidum* is oviparous, laying 4–20 eggs. Eggs incubated in captivity should be placed in a bowl containing a 1:1 mix of sand and peat inside another, larger bowl containing water at 25°C. The bowl should be covered with bark and maintained at an 80%–90% relative humidity (Stemmler 1971b; Pitman 1973; Spawls and Branch 1995).

Eggs are laid in May or in November–December. The incubation period is usually 50–64 days, and a 2-cm embryo observed in a newly laid egg indicates that development begins inside the mother (Stemmler 1971b). There seems to be a correlation between birthing time and the occurrence of rain, with most young born when the ground is wet. This suggests that birth can possibly be delayed a few weeks during unfavorable weather (Duff-Mackay 1965).

Neonates are 11–15 cm in total body length. Little is known about the growth rate, with estimates based primarily on the number of shed skins (Duff-Mackay 1965). During August and September, large numbers of subadults and juveniles are present in African populations. Of 218 specimens collected, only 35 (16%) were adults, 70 (32%) were subadult, and 113 (52%) were juveniles under 25 cm (Ionides and Pitman 1965a). Based on February collections which were mostly adults, Ionides and Pitman (1965a) suggested that *E. pyramidum* probably grows to adult size within 6 months.

An attempted mating between a male *E. coloratus* and female *E. pyramidum leakeyi* was reported by Stemmler (1971b). On 5 November, the male crawled along the back of the female in a series of jerky motions. His chin tip stroked the female's sides as he moved his tail under hers to copulate. In December, 1 fertilized egg with a clearly discernible yolk was laid on a stone. The egg was placed in a sand-peat substrate, but by 1 January was dead. On 2 January, 3 more fertile eggs were laid with very thin, shiny, and smooth shells. They had a milky transparency through which a fine network of blood vessels could be seen but these eggs also failed to hatch.

Echis Bite and Venom

Because of the unsettled taxonomy, we discuss the venom and bites of all *Echis* species together in this section.

Epidemiology—*Echis* envenomation probably results in more deaths than from the bites of any other snake, and is recognized as an important medical problem in many tropical rural areas (Warrell et al. 1977). In Nigeria, *Echis* bites occur at a rate of 120 bites per 100,000 people per year, and 10% of all snakebite patients in hospitals from the Nigerian savanna were victims of saw-scaled bite (Warrell and Arnett 1976). In Israel, *Echis* is the second most common cause of snakebite (Benbassat and Shalev 1993). However, worldwide the total number of bites is underreported due to limited medical access and the frequent use of "local cures." For example, in Kenya, Rendille herdsman treat bites by opening the wound to cause bleeding, and the victim is fed burnt milk and goat hide as an emetic for 3–4 days. Surprisingly, few fatalities result (Duff-Mackay 1965).

Most bites occur after dark due to the nocturnal habits (Pitman 1973; Warrell and Arnett 1976). In Nigeria, bite incidence increases in February and peaks in May and June, corresponding to the peak farming and field work periods (Pitman 1973). In India, 91% of bites occur from May through September, peaking in July and August, with mortality higher during December and January. This is correlated with the dry season when prey is scarce and the venom more concentrated (Bhat 1974).

Eighty-one percent of bites occur on the lower limbs with 77% on the ankle or foot (Warrell and Arnett 1976; Warrell et al. 1977). Fang marks, usually two, are often visible with the persistent oozing of blood, and are between 0.5 and 1.1 cm apart (for a comprehensive review of bite frquency, see Warrell and Arnett 1976).

Yield—The venom apparatus is typical of most viperids. However, there are also similarities in the skeletal muscles associated with the main venom gland with *Naja naja,* and the main gland epithelium is similar to that of *Naja naja* and *Hypnale hypnale* (Gopalakrishnakone 1985).

Yield reports include 20–35˚ mg dried venom from specimens 41–56 cm long (Minton 1974; U.S. Navy 1991), 6–48 mg (mean 16 mg) from Iranian snakes (Latifi 1991), and 13–35 mg of dried venom from animals collected from various localities (Boquet 1967b). Yield varies seasonally, with maximum yields observed during summer (40 mg), followed by spring (33 mg), winter (22 mg), and fall (18 mg). Yield differs between males and females, with means of 53 mg and 35 mg respectively (Latifi 1984).

Toxicity—LD_{50}s vary among *Echis* species, individual snakes, geographic variants, season, test animals, milking sequence, and injection modality (i.v, s.c., i.p.).

Intravenous LD_{50}s in mice are reported as 2.30 mg/kg (U.S. Navy 1991), 24.1 mg/kg (Christensen 1955), 0.44–0.48 mg/kg (Cloudsley-Thompson 1988a), and 0.031 mg per animal for freeze-dried venom (Boquet 1967b). Subcutaneous LD_{50}s in mice are reported as 0.007–0.077 mg (Cloudsley-Thompson 1988a); and freeze-dried i.p.-injected mice as 0.031 mg per animal (Boquet 1967b). Human lethal dose is estimated at 3–5 mg (Minton 1967a). LD_{50}s vary between males and females (Latifi 1984). Male LD_{50}s range from 4.6 to 9.2 μg/mouse (mean 6.6) and female LD_{50}s from 2.4 to 3.8 μg/mouse (mean 3.0). LD_{50} values do not vary greatly between left and right fangs. LD_{50} values vary according to milking sequence. Latifi (1984) reported LD_{50} values declined in subsequent milkings from venom taken from the same snake during the first and second milkings.

Content—Content varies genetically both intra- and interspecifically and this variability may cause contradictory biochemical and/or physiological test results (Schaeffer 1987; Paine et al. 1992). Several active enzymes are found in whole venom including: L-amino acid oxidase, phospholipase A, phosphotases, cholinesterases, hyaluronidase, and a coagulation accelerator and inhibitor (Boquet 1967b).

Exopeptidase activity and bradykinin release was found by Tu and Toom (1967) but Diniz (1967) reported that *Echis* venom was a poor liberator of bradykinin in comparison to North American pit vipers. Significantly higher proteinase activity levels were found in *E. carinatus* than from five other viperine species (*Bitis arietans, B. gabonica, E. coloratus,* and two subspecies of *D. russelii*).

Master and Rao (1963) found one toxic principle that has seven components exhibiting proteolytic and clotting activities. Rosenfeld et al. (1967) found two fractions in the venom of *Echis* spp. Fraction I was an anticoagulant via fibrinogenolysis whereas fraction II was a coagulant. Other proteins found are hemorrhagin, echistatin, and ecarin (Taylor and Mallick 1935; Nahas et al. 1964; Minton 1974; Franza et al. 1975; Kornalik and Blombach 1975; Warrell et al. 1977; Gans et al. 1988).

The presence of copper is reported from the venom of *E. coloratus* and this element has been linked to the neurotoxic protein occurring in the venom of *Vipera xanthina* (Gitter et al. 1960). A neurotoxic fraction antigenetically related to *Naja* venom has also been isolated from the venom of *E. coloratus* (Detrait and Saint-Girons 1979).

Variation in dried venom appearance from *E. pyramidum* seems correlated to snake length. Milked snakes 30 cm and under produced almost colorless pale yellow venom, while snakes 45 cm and larger produced deep orange-yellow dried venom. Darker dried venom has also been reported from both large and small, captive and penned specimens (Duff-Mackay 1965). The significance of these color differences is unknown but probably reflects L-amino acid oxidase concentration (Minton 1977, pers. comm.).

Symptoms and Physiological Effects—In general, venom of *Echis* causes massive clotting, convulsions, and death within minutes in natural prey. However, these effects are rare in humans since injected dosages are smaller in relation to body size. No significant correlation between snake length and symptomology signs occurs in humans (Warrell et al. 1977).

Lethal envenomation by *E. coloratus* in lab animals causes bleeding, hypotension, afibrinogenemia, and increased prothrombin time and fibrin degradation products within 4–6 hours. Sublethal doses produce similar hemostatic disturbances which peak after 24 hours and gradually normalize after 10 days (Benbassat and Shalev 1993).

Mice i.v.-injected with venom from *E. coloratus* suffered hemorrhage, nervous system disturbance, anemia, thrombocytopenia, and afibrinogenemia (Gitter et al. 1960). Intraperitoneal injection of lethal doses from *E. carinatus* in mice preterminally caused immobility, breathing difficulty, general weakness, hind leg paresis, and jerky uncoordinated movements of the forelimbs (Boquet 1967b). Gaertner et al. (1962) found that venom from *E. coloratus* caused *in vitro* destruction of embryonic mouse and chick cells.

White rats exhibited profound coagulation defects following sublethal s.c.-injection of venom from *E. carinatus* (Kornalik and Pudlak 1971), but normal fibrinogen levels were restored within 5–6 days. A moderate decrease in thrombocyte levels occurred which was restored faster than any other coagulant parameter. Effective venom doses differed suggesting variability in venom concentration at any one time. It was shown that after an hour, rats treated with hydrocortisone withstood venom from *E. carinatus* better than untreated rats (Christensen 1967b).

Lethal i.v. injections of venom from *E. coloratus* in guinea pigs produced intravascular clotting, hemorrhage, afibrinogenemia, Factor V deficiency, and thrombocytopenia (Rechnic et al. 1962; de Vries et al. 1963). Injections of heparin after venom injection caused rapid fibrinogen regeneration, inhibited fibrinolysis, and decreased thrombocytopenia (de Vries et al. 1963).

Heparin injected rabbits showed no coagulative or hemorrhagic defect from venom of *E. carinatus*. A minimal lethal dose (MLD) of 0.1 mg in rabbits could be increased 20 times if preceded by 10 mg of heparin (Ahuja et al. 1946). At 30 times the MLD, all rabbits died regardless of heparin dose (Forbes et al. 1966). Envenomated rabbits and monkeys experienced toxicoglomerulitis, edema, and degenerative neuronal changes near cardiac and respiratory centers of the brain stem (Sant 1978).

In dogs, venom from *E. pyramidum* caused hypercoagulabilty within 5 minutes, followed by incoagulability after 15 minutes (Mohammed et al. 1969d). Clots introduced in the inferior vena cava of dogs were dissolved with venom from *E. coloratus* within 24 hours. Localized activation of the fibrinolytic system is credited with this phenomenon, so *E. coloratus* venom may be of therapeutic value in thrombosis (Djaldetti et al. 1966).

Following introduction of venom from *E. carinatus,* Rhesus monkeys exhibited an initial procoagulant effect which led to disseminated intravascular coagulation (Chugh et al. 1981). However, the main effects in monkeys were local and visceral hemorrhaging. Hemostatic profiles of monkeys injected with sublethal doses showed peak abnormalities within 24 hours and full recovery within 10–20 days. The severity of the hemostatic disturbance was dependent on dose and concentration of various active principles. Plasma fibrinolytic activity was not affected by dose (Taylor and Mallick 1935).

There are many reports of human envenomation by *Echis*. In general, initial symptoms in humans are pain, swelling, and bleeding at the bite site. Of 115 patients bitten by *E. ocellatus* in Nigeria, Warrell et al. (1977) reported pain occurred within 10 minutes in 80% of cases, with 77% reporting pain in regional lymph nodes. Local swelling was noticed within 30 minutes by 60% of the victims. Swelling was the most common sign of envenomation, occurring in 100% of cases and lasting up to 1 week. Spontaneous bleeding from the bite site and sites distant from the bite occurred in 57% and 68%, respectively. Bleeding from fang punctures lasted less than 10 minutes in 80% but continued from 3 to 18 hours in the remainder of cases, eventually stopping spontaneously or after antivenin administration. Bleeding from the gingival sulcus was reported in 57%. Thirty-two percent reported blood in the sputum and 31% bled persistently from remote wounds.

Echis coloratus and *E. carinatus* envenomations differ in bleeding duration. Benign clinical bleeding following hospital admission lasted several days in only 5% of *E. coloratus* patients, with bleeding declining despite persistent hemostatic failure and lack of treatment. The risk of anemia and thrombocytopenia was correlated with hemostatic defect duration. This is in sharp contrast to bites by *E. carinatus* where overt bleeding continued in 57% of untreated patients, with 10%–20% mortality, due mostly to cerebral hemorrhage (Benbassat and Shalev 1993).

Venom from *E. carinatus* may cause spontaneous sublingual, subconjunctival, intramuscular, subcutaneous, and retroperitoneal hematomas. In addition, subarachnoid hemorrhage was detected in three patients. Liver and capillaries are the main affected systemic sites (Mole and Everard 1947).

Venom from *E. coloratus* causes other local and systemic signs of envenomation including fang wounds (51%), fever (72%), enlarged and tender lymph nodes draining bitten area (50%), tachycardia greater than 90 bpm (48%), vomiting (28%), headache (17%), dizziness/faintness (17%), drowsiness (14%), blistering (13%), localized necrosis (11%), local bruising (10%), altered consciousness (8%), abdominal pain (7%), and hypotension (4%). In addition, 28% of all newly admitted patients had submucosal and general microhematuria, and 21% had a platelet count <100,000/μl (Benbassat and Shalev 1993). Venom from *E. ocellatus* causes necrosis at the bite site from 1 to 13 days (mean 6 days) postbite, destroying skin, subcutaneous tissue, and blood vessels to the muscle (Warrell et al. 1977). Cortical necrosis, hematuria, glomerulitis, and hepatotoxicity have has been reported (Sant 1978).

Venom from *E. coloratus* causes transient renal damage evidenced by oliguria and albuminuria. Cylinduria in the absence of hypotension can be ascribed to the presence of microthrombi in the glomerular capillaries (Rechnic et al. 1962; Fairnaru et al. 1974a, b). Systolic pressures lower than 60 mm Hg have been reported and acute renal failure resulting from hypotension was diagnosed as the cause of death in a case study of *E. coloratus* envenomation (Oram et al. 1963; Schulchynska-Castel et al. 1986).

Venom from *E. carinatus* may mobilize glycogen from the liver and muscles causing hypoglycemia (Mohamed et al. 1963).

Venom from *E. coloratus* has a direct toxic effect on heart myocardium causing electrocardiographic changes (Fairnaru et al. 1974a, b).

Symptomatic neurological disturbance is minimal despite a neurotoxic element in the venom (Boquet 1967b). Breakdown of the blood-brain barrier by venom from *E. coloratus* can simulate neurotoxic effects (Sandbank and Djaldetti 1966).

Warrell et al. (1977) reported two patients envenomated by *E. ocellatus* suffered convulsions and upper motor neuron lesions, but no cranial nerve lesions or respiratory paralysis were observed. The exact nature of neurological effects is still not fully understood. Venom from *E. carinatus* may cause complete or partial blindness due to optic atrophy following hemorrhage or prolonged spasms of the retinal vessels (Davenport and Budden 1953; Guttman-Friedman 1956).

Hematological Activity—The principle physiological response to envenomation by *Echis* is transient hemostatic failure (Warrell et al. 1977). Venom from all *Echis* species produce this response through combined effects of procoagulant, fibrinolytic, antiplatelet, and hemorrhagic components (Benbassat and Shalev 1993). Hemostatic failure is less severe with *E. coloratus* bites than those of *E. carinatus* (Coppola and Hogan 1992). Intraspecific variation in hematological activity occurs primarily with thromboplastic coagulant activity *in vitro,* and defibrinogen capacity *in vivo* (Taborska 1971).

Coagulant or anticoagulant effects are proportional to venom dose. Low doses increased prothrombin conversion to thrombin, confirming coagulant activity. At high doses fibrinogen was destroyed and clots could not form even after thrombin administration. Although fibrinogenolysis and fibrinolysis are major factors involved in coagulation failure, disruption of other factors, especially in early stages of coagulation, are important. However, these events are difficult to assess as a consequence of rapid venom activity at high concentrations (Boquet 1967b; Rechnic et al. 1967).

Clinical assessment of human envenomation by *Echis* reveals hypofibrinogenemia, prolonged prothrombin and clotting time, and increased circulation of fibrinogen split products (Benbassat and Shalev 1993). Intravascular and intracapillary clots form in guinea pigs within 5 minutes following injection of a lethal dose of *E. coloratus* venom. Sublethal doses produce defibrination and thrombocytopenia (Rechnic et al. 1962).

In envenomation by *E. ocellatus,* hemostatic defect can be detected as early as 75 minutes and as late as 27 hours postbite. This delay may be due to delayed venom absorbtion (Warrell et al. 1977), and delayed onset of symptoms obviates the rule of thumb that if envenomation signs are not apparent within 1 hour, clinically significant envenomation will not occur (Benbassat and Shalev 1993).

Thrombinlike activity of venom causes hypercoagulability, usually manifested as disseminated intravascular coagulation (Minton 1974). However, venom of *E. coloratus* may not cause disseminated intravascular coagulation in humans (Dvilansky and Biran 1973). Following such coagulation, incoagulability resulting from fibrinogen degradation, fibrinolysis, hypofibrinogenemia, afibrinogenemia, and prolonged clotting time leads to massive hemorrhaging (Coppola and Hogan 1992; Russell 1983; Warrell et al. 1977). Spontaneous hemorrhaging may begin from 1 to 36 hours following envenomation. Evidence of coagulopathy was observed in one-third of patients in Israel and 93% of Nigerian patients bitten by *Echis* (Minton 1994). This is clinically the most important effect of *Echis* venom and is the cause of most deaths (Warrell et al. 1977). Demonstration of noncoagulating blood should be the number one diagnostic test given in suspected cases of envenomation by *Echis* (Reid et al. 1963). Patients with evidence of intercranial hemorrhage have died on days 11 or 12 postbite (Minton 1974).

The procoagulant enzyme "ecarin," an acidic glycoprotein, exhibits strong thromboplastin activity and is able to convert prothrombin into thrombin, even in the absence of calcium (Nahas et al. 1964; Weiss et al. 1973; Franza et al. 1975; Kornalik and Blomback 1975; Morita et al. 1976). Of 19 snake venoms investigated, only two genera, *Echis* and *Notechis* (Australian tiger snake), were able to convert prothrombin to thrombin (Jobin and Esnouf 1966). However, ecarin does not aggregate platelets (Minton 1994).

Venom from *Echis* varies in its ability to convert prothrombin to thrombin. Venom extracted from *E. pyramidum* was a poor converter of prothrombin compared to *E. carinatus* (Kamiguti et al. 1988). The ability of *Echis* venom to convert prothrombin to thrombin has been used in medical and physiological research. A clotting assay using *Echis* venom was used to measure prothrombin content (Franza and Aronnson 1976), and venom has been used to initiate intravenous formation of microemboli for modeling pulmonary microembolisms (Schaeffer et al. 1987).

Echis venom does not contain fibrinolytic components and the fibrinolytic response results from activation of the animal's own fibrinolytic system (Rechnic et al. 1962; Moav et al. 1963; de Vries et al. 1963; Warrell et al. 1977). Afibrinogenemia results from two actions: lysis of fibrin clots formed early by procoagulant action and consumption coagulopathy following secondary fibrinolysis (Gitter and de Vries 1967; Yatziv et al. 1974). *Echis coloratus* venom also destroys fibrin-stabilizing factors and impairs platelet aggregation (Djaldetti et al. 1965, 1966, Biran et al. 1972, 1973).

Hemorrhage is exacerbated by hemorrhagin, another protein fraction isolated from *Echis* venom. Hemorrhagin causes vascular tissue damage resulting in leakage across endothelial cells of cerebral capillaries (Taylor and Mallick 1935; Sandbank et al. 1974). Additional factors contributing to hemorrhage are depletion of clotting factors II, V, VIII, and XIII, and the presence of echistatin, a potent platelet aggregation inhibitor (Warrell et al. 1977; Gans et al. 1988).

Systemic hemorrhaging results in bleeding from the bite and from sites and wounds distant from the bite. The most common site of hemorrhage is the gingival sulcus of the gums. Hemorrhage results in hematuria and ecchymosis, and commonly occurs in the subarachnoid space, cerebrum, and uterus. It is also evidenced in the gastrointestinal tract by blood in the vomitus, sputum, and stools. (Hall 1962; Warrell et al. 1977; Coppola and Hogan 1992). Despite microscopic perivascular hemorrhages in the lungs, the vessels themselves remain intact (Gitter et al. 1960).

Case Histories—Warrell et al. (1977) described a case of a 12-year-old boy bitten on the left foot by a 54-cm male *E. ocellatus* while plowing a field. He experienced immediate pain, noticeable swelling after 30 minutes, and bleeding from the gums and nose within 1 hour. A local herbal medicine was chewed during the 19-km journey to the hospital. Upon arriving, he was confused and drowsy, and had shallow respiration, an impalpable arterial pulse, and nonrecordable blood pressure. He was still bleeding from the gums and nose and the left leg was swollen to the groin. His blood was incoagulable. He was given 40 ml of Behringwerke antivenin i.v. Tilting his head downward resulted in a blood pressure of 50/30 mm Hg; however, blood pressure dropped again despite further infusion of 500 ml of Dextran. Three hours after a blood transfusion of 600 ml his blood pressure stabilized at 95/50 mm Hg. The following day, blood pressure dropped slightly to 90/50 mm Hg, and blood blisters appeared on the top of the left foot. Blood was still incoagulable (grade 5 clot quality), but no signs of bleeding were present. Other blood parameters were platelets 30,500/mm^3, fibrinogen less than 5% normal, and FDP 80 μg/ml, packed cell volume (PCV) 20%. Evidence of microangiopathic hemolysis was indicated by the presence of numerous poikilocytes and schizocytes. A 20-ml i.v. injection of SAIMR *Echis* antivenin was given along with one unit of transfused blood. Two days after the bite, his blood pressure improved markedly (115/70 mmHg), blood was coagulable (clot quality grade 1), platelets were 50,000/ mm^3, FDP 40 μg/ml, and PCV 24%.

Although the prognosis was good at this point, between days 3–13 several pathological events transpired. His kidneys failed, yielding a urine volume of 150–300 ml/day that contained granular casts, red blood cells, and albumen. Transient nose bleeds occurred on days 5, 7, and 8, with bloody stools on day 8. The bitten limb was extensively bruised, with new blisters erupting daily during the first 7 days. On day 8 he became ill with generalized edema, breathlessness, and bilateral joint pain. On day 10, blood urea rose to 330 mg/100 ml and serum potassium rose to 6.9 mEq/L. A distinctly marked area of putrified necrotic tissue at the bite site appeared by day 12. During this time his platelet count rose from 65,000 on day 3 to 138,500 by day 7 and to 404,000 on day 12; PCV also rose to 35% by day 12. Blood transfusions were given on days 5 and 8, and the patient was able to get out of bed by day 14. However, this resulted in sudden venous hemorrhage of about 1 L from the necrotic area of his left foot. A tourniquet was temporarily applied and the area was further debrided. By day 20, general improvement was steady and split skin grafts were applied to the left foot. On day 27, blood urea was 20 mg/100ml, platelets were 218,000/mm^3, and hematocrit was 32%. A followup examination after 6 months showed full recovery.

A fatal case history is provided by Warrell et al. (1977) in which a 50-year-old Nigerian man was bitten on the right ankle by a 40-cm *E. ocellatus*. Pain occurred immediately, and after 2 hours swelling had progressed to midcalf. Herbal medicine was drunk and applied locally. The next day there was considerable bilateral joint pain and stiffness, and blood was present in the urine. This was followed by abdominal pain and more swelling. Six days later, the victim was transferred 80 km to a hospital. At this time he was grossly anemic and would sweat and faint with every attempt at sitting up. His blood pressure was 90/10 mm Hg and pulse was 122 and regular. The abdomen was swollen and tender, and a drop of blood at the urethral meatus was the only sign of bleeding. The right ankle was minimally swollen and tender, without necrosis. His blood was incoagulable (clot quality grade 5) and hemoglobin was 2.5 g/100 ml. Antivenin was not available but an i.v. infusion of dextrose saline was started. In addition, 100 mg hydrocortisone i.v. and 10 mg vitamin K i.m. were administered. The patient died 7 hours after admission before a blood transfusion could be started. An autopsy revealed the following: extensive internal bleeding including perivascular hemorrhage around the large vessels of the thoracic inlet and aortic arch; a hemoperitoneum; an extrapleural hemorrhage of approximately 1.5 L displaced the mediastinum and compressed the right lung; a large retroperitoneal hemorrhage tracking along the pelvi-calyceal system; and hemorrhage into the adventitia of the bladder. There were bilateral intercostal hemorrhages and massive bleeding into the sheath of the rectus abdominus and other muscles of the pelvic floor. No hemorrhage was found within the brain or its meninges or in the digestive tract.

Fairnaru et al. (1974a, b) reported on a 15-year-old boy bitten on the index finger by *E. coloratus* in the Judean Desert. Upon hospital admission 90 minutes postbite, severe pain radiated from the bite site to the shoulder. Within 6 hours, the boy was pale, nauseous, vomiting, and the entire limb was markedly swollen. Blood pressure, pulse, and temperature, heart, lungs, and abdomen were normal. No signs of spontaneous hemorrhage or bleeding were observed at the bite site but venous blood failed to clot. An EKG showed heart irregularities. Intravenous administration of glucose, hydrocortisone, and penicillin were started immediately along with 20 ml i.m. antivenin of *D. palaestinae*. After 12 hours, afibrinogenemia was detected, along with very low Factor V activity, low prothrombin time and platelet count. By day 2, blood glucose and urea, serum electrolytes, and hemoglobin and leukocytes were within normal limits. Serum protein, liver function, and urinary excretion of catecholamines were all within normal limits. After 42 hours, the patient finally urinated. The urine contained protein, numerous granular casts, and a few leukocytes. During the next 24 hours, 500 ml of urine were produced containing 4.8 g protein/L. Treatment during the first day included 3000 ml of 5% i.v. glucose, normal saline, 1 g hydrocortisone, and four M units of crystalline penicillin. Fluid intake during the first day was *ad libitum*. During the second day blood urea elevated to 64 mg/100ml. An i.v. of 0.4 g hydrocortisone and i.m. of penicillin were given. A small clot which dissolved after 20 minutes was observed from blood drawn after 36 hours but, except for oozing of blood from the puncture sites, no subcutaneous bleeding was observed. Granular casts and protein disappeared from urine over the next 5 days and urine output increased. By day 5, fibrinogen, prothrombin time, clotting time, and Factor V activity were approximately normal; EKG tracings showed nodal rhythm alternating with sinus rhythm, and T waves became inverted in leads II and III. Edema of the affected arm remained unabated for 7 days, after which, over the next 4 days, it normalized. The patient was released after 11 days.

Development of hypersensitivity in the form of eye irritation and incessant sneezing by workers constantly exposed to dried venom dust has been reported (Duff-Mackay 1965). Venom dust entering a small cut in a workers forearm produced bruising of the limb that lasted 24 hours. Further exposures by workers resulted in a reduced effect.

Treatment and Mortality—Bleeding from sites distant from the bite is the most important sign of systemic poisoning (Warrell et al. 1977). Tourniquets greatly increase tissue damage (Spawls and Branch 1995). In individuals with bleeding complications, *Echis coloratus* antivenin use is favored in adults not previously treated with antivenin (Benbassat and Shalev 1993).

Several antivenins are available (Theakston and Warrell 1991), however, antivenin effectiveness may vary and is dependent on the venom source (Nahas et al. 1964; Taborska 1971; Kornalik and Taborska 1973; Latifi 1973). Comparisons between SAIMR (South African Institute for Medical Research) and Beringwerke (North and West African polyvalent antivenin), showed the former to be more effective, restoring coagulability and reducing necrosis with only 15.2 ml compared to 37.9 ml (Warrell et al. 1977). These findings imply that antigenic differences exist among individuals from different populations and, therefore, antivenins should be made from pooled venom sources. Monovalent antivenins are far more effective in neutralizing venom of *Echis* than polyvalent ones (Latifi 1984), and Porath et al. (1992), in a risk assessment of patients envenomated by *E. coloratus*, strongly recommended using monovalent antivenin.

Antigenicity of saw-scaled viper venom can be increased from 45 LD_{50}s to 500 LD_{50}s using venom-adjuvant mixtures (Moroz et al. 1966b).

Venom from *Echis* contains at least two antigens common to several elapid species, consequently antivenin from two species of *Naja* (*N. haje*, *N. nigricollis*) neutralized moderate amounts of venom (Boquet et al. 1967b; Mohamed et al. 1973). Antivenin prepared from saw-scaled viper venom can counteract *Cerastes* venom (Schwick and Dichgiesser 1963). However, New World crotalid polyvalent antivenin has little effect cross-neutralizing venom from *Echis* (Minton 1976).

The average duration of hemostatic failure in patients given antivenin was 2.5 days, compared to 3.6 days in untreated or treatment delayed patients (Schulchynska-Castel et al. 1986). Mortality rates of 10%–20% in untreated patients decreased to 2.8%–3.4% in treated patients (Bhat 1974; Warrell et al. 1977), and blood coagulability was restored 2–39 hours (mean 12 hours) after the first antivenin dose. The antivenin mode of action is primarily neutralization of the procoagulant, with little effect on the local swelling, necrosis, or blistering (Reid et al. 1963).

Hemorrhagin in *Echis* venom is very strong and antihemorrhagin serums cannot reverse blood vessel damage, they can only prevent further damage (Gitter and de Vries 1967). Administering hydrocortisone to protect blood vessels has been recommended (de Vries and Cohen 1969).

Heparin has been used in patients where antivenin and/or fibrinogen therapy fail to correct coagulation defects (Weiss et al. 1973). However, heparin has not been recommended for bites by *E. coloratus* (Benbassat and Shalev 1993) and its use is now questionable (Minton 1977, pers. comm.). Fresh frozen plasma is also not recommended. Patients receiving fresh frozen plasma have higher bleeding rates, larger drops in hemoglobin during follow-up, larger increases in blood urea, and more prolonged hospitalizations (Porath et al. 1992).

Benbassat and Shalev (1993) provide a useful "decision tree" for treating *E. coloratus* envenomated victims. For a detailed accounting of clinical analysis, Pitman (1973) and Warrell and Arnett (1976) should be reviewed.

Alternative methods of protecting against envenomation by *Echis* have been investigated (Theakston et al. 1985). Incorporating snake venom into stabilized liposomes and injecting these into sheep resulted in an intense, sustained, and protective antibody response. This technique has not been investigated in humans. Iddon et al. (1985) developed a monoclonal antibody against the hemorrhagic principle of Nigerian *Echis* which neutralized both its hemorrhagic and necrotizing activities. However, no significant neutralization of procoagulant or defibrinogenating agents occurred. Monoclonal antibodies had no effect on venoms from Middle Eastern or Asian *Echis*.

Mortality rates from *Echis* bites vary from 5% to 33% (Hall 1962; Coppola and Hogan 1992). Hospitals and dispensaries in northeastern and Benue Plateau states of Nigeria treat over 150 snakebite patients per year and have an overall *Echis* mortality rate between 7% and 15% (Warrell et al. 1977; Theakston and Reid 1982). Bites by *Echis* are a major cause of snakebite mortality in India with mortality rates for untreated envenomation exceeding 20% (Sarkar and Devi 1967; Bhat 1974). Morbidity rates for *E. coloratus* are less than those of other *Echis* species. This may be related to differences in venom toxicity and characteristics of the population at risk. In addition, many bites by *E. coloratus* occur in Israel where medical attention is readily available, in contrast to bites by other *Echis* species which occur in less developed nations (Benbassat and Shalev 1993).

Remarks

An array of biotic associates have been found under rocks with *Echis:* snakes (*Amphiesma stolatus, Boiga trigonata, Daboia russelii* and *Naja naja*), lizards (*Calotes versicolor, Riopa guentheri, Ophisops bedddomei, Hemidactylus brooki, H. maculatus, H. frenatus,* and *H. albofasciatus*), and frogs (*Rana hexadactyla, R. tigerina, R. limnocharis, R. cyanophlyctis, R. breviceps, Microhyla ornata,* and *Rhacophorus maculatus*). In addition, arthropods were found including red and black scorpions (*Buthus tamulus* and *Heterometrus gravimanus* respectively), other arachnids, centipedes, millipedes, crabs, earwigs, and beetles. Many of these are known prey for *Echis* (Whitaker 1973b).

Batesian mimicry with the egg-eating snake, *Dasypeltis scabra,* has been reported (Corkill 1935; Pitman 1973). The common cat-eyed snake, *Boiga trigonata,* in Pakistan, and Soosan vipers and leopard vipers, *Telescopus* spp., in the Middle East are also *Echis* mimics, although the former only weakly so (Minton, 1977, pers. comm.). Similarities in color, shape, keeled scales, and scale stridulation suggests Mullerian mimicry may exist between *E. carinatus* and *Cerastes* spp. (Cloudsley-Thompson 1988a).

Saw-scaled vipers have few natural enemies. The red cobra, *Naja nigricollis pallidus,* a known snake eater, seems to be the main one in Africa (Duff-Mackay 1965). In areas where *E. pyramidum* are abundant, puff adders, *Bitis arietans,* are rare, and vice-versa, suggesting some type of competitive exclusion (Pitman 1973).

Echis pyramidum does not do as well in captivity as *E. carinatus*. Stemmler (1971a) kept specimens at temperatures varying from 14°C to 33°C depending on season, with no nocturnal heat provided at any time of year.

In captivity, Goode (1979b) recommended keeping *E. coloratus* offspring on "number 16" sand with no access to water.

Chapter 9
Eristicophis

The genus *Eristicophis* contains a comparatively rare desert viper known as Macmahon's viper, leaf-nosed viper, or Asian sand viper. This monotypic genus is regarded as a most specialized derivative of the Eurasian viperid lineage (Marx and Rabb 1965). Phylogenetic analysis of Palearctic viperids distinguished *Eristicophis* as a separate genus from *Pseudocerastes* and *Vipera* based on differences in osteology, head and body scalation, and range of locomotion (Marx and Rabb 1965). Together with *Pseudocerastes*, it forms a monophyletic group, either with *Daboia* or with *Macrovipera* as its sister taxon (Herrmann et al. 1992; Herrmann and Joger 1997).

A particular oddity is the premaxilla which has a concave, spatulate, dorsoposterior process with an expanded posterior "paddle." This osteological character is not found in any other genus in the family.

Eristicophis macmahonii
Alcock and Finn 1897

Recognition
(Plate 9.1)

Head—Large, broad, flat, and wedge-shaped; distinct from the neck; snout broad and short; eyes moderate; pupils vertical; crown covered with small scales; nostrils small and slitlike; rostral broad, being wider than high, strongly concave, and bordered dorsally and laterally by 4 greatly enlarged nasorostral scales which are arranged in a butterfly shape; supralabials 14–16, separated from suboculars by 3–4 small scales; sublabials 16–19; circumorbital ring with 16–25 scales.

Body—Dorsoventrally slightly depressed, moderately to markedly stout; tail short and prehensile, tapering abruptly from the vent; skin soft and loose; dorsal scales short, keeled, in 23–29 vertical rows at midbody, dorsal scale rows arranged in a straight regular ring pattern; ventrals 140–144 in males, 142–148 in females, with lateral keels; subcaudals 33–36 for males, 29–31 for females, without keels; tail prehensile.

Size—A rather small viper less than 1 meter long, with males ranging from 22 to 40 cm and females from 28 to 72 cm (Marx and Rabb 1965).

Pattern—Head with white stripe posteriorly from above eye to corner of the mouth; crown may have scattered dark flecks; labials and throat white; dorsally from reddish to yellowish brown; uniform or with a regular series of 20–25 dorsolateral dark spots bordered partly or completely with white scales, becoming more distinct posteriorly; white border areas often extend over back as bands; venter white; tip of tail yellow with distinct crossbands. Juveniles with a series of about 30 dark brown crossbands (Wall 1913; Marx and Rabb 1965; Minton 1966; Joger 1984; U.S. Navy 1991).

Taxonomy and Distribution

The genus was established from the species *Eristicophis macmahonii* collected in the desert south of Helmand River, Baluchistan.

Macmahon's vipers are reported from Pakistan, Afghanistan, eastern and northwestern Baluchistan, and southern Iran. In India, they are found in the Rajasthan Desert (Werman 1986). They are restricted to the Dast-i Margo Desert and adjacent dune areas, from Seistan in the extreme east of Iran into Afghanistan south of the Helmand River. Also in northwest Baluchistan, between the Chagai Hills and Siahan Range, east to Nushki (Joger 1984).

Habitat

Eristicophis macmahonii is found associated with sand dunes and fine loose sand (shifting dune) habitats (Marx and Rabb 1965; Joger 1984; Phelps 1989; U.S. Navy 1991). It does not occur above 1300 m (Joger 1984).

Food and Feeding

Macmahon's vipers feed on small rodents, lizards, and occasionally birds. Mice are held until dead or nearly so (Mahendra 1984).

Behavior

Eristicophis macmahonii utilizes rectilinear and serpentine locomotory techniques. Sidewinding is used when moving over loose sands or when alarmed (Khan 1990). It occasionally climbs into bushes using its prehensile tail (Marx and Rabb 1965).

These snakes are primarily nocturnal but they also may be crepuscular (Minton 1966; Khan 1983, 1990). In western Pakistan, specimens have been collected from mid-April through August (Minton 1966).

When disturbed *E. macmahonii* quickly buries itself in sand. It can appear to sink into loose sand by a rocking or peristaltic motion using the laterally keeled scales. After burrowing, it usually shakes and rotates its head along its longitudinal axis, covering its head yet leaving the snout and eyes free of sand. It is believed that the enlarged nasorostral scales are used to prevent sand from entering the nostrils. J.A. Anderson, in a personal communication to Minton (1966), wrote, "Every scale of *Eristicophis* is designed for sand burrowing, and even on a hard floor it tries to burrow as though it knows no resistant surface."

They are also reported to be bad-tempered, hissing very loudly and deeply, raising the anterior part of the body in a loop to strike vigorously (Khan 1983, 1990; Mahendra 1984).

Reproduction and Development

Reproductive data for *E. macmahonii* is based on captive specimens. Up to a dozen eggs are laid. Incubation is between 6 and 8 weeks, after which 15-cm hatchlings emerge (Mehrtens 1987).

Bite and Venom

No data regarding envenomation statistics are available.

Eristicophis macmahonii is reported as a potentially dangerous snake with venom similar to that of *Echis* (U.S. Navy 1991).

An LD_{50} of 7.5 µg per 20 g mouse is reported (Minton 1966, 1976).

Two phospholipase A_2 isoenzymes and two polypeptides, "eristocophin I and II," have been isolated (Siddiqi et al. 1991, 1992). These polypeptides inhibit platelet aggregation and contain adhesive proteins. Eristocophin I contains the tripeptide Arg-Gly-Asp which is known to inhibit fibrinogen interaction with platelet receptors.

Human symptoms include pain, local swelling, and ulceration at the site of the bite (Khan 1990). The venom is strongly hemotoxic (Minton 1966, 1976).

The only available accounts of envenomation by *E. macmahonii* are anecdotal reports by Shaw (1924). His first case history described a woman bitten on the right instep. Several hours post envenomation (following delays due to "local treatments" and transport time) she was brought to the location where Shaw was stationed. After several hours, she was conscious but weak, feverish, and unable to sit up. The injured limb was inflamed, swollen, and immobile. A bronze-colored mark about the size of an "ordinary man's ring" surrounded the bite. She complained of being thirsty but was unable to swallow. Her eyes were partially closed and could not be opened. Her abdomen became "inflated and drumlike," accompanied by pain and pressure under the breasts. Although the patient was hurried to the nearest hospital by train, she died shortly after admission.

A second case history involved a young man in Shaw's unit who was bitten on the left hand while he rested it on a wall. The bite was located just under the nail on the left thumb. Within minutes of being bitten, the "whole of the fleshy portion of the underpart of the thumb" was cut away. A ligature was applied above the elbow and as much blood as possible was drained from the wound. The thumb was then placed in a cup of vinegar and salt. The arm swelled and remained swollen for 3 days. Shaw reported that the pain was so severe the man constantly requested that he be shot to rid him of the agony. After day 3, swelling subsided but the hand and wrist broke out in ulcers. No further details are given.

Shaw's third case history involved a sleeping 15-year-old boy bitten on the abdomen just superior to the root of the penis. The snake was found and killed in the boy's blankets. At least 4.5 hours passed before he was brought to the station. At that time the patient was comatose. Whisky was "administered internally" and the wound was incised. Artificial respiration was attempted but the patient's condition continued to worsen. The stomach became distended and the testicles, penis, and both thighs became abnormally swollen. Again, a bronze-colored mark around the wound was present and blood-colored serum oozed from the wound and penis. No further details are given.

Remarks

Although rarely seen in captivity, conditions similar to housing *Cerastes* seem suitable. In captivity, drinking water should be offered weekly and not left in the cage. A captive male specimen lived for more than 9 years (Mehrtens 1987).

Chapter 10
Macrovipera

The genus *Macrovipera* contains big snakes that are responsible for a number of bites in western Asia and North Africa each year. The members of this genus can be bad-tempered and dangerous. Bites are painful and much venom can be injected.

Recognition

Head—Broad, flat, distinct from neck; dorsally covered with small, irregular, keeled scales, which usually are 6–11 across; supraoculars fragmented or semi-divided. A lot of variation occurs in differnt scale characteristics. During preparation of this book, six specimens (of *M. lebetina* and *M. mauritanica*) in the Smithsonian Institution collection were examined. Of these, four (from Algeria, India, and Iran) had nasal plates clearly sutured, two (from Cyprus and Turkey) did not. It is worth noting that these specimens also possessed variable supraoculars, with one from India exhibiting an apparently complete and unfragmented plate, two (from Iran and Cyprus) with a single large and several small components, and three (from Algeria, India, and Turkey) with the supraocular highly fragmented. These characteristics are to some extent geographical (see Taxonomy and Distribution for *M. lebetina*); snout broad; eye moderate with vertically elliptical pupil.

Body—Large and reaches more than 1.5 m (except in *M. schweizeri*); 23 to 27 midbody scale rows; keeled scales.

Taxonomy and Distribution

Type species is *Coluber lebetinus* Linnaeus (1758) by original designation. The genus *Macrovipera* was created in 1927 specifically to accommodate the species *lebetina*. Primary differentiating morphological characteristics of the genus, besides biochemical ones, are size, pattern, and scalation (Herrmann et al. 1992). The four species currently included in *Macrovipera* are *lebetina, mauritanica, deserti,* and *schweizeri,* and all have earlier been treated as subspecies of *lebetina*. They are all characterized by large size, oviparity, and adaptation to arid and dry habitats.

The geographic range of the genus is from northern Africa (Morocco, Algeria, and Tunis) east to Pakistan, India, and Kashmir, north to Greece (Milos Archipelago in the Aegean Sea), Armenia, and Russia (Dagestan) and south to a single old record from Yemen. Fossil species are found in Greece, Austria, Hungary, and the Ukraine.

Macrovipera deserti (Anderson 1892)

Called the desert viper.

Recognition

Head—Broad, triangular, distinct from neck, snout blunt and rounded when viewed from above, canthus rostralis is not pronounced, eyes relatively small, dorsal scales small, numerous, with prominent keels; supraocular broken into several small plates, rostral approximately as high as broad, in contact with three scales on the snout, 6–10 interocular scales, (including remnants of supraoculars), fragmented supraoculars, row of scales along the head margin, and scales forward on the head are of the same size or somewhat larger than those on top of the head and lack the keel, circumorbital scales 12–18 (on average 15) on each side, including supraocular remnants, 2–3 rows of scales between eye and supralabials, which number 11–12, nostril large, located in a large scale representing the fused nasal and nasorostral. Supraoculars fragmented. Gular plates unkeeled.

Body—Dorsals strongly keeled except outer row adjoining ventrals, 27 rows at midbody, ventrals 164–170, with 44–51 paired subcaudals, anal single. Tail is rather long.

Size—A large viper, reaching 160 cm (Schwarz 1936).

Pattern—Body dorsal ground color is often grayish to yellowish, patterning darker consisting of 23 to 26 alternate spots that tend to form a broad blotchy pattern along the back. The pattern tends to fade with age, and adults often have a very weak pattern. Occasionally specimens without any pattern can be seen.

Taxonomy and Distribution

Originally described by Anderson in 1892, as *Vipera lebetina* var. *deserti,* and treated as a subspecies of *Vipera mauritanica* by Kramer and Schnurrenberger (1959) (*V. m. deserti*). In 1992 Herrmann et al. placed it as full species in the genus *Macrovipera*. It occurs in Tunisia and Libya, and inhabits the southern foothill region of the Atlas Mountains in Algeria.

Habitat

Macrovipera deserti is a snake of rocky, semiarid, vegetated areas in mountain regions. It is nocturnal and spends daylight hours under stones and bolders, in crevices, rodent burrows, or under big grass tussocks (Halfa grass).

Food and Feeding

Feeds on a wide variety of prey, largely in proportion to availability. Small mammals and birds are preferred by adults, lizards by young. Birds, including adults, chicks, and eggs, are favored. Adults will climb shrubs and trees to forage.

Behavior

No information is available.

Reproduction and Development

Macrovipera deserti is an oviparous species.

Bite and Venom

No bites seem to have been recorded in literature from this species.

Macrovipera lebetina (Linnaeus 1758)

Called the blunt-nosed viper, Lebetine viper or Levant viper, the latter names referring, of course, to its range which includes both the eastern Mediterranean mainland and nearby islands as well as a small area in northwest Africa. It is often referred to as Kufi or Kufi viper, from its Arabic name, or Gjursa in Russian-speaking areas.

Recognition
(Plates 10.1 and 10.2)

Head — Broad, triangular, distinct from neck, snout blunt and rounded when viewed from above, margins sharp and clear, with sides vertical, meeting the top at nearly a right angle, eyes relatively small and elongate, set forward; dorsal scales small, numerous, with prominent keels, supraocular fragmented, somewhat variable, broken into several small plates, rostral approximately as high as broad, in contact with 3 scales on the snout, 7–11 interocular scales (including remnants of supraoculars), supraoculars, row of scales along the head margin, and scales forward on the head larger than those on top of the head and lack the keel, circumorbital scales 11–18 including supraocular remnants, 2–3 rows of scales between eye and supralabials, which number 9–11, nostril large, located in a large scale representing the fused nasal and nasorostral, which is the normal state, but some variation occurs. Nikolsky (1916) reported a single plate with an indistinct suture, Joger (1984) illustrated the nostril in a clearly divided plate. Gular plates unkeeled.

Body—Dorsals strongly keeled except outer row adjoining ventrals, 25 rows at midbody, ventrals 146–177, with 35–51 paired subcaudals, anal single.

Size—A large viper, reaching 214 cm with females and males of similar size. A male and female in copula, observed in Armenia, measured more than 200 cm each (Orlov pers. obs.). Size varies with population. The population on Cyprus (*M. l. lebetina*) is somewhat smaller than mainland forms.

Pattern—Given the relatively large geographic area occupied by this species, the pattern is less variable than might be expected. Head normally unicolored, but can occasionally be marked with a dark inverted V, with oblique lines converging to the neck behind the V, usually with weak dark lines running from eye to angle of the jaw, frequently a dark blotch or spot below the eye. Body dorsal ground color grayish (very pale in some specimens from Cyprus), gray, beige, pinkish, brown, khaki, or olive, patterning darker gray, bluish, rust, or brown, sometimes indistinct, usually a middorsal row or double row of large dark spots, when dorsal pattern is double-rowed, the rows may alternate or oppose, yielding patterns ranging from saddled through alternate spots to continuous zigzag striping, spots or blotches may be red, brick, olive, or yellow in addition to more common brown, dark gray, or blackish, some specimens unicolored or sparsely patterned, having light gray, pinkish, or beige dorsal ground, often a row of spots lower on dorsum alternating with or adjoining middorsal row along each side, these are usually colored as the middorsal markings, but sometimes differ. Northeastern subspecies (*M. l. turanica*) commonly shows orange or gold crossbars on a pale gray or silver ground.

The venter is usually white or yellowish, often grayish in females. Frequently speckled with brown or marked with black and white spots. Underside of the tail tip light, often yellow or pinkish (Nikolsky 1916; Steward 1971; Isemonger 1983; Joger 1984; Nilson and Andrén 1988b; U.S. Navy 1991; Latifi 1991; Herrmann et al. 1992).

Taxonomy and Distribution

Originally described by Linnaeus in 1758, from specimens of undetermined locality. Type later restricted to Cyprus (Mertens and Müller 1928). In the definition of this species employed here, the range is from North Africa to Pakistan and Kashmir, and from the Gulf of Oman to the Caspian Sea and Dagestan (Russia).

We are incorporating as subspecies five distinct populations (discussed individually below). It is anticipated the new generic standing will be widely accepted, and certain populations treated here as subspecies may be elevated to valid species status.

Macrovipera lebetina lebetina: the nominate subspecies is restricted in range to the island of Cyprus. This form has usually 146–163 ventrals, and is the only venomous snake on Cyprus.

Macrovipera lebetina transmediterranea: A small subspecies from costal mountain areas of Algeria and Tunis in northern Africa. It may be sympatric with *M. mauritanica* and/or *M. deserti*.

Macrovipera lebetina obtusa: widely distributed, from central Turkey through

Syria, Lebanon, Iraq, northern Jordan, the Caucasus region including Armenia, Azerbaijan and Dagestan, and Iran. The populations occurring in southern Afghanistan, Pakistan and northern India (Kashmir) are currently referred to this subspecies, but the name *peilei* is available for this eastern group (the eastern populations normally have semidivided supraoculars). In Pakistan, *M. lebetina* is confined to western highlands, and the range does not overlap that of *D. russelii* which inhabits the Indus River valley (Khan 1983). This eastern form tends to be relatively dark in coloration and has high ventral (usually 170–175) and dorsal (usually 25, rarely 27 rows at midbody) scale counts. Another name is *euphratica* that is given for the populations inhabiting the Euphrates basin of Turkey, Iraq, and Syria. It has been synonymized with *M. l. obtusa* in several publications (e.g., Joger 1984), because of a lack of geographical foundation for the pattern differences on which the description is based. This form may be somewhat darker in color, with indistinct dark crossbands, and grows to a relatively large size. The type locality for *euphratica* is in southern Turkey (Birecik).

Macrovipera lebetina cernovi: A northern counterpart of *M. l. obtusa*. It inhabits northeast Iran, southern Turkmenistan, and parts of northern Afghanistan. This taxon normally has semidivided supraoculars

Macrovipera lebetina turanica: A northeastern subspecies of *M. lebetina,* characterized by a dark ground color with a lighter, orange dorsal zigzag pattern. It inhabits eastern Turkmenistan, Uzbekistan, Tadshikistan, southwestern Kazakhstan, and parts of northern Afghanistan. Also, this taxon normally has semidivided supraoculars.

Habitat

In general, *M. lebetina* is a snake of rocky, semiarid, vegetated areas or in sandy steppe habitats. It is not often found in either well-watered or very dry desert regions. Mehrtens' (1987) description of its habitat as "brushy, overgrown, rocky hillsides, ravines, and valleys," is a concise summary of accounts in the literature. While it is apparently neither a high mountain nor low desert inhabitant, it prefers some altitude and is found generally up to 1000 m in the central and western parts of its distribution. Joger (1984) suggested that it is never found at high elevations—above about 1500 m in the north and 2000 m in the south of its range. However, Nikolsky (1916) recounted receiving specimens taken at over 2400 m in "the Pyandzh River area." Coastal and island populations inhabit lower sites but the habitat (dry shrubby ground) is generally similar to that of mainland forms.

Due to the arid nature of the habitat and the preference for some vegetation, standing water is rare but locally present where *M. lebetina* is found. Hibernation sites are selected to avoid flooding which might interfere with the annual cycle. There are reports of individuals observed near ponded water and even actively swimming (Nikolsky 1916).

In Armenia, within walking distance of a tourist hotel in the city of Erevan, an individual was observed on a rocky hillside. Vegetation included grassland and shrubs interspersed with bare soil and rock, including stone ruins and some patches of second growth forest. The hillside was found sloped to a running stream (Anton 1987).

This species is frequently found associated with human landscapes. It is common in and around agricultural fields, pastures, farmyards, gardens, and villages (Nikolsky 1916; Anton 1987).

Food and Feeding

Feeds on a wide variety of prey, largely in proportion to availability. Small mammals and birds are preferred by adults, lizards by young. Birds, including adults, chicks, and eggs, are favored. Adults will climb shrubs and trees to forage. In some areas, *M. lebetina* is found in association with dense populations of partridges and quail and will raid henhouses for chicks and eggs. Where hares or gerbils are abundant its diet reflects this distribution of prey. It is reported to have been observed feeding in the trees containing a nesting colony of herons (Nikolsky 1916; Anton 1987).

In captivity *M. lebetina* has a broad diet and is easy to feed. In fact, they feed so readily they often become unhealthily obese (Mehrtens 1987). Rabbits are reported to be bitten but not eaten, undoubtedly due to size. There is one report of a captive consuming a grass snake (Nikolsky 1916).

Behavior

During the warm season, *M. lebetina* is active in the evening and night. It is found abroad during daylight hours only in cool weather. Its aggression level also seems related to time of day. It is "normally placid during day . . . occasionally aggressive at night" (U.S. Navy 1991). An earlier author (Nikolsky 1916) reported a similar observation, that the snake is active at night and usually bites only after sunset. One specimen was taken from under a rock during the day (Anton 1987).

Hibernation time varies, but often begins in late October. Rodent holes are favored and are selected to avoid late autumn or early spring flooding (Nikolsky 1916). In many areas, *M. lebetina* emerges from hibernation in April, reaches peak numbers abroad in May, and is rarely observed during summer, becoming common again in September and October. This pattern is due to nocturnal behavior. In Iran and Turkmenistan the species is very active during the summer nights (Nilson pers. obs.).

In general demeanor, *M. lebetina* is sluggish and ungainly. However, it is capable of quick effective response in the form of rapid strikes accompanied by loud hissing when provoked. It is considered a dangerous and aggressive snake, although the relatively large number of bites recorded may be due primarily to its presence around human habitations.

Reproduction and Development

Macrovipera lebetina lebetina on Cyprus is oviparous (Gumprecht and Lauten, 1997). *Macrovipera lebetina cernovi* is oviparous as well (Nilson pers. obs.). About *M. l. obtusa* there are different opinions in literature although the taxon must be considered oviparous as well. In a number of captive bredings, *M. l. obtusa* has always laid eggs. In the wild, a maximum of about 35 eggs per clutch is reported (Latifi 1991) for *M. l. obtusa*. According to unverified literature reports, some populations consistently lay eggs, others consistently bear living young, and some may reproduce either way depending on local conditions. According to Mehrtens (1987), *M. l. obtusa* lays eggs in the northern portion of the range and bears living young in the south. In the Caucasus region (at the northern edge of the range) *M. l. obtusa* are live bearers (Terentyev and Chernov 1949). This reproductive flexibility is maintained by staged embryogenesis that may include periods of diapause during development (Nadjafov and Iskenderov 1994). In oviparous populations, eggs may be laid with just a few weeks required for hatching (Mehrtens 1987). It seems that *M. l. obtusa* in general is an oviparous taxon, with variable hatching time depending on the degree of development of the young when the eggs are deposited.

Bite and Venom

Epidemiology—This is the most common and widespread venomous snake over much of its range and bites are correspondingly common. Field workers and shepherds tending crops or animals after dark were the most frequent victims decades ago (Nikolsky 1916). In Iran, it is among the most commonly reported cause of venomous snakebite (Latifi 1991). Its ability to survive, indeed to prosper, in human landscapes contributes to its current important status as a cause of human envenomation. People often time their activities to take advantage of the snake's quiescent periods. (Leviton et al. 1992).

Yield—Venom yield is relatively high, with a maximum yield per individual of 105 mg and an average of 47 mg (Latifi 1991).

Toxicity—This group is acknowledged in the general literature as dangerously venomous. Steward (1971) stated that the venom is potent and a bite should be considered dangerous. It has been known to kill humans, horses, cows, and camels. It must be remembered, however, that this is a large snake with long fangs, and the severity of the bite may be due as much to physical trauma as the chemical nature of the venom.

The venom is potent but variable. Intravenous mouse LD_{50}s ranged over an order of magnitude from 0.64 mg/kg (Hassan and El-Hawary 1975) to 7.7 mg/kg (Latifi et al. 1973). Most detailed figures reported in Latifi (1984) for i.v. mouse toxicity yield calculated LD_{50} values less than 1 mg/kg. The venom is apparently somewhat more potent than *V. aspis* and substantially less potent than *V. xanthina* (=*palaestinae*) (Khole 1991).

Content—As may be expected from the symptomology of the bite, the venom components that are most potent are the hemorrhagic, necrotic, and proteolytic fractions (Joger 1984). There is very high amino acid oxidase activity (Zeller 1948). Proteolytic activity is high, approaching the level of crotalids, which have among the most active protease systems known (Geiger and Kortmann 1977).

Venom of this species is among the most powerful of those tested for L-amino acid oxidase activity. In assay experiments designed to compare the activities of elapids, crotalins and viperins, only *V. aspis* had a higher oxidase activity (Zeller 1948).

Symptoms and Physiological Effects—In humans, bites are very painful, and swelling of the bitten extremity is rapid and severe (Nikolsky 1916; Reymond 1956; Latifi 1991). Breathing quickly becomes rapid, pulse rate increases, and there is extreme thirst. Characteristic symptoms include swelling, local hemorrhage, red spots on the skin, and reduced blood-clotting capability, all within the first few hours after the bite. Necrosis of the bitten area follows. Post-treatment deformity may continue for relatively long periods.

Hematological Effects—Hemorrhagic symptoms following *M. lebetina* bite are due largely to venom components which act directly on the blood vessel walls (Efrati 1969).

Macrovipera mauritanica (Duméril and Bibron 1848)

Called the Moorish viper, referring to its range,
which includes Morocco and coastal areas of Algeria.

Recognition
(Plate 10.3)

Head—Broad, triangular, distinct from neck, snout blunt and rounded when viewed from above, margins sharp and clear, with sides vertical, meeting the top at nearly a right angle, eyes relatively small and elongate, set forward, dorsal scales small, numerous, with prominent keels, supraocular broken into several small plates, rostral approximately as high as broad, in contact with 3 scales on the snout, 6–9 (on average 7) interocular scales, (including remnants of supraoculars), fragmented supraoculars, row of scales along the head margin, and scales forward on the head are of the same size or somewhat larger than those on top of the head and lack the keel, circumorbital scales 11–17 (on average 14 on) each side, including supraocular remnants, 2–3 rows of scales between eye and supralabials, which number 9–12, nostril large, located in a large scale representing the fused nasal and nasorostral. Supraoculars fragmented. Gular plates unkeeled.

Body—Dorsals strongly keeled except outer row adjoining ventrals, 27 rows at midbody, ventrals 157–174, with 40–51 paired subcaudals, anal single.

Size—A large viper, reaching 180 cm. In two individuals of each sex measured, tale length was 12.6% and 13.2% of total in males, and 10.1% and 13.1% in females.

Pattern—Body dorsal ground color brownish-gray, patterning darker consisting of 27 to 30 alternate spots that tend to form a continuous broad zigzag striping along the back. A pale, reddish morph with very weakly developed pattern can be found as well.

Taxonomy and Distribution

Originally described by Duméril and Bibron (1848) as *Echidna mauritanica*, and transferred to *Vipera* by Strauch (1869). Boulenger (1896) included it in *Vipera lebetina* and Schwarz (1936) regarded it as a subspecies of the latter (*V. l. mauritanica*). In 1992 Herrmann et al. placed it as full species in the genus *Macrovipera*. Its range is restricted to the northwestern corner of Africa.

Habitat

In general, *M. mauritanica* is a snake of rocky, semiarid, vegetated areas. Rocky ravines with some water is a favorable habitat. Due to the arid nature of the habitat and the preference for some vegetation, standing water is rare but locally present where *M. mauritanica* is found.

Hibernation sites are selected to avoid flooding which might interfere with the annual cycle.

Food and Feeding

Feeds on a wide variety of prey, largely in proportion to availability. Small mammals and birds are preferred by adults, lizards by young. Birds, including adults, chicks, and eggs, are favored. Adults will climb shrubs and trees to forage. One successful breeder fed his captive *M. mauritanica* both live and dead guinea pigs, rats, mice, gerbils, hamsters, small chickens, and other birds (Weima 1990).

Behavior

The species is terrestrial and an ambusher that prefers to be still much of the time. It is mainly nocturnal, and displays a more active and perhaps aggressive behavior during nighttime. In general demeanor, *M. mauritanica* is sluggish and ungainly. However, it is capable of quick effective response in the form of rapid strikes accompanied by loud hissing when provoked. Captive specimens hiss loudly when startled and may spit venom (Weima 1990).

Reproduction and Development

Macrovipera mauritanica is an oviparous species. Detailed accounts of breeding are provided by Weima (1990). The female was mated around 10 May, and ate heavily during pregnancy. She laid 40 eggs on 12 and 13 July. Hatching began on 31 August, and all 40 eggs hatched successfully (although the last was so weak as to require euthanasia). Newborn young from this clutch averaged 14 grams in weight and 20 cm in total length. Sex ratio was approximately 1:1.

Bite and Venom

Toxicity—This group is acknowledged in the general literature as dangerously venomous. Steward (1971) stated that the venom is potent and a bite should be considered dangerous. It has been known to kill humans, horses, cows and camels. It must be remembered, however, that this is a large snake with long fangs, and the severity of the bite may be due as much to physical trauma as the chemical nature of the venom. The venom is potent but variable. Intravenous mouse LD50s ranged over an order of magnitude, from 0.64 mg/kg (Hassan and El-Hawary 1975).

Hematological Effects—Hemorrhagic symptoms following *M. mauritanica* bite are due largely to venom components which act directly on the blood vessel walls (Efrati 1969). However, there is other damage to the blood, including weakening and lysis of red blood cells within the vessels (Balozet 1957). This effect occurs in guinea pigs exposed to relatively high doses (Balozet 1962).

Macrovipera schweizeri (Werner 1935)

Called the Milos viper, referring to its range which includes Milos and three adjacent small islands in the western Cyclades in the Aegean Sea, Greece.

Recognition
(Plate 10.4)

Head—Broad, triangular, distinct from neck, snout blunt and rounded when viewed from above, margins clear, with sides vertical, eyes normal and round, dorsal scales small, numerous, with prominent keels, supraocular broken into several small plates, rostral approximately as high as broad, in contact with three scales on the snout, 8–9 (on average 8.3) interocular scales, (including remnants of supraoculars), row of scales along the head margin, and scales forward on the head larger than those on top of the head and lack the keel, 14–16 (on average 15) circumorbital scales including supraocular remnants, 2–3 rows of scales between eye and supralabials, which number 10–11, nostril large, located in a large scale representing the fused nasal and nasorostral. Gular plates unkeeled.

Body—Dorsals strongly keeled except outer row adjoining ventrals, 23 rows at midbody, ventrals 148–160, with 33–47 paired subcaudals, anal single.

Size—A maximum size of almost 98.5 cm total length, but average size is around 50 to 70 cm. The two sexes are of equal size.

Pattern—The pattern is most variable. Head mostly unicolored with no or little pattern. Body dorsal ground color light gray, brown, pink, or yellowish. Patterning darker gray, dark brown, olive, or reddish-brown. Almost blackish specimens occur as well as unicolored red or brownish red (Steward 1971; Brodmann 1987; Herrmann et al. 1992).

Taxonomy and Distribution

The populations on the Cyclades Islands off the Greek mainland coast were originally described by Werner (1935) as a subspecies of *Vipera lebetina*. Because of many unique characters, these viper populations on the western Cyclades were raised to species level (Nilson and Andrén 1988b), and transferred to the genus *Macrovipera*, which was reestablished for the *Vipera lebetina* complex (Herrmann et al. 1992).

Macrovipera schweizeri is found only on the Cyclades Islands of Greece (Milos, Kimolos, Polyagos, and Sifnos). Two subspecies:
Macrovipera schweizeri schweizeri on Milos, Kimolos, and Polyagos; and
Macrovipera schweizeri siphnensis on Sifnos.

This is the smallest species in *Macrovipera*, and it is highly variable in color. Further, it has only 23 midbody dorsal scale rows vs. 25 or 27 in other *Macrovipera*. The form from Sifnos is larger and has a different color pattern than the nominate subspecies (Cattaneo 1989).

Habitat

In general, *M. schweizeri* is a snake of rocky hills and vegetated areas. However, it is not rare in and around agricultural fields, pastures, farmyards, and gardens (Nilson et al. 1999). Typical habitat is maccia areas with a mixture of open terrain and large bushes.

Food and Feeding

Macrovipera schweizeri is a bird specialist (Cattaneo 1989; Nilson et al. 1999). Twice a year, during passerine bird migration, food is available in large numbers and these are the main feeding periods. Altogether 58 species of passerine bird species have been recorded from Milos. Natural rodents are absent, but *Mus musculus* and *Rattus rattus* have been introduced by man, and are preyed upon to some extent. However, big *Rattus* are of the same size as most snakes. During the spring the vipers are ambushing around water holes and streams where migratory birds rest, and in the fall adult vipers will climb shrubs and trees in the evening to forage on resting birds (Nilson et al. 1999).

Behavior

Hibernation begins in October–November, but specimens can emerge anytime during hibernation in sunny weather. Hibernacula are on south facing slopes and the snakes can use communal dens. Normally *M. schweizeri* emerges from hibernation in March or April, but this can vary much between years depending on climate. It is nocturnal during summer, becoming common again in September and October (Nilson et al. 1999).

Reproduction and Development

The mating period is May 10–20 on Milos, and the eggs can be laid in communal nests under rocks. Individual cluch size is around ten eggs.

Bite and Venom

It is considered a dangerous snake, but no fatal cases of envenomation have been recorded by the hospitals and doctors on Milos, Sifnos, and Kimolos (Nilson et al. 1999).

Chapter 11
Montatheris

The single species was originally described as *Vipera hindii*, and has currently been included in *Bitis* and *Atheris*. Broadley (1996) transferred it to the monotypical genus *Montatheris*.

Montatheris hindii (Boulenger 1910)

Commonly called the Kenya mountain viper.

Recognition

Head—Elongate, not greatly distinct from neck; covered with small keeled scales. Eye small, set forward on snout, with vertical pupil.

Body—Dorsal scales strongly keeled, 24–27 midbody dorsal rows.

Size—Average adult total length 20–30 cm, maximum approximately 35 cm.

Pattern—Dorsal ground color brown or gray, paired rows of black triangular patches edged with whitish, cream, or yellow. Venter grayish, speckled with dark gray. Head with a conspicuous V or forward-facing arrow, broad dark stripe edged with lighter color along jawline. Labials whitish (Spawls and Branch 1995).

Taxonomy and Distribution

Found only in Kenya, with one population on Mount Kenya and another in the Aberdare Mountains (Spawls and Branch 1995).

Habitat

Occurs at high altitudes in treeless moorland. A terrestrial species that favors bunch grass clumps for cover (Andrén 1976; Spawls and Branch 1995).

Food and Feeding

Eats lizards (including chameleons and skinks), frogs, and small mammals (Spawls and Branch 1995).

Behavior

Because of low night temperatures in its high altitude habitat, restricted to diurnal activity. Basks in the sun, forages only when warm. On cool or cloudy days, does not emerge from shelter. Irritable and prone to strike when threatened (Spawls and Branch 1995).

Reproduction and Development

One female collected in the wild produced a litter of 2 young in late January (Spawls and Branch 1995), and a second wild-caught female gave birth to 3 young in May 8 (Andrén 1976). Juveniles have been collected in February. Newborns are between 10 and 13 cm total length.

Bite and Venom

No information is available.

Remarks

Birds of prey have been observed feeding on these snakes (Spawls and Branch 1995).

Chapter 12
Proatheris

The single species was originally described as *Vipera superciliaris*. According to Stevens (1973), the species was placed in *Vipera* prior to 1961. Moved to *Bitis* by Kramer (1961a) and put in *Atheris* by Marx and Rabb (1965) based on skull characteristics. *Atheris* is commonly held to be reserved for prehensile-tailed arboreal taxa and *P. superciliaris* was placed in this genus because its tail is partially prehensile, and subcaudals tend to fuse as they do in arboreal species. Broadley (1996) transferred it into the monotypical genus *Proatheris*.

Proatheris superciliaris (Peters 1854)

Commonly called the lowland viper and swamp adder.

Recognition
(Plate 12.1)

Head—Triangular and broad ("elongate" according to Branch 1988); distance between snout tip and commissure of jaws 30% (juvenile) to 60% (adult) greater than greatest width of head; head very distinct from the neck; head rather depressed, flat on top; parietal region slightly domed; snout smoothly rounded; canthus rostralis distinct; eyes directed laterally; nares open laterally, directed posterodorsally; eye moderate size, longer than deep, separated from lip by 1 time or less its own vertical diameter; pupil vertically elliptic; rostral 1.5 times wider than deep, visible from above; nostril between 2 nasal shields; large anterior nasal shield in broad contact with rostral; first supralabial separated from its pair by 1–4 small scales behind rostral; posterior nasal shield smaller, separated by at least 1 scale from supralabials and by 1 or 2 scales from circumoculars; small juxtaposed scales cover top of head with exception of large supraocular shields; supraoculars almost twice as long as wide, separated by almost their length from rostral and from each other by 4–6 small scales; supralabials usually 8–9, rarely 10, exceptionally 11 with third (rarely fourth) supralabial in contact with first 3; rarely 4 infralabials; subtriangular mental well separated from sublinguals by common median suture of first infralabials; scales of lower jaw, rostral, nasals, loreals, supralabials, circumoculars, and lowermost row of temporals smooth; scales on top of head strongly keeled with exception of smoothly tubercular internasal scales; supraoculars enlarged and rugose with obliquely arranged keels.

Body—Dorsal scales imbricate with strong median keel, numbering 26–30 (commonly 27–29) at midbody; 17–19 dorsal rows precloacal; dorsal scale reduction following a standard pattern on posterior body: 27–29 at midbody, rows 4 and 5 fuse repeatedly to yield a reduced number (17–19) in the precloacal area; scale rows 3 and 4 occasionally fuse instead of 4 and 5; scales of outermost dorsal scale row almost as wide as long and not or only weakly keeled; scales of dorsal row 2 not as wide as row 1 but wider than rest of dorsum; 131–159 smooth ventrals; 32–45 (common maximum 43) smooth paired subcaudals, frequently with up to 5 distal pairs fused to single plates; anal entire. Some morphometric dimorphism with female ventral counts statistically significantly higher (140–150 vs. 131–144 males), females more variable dorsal count, averaging higher (avg. 28 vs. 27 males)

Size—Branch (1988) called this species "robust," Sweeney (1961) reported it as "slender." Adults average 40–50 cm with maximum just over 60 cm, tail short, tapering in males, distinct in females. Morgan (1988) provided a biometrics table for 5 adults, 14 neonates. In this series, one adult male was 40 cm snout-to-vent length, 8 cm tail length, with a weight of 68 g, four females ranged from 39 to 51 cm snout-to-vent length with a range of tail lengths of 6–7 cm and a range of total lengths of 46–59 cm, weight 44–98 grams. Neonates ranged 13–14 cm snout to vent, 2–3 cm tail, 15–16 cm total, 2.0–3.0 g. Stevens (1973) worked with a separate series of specimens from Mozambique. A female at 60 cm was largest in the series, the record from Mozambique is 61 cm. The largest known male is the type specimen at 59 cm.

Pattern—Head dark with grayish white oblique broad line from nostril to upper labials continues to lower labials, similar line from behind eye to lower labials, top of head has inverted V; pair of large, roundish, black markings covering parietal region sometimes fused anteromesially with themselves and a smaller roundish marking in the postfrontal region; parietal marking not flanked laterally with associated yellow bar and lateral element; 3 black chevrons cover anterior head, apex of posterior one in interorbit, passing obliquely behind eye onto posterior labials, apex of middle one in prefrontal region, passing obliquely back through eye and posterior lobes onto labials below and behind eye, apex of anterior line in internasal, passing through edge of rostral and first 2 supralabials·onto lower jaw, extending posteriorly through side of mental, terminating approximately opposite first ventral plate; chevrons generally broken mesially but sometimes entire and/or fused with adjacent pattern elements; small discrete internasal spot in some specimens; sides of head creamy white, outermost dorsal scale row very pale, merging into white coloration of venter.

Dorsum dark with a series of fairly regular paler dashes down each side, each dash about 1.5 to 3 scales long, separated by 2 times the dash length from the next; a series of lighter but more variable marks along the midline; 29–39 black vertebral blotches, 3 scales long × 10 scales wide, separated from each other by less than 1 scale length; blotches become indistinct on tail; blotches bounded laterally by short longitudinal yellow bars of less than 2 scales length on upper half of 8 and 9 dorsal rows; black semicircular mark 2.5 scales long × 5 deep associated with each vertebral blotch lies on flanks immediately below, contiguous with longitudinal yellow dorsolateral bars; oblique series of 2–4 small black blotches present on lower flanks between and below lateral marking, extend onto edges of ventral plates; small black spot usually present above and between dorsolateral yellow bars.

Venter grayish white with spots down the middle, some coalesced into lines; each side of midbelly line has numerous large blackish spots which resemble the spots on dominoes; subcaudals straw yellow to bright orange, extending onto tail dorsum in some young; black longitudinal band arising just behind sublinguals on midline of venter, extends posteriorly to between fifth plate and midbody before breaking up and diffusing; outer third of venter discretely spotted with black; diffuse dark markings usually present on midline to basal half of tail. No sexual differences occur in dorsal pattern (Sweeney 1961; Stevens 1973; Branch 1988; Marais 1992).

Taxonomy and Distribution

Described by Peters in 1854 from single Mozambique specimen (facsimile in Bauer et al. 1995). Capture location given as "Terra Querimba," on the mainland opposite Ilha Quirimba, Mozambique.

Proatheris superciliaris has a limited distribution in East Africa. There is a specimen in the British Museum collected in 1937 from Nchisi Island, Lake Chilwa (Sweeney 1961). More recently, the range has been extended to the Liwonde region of the upper Shire River. In this inland region, reported to occur at "moderate" altitudes. Also present in Mozambique on lower Zambesi River and coast north of Quelimane, two specimens from north end of Lake Nyasa in Tanganyika. Coborn (1991) gave a range summary as "southern Tanzania, Malawi, and Mozambique." Marais (1992) similarly reported it from the Mozambique Plain from Beira northward. According to Branch (1988), it inhabits the Mozambique Plain from Beira north to Quissanga, and follows the Zambesi River floodplain to Lake Malawi and Tanzania.

Habitat

Presence is closely determined by suitable soil texture and moisture, which determine prey availability (Stevens 1973). It is not found in saturated or dry unstable soils since rodent populations are low due to burrowing restrictions. All populations collected were found on grazing land, burned by local herders in dry season and grazed the rest of year. They are particularly abundant where surface soil is removed to expose sandy "bunkers" with rodent burrows at the periphery. Branch (1988) found them exclusively in low-lying marshes and floodplains.

Behavior

In the wild, *P. superciliaris* is crepuscular and in captivity there is a regular diurnal activity pattern. Major active periods are in morning and in late afternoon/evening. Individuals hide at night (Stevens 1973).

Individuals in captivity are lethargic except when food is introduced. Snakes emerge to bask with onset of artificial lighting in the morning, then disperse after a few hours, returning to bask if the temperature outside the cage is low. When not basking, they coil in very characteristic fashion, similar to pygmy rattlesnakes, with the tail outside the coil and the head in the center propped over the body and held motionless at an angle of about 70°. Freshly caught or imported specimens are "short-tempered," striking aggressively. They become placid after a few months, rarely striking, but retreat quickly. Hatchlings are aggressive at birth, soon quieting down. Subadults climb on limbs to bask, adults do so only rarely. Interestingly, the tail is fully prehensile in young individuals and subadults can support their own weight. This is not so in adults, although the tail is employed as an aid when climbing (Stevens 1973).

In the wild, during prebreeding season (October to December), snakes bask at mouths of unused rodent burrows in open grasslands on floodplains. Oddly, at this time of year there are invariably two females per burrow. No males are found during this season, and after breeding the female "pairs" break up and only one female is found per burrow (Stevens 1973).

In captivity, individuals shed every 5–8 weeks, with sloughing remarkably quick, clean, and free of the trouble often associated with shedding in captive animals. There is abundant oil beneath old skin, with a discernible and distinct odor in freshly shed animals. There is apparently no tendency to soak prior to shedding (Stevens 1973).

Reproduction and Development

Stevens (1973) described the hemipenis as deeply bifurcate, forking at the level of the second subcaudal, extending posteriorly to between levels of subcaudals 7–10. The base of the organ is transversely folded, with the base of each fork densely spined; the distal half with shorter broader spines which merge into shallow calyces. Deeply divided sulci extend to the base of the short blunt terminal awn at the tip of the organ.

In April 1988, 7 infertile ova were found in a recently captured individual from Lake Chilwa, Malawi. One freshly caught female bore 8 young (3 stillborn) on 21 December, another 10 young on 10 January. All babies shed within hours of birth (Morgan 1988). In another report (Marais 1992) 3–8 young were born in summer, with the neonate length 13–15 cm. In contrast, captured females bred in December and November in 2 succeeding years (Stevens 1973). No mature male was present between birth events, indicating that viable sperm are maintained within the female for a minimum of more than 1 year. Embryos 2 months prior to parturition are 40–45 mm long within eggs averaging 22×14 mm. Birth is rapid, with young breaking the membrane within a few minutes with the egg tooth, then molting within 24 hours. The yolk persists at birth, but prior to birth most is retracted into the abdomen. The small external amount at the umbilicus is quickly absorbed. The umbilical scar covers 3 scales between the ninth and fifteenth ventral scute anterior to the cloacal plate. Scars persist as an unpigmented "dash" mark on abdomen of adults. The largest individuals may have more than 10 babies, but this is exceptional.

Total length at birth is approximately 13–16 cm. Growth is rapid until the yolk is absorbed, then slows to a constant rate for the first 1.5 years. In captivity, growth

Plate 1.1 *Azemiops feae*—John H. Tashjian, courtesy Dallas Zoo.

Plate 2.1 *Causus maculatus*—John H. Tashjian, courtesy Houston Zoo.

Plate 2.2 *Causus resimus*—John H. Tashjian, courtesy California Academy of Sciences.

Plate 2.3 *Causus rhombeatus*—courtesy R.D. Bartlett.

Plate 2.4 *Causus rhombeatus*—courtesy R.D. Bartlett.

Plate 4.1 *Atheris ceratophora*—John H. Tashjian, courtesy San Antonio Zoo.

Plate 4.2 *Atheris chlorechis*—John H. Tashjian, courtesy Houston Zoo.

Plate 4.3 *Atheris desaixi*—John H. Tashjian, courtesy Dallas Zoo.

Plate 4.4 *Atheris hispida*—John H. Tashjian, courtesy Dallas Zoo.

Plate 4.5 *Atheris nitschei*—courtesy R.D. Bartlett.

Plate 4.6 *Atheris nitschei*—courtesy R.D. Bartlett.

Plate 4.7 *Atheris squamigera*—John H. Tashjian, courtesy Fort Worth Zoo.

Plate 4.8 *Atheris squamigera*—John H. Tashjian, courtesy Abilene Zoo.

Plate 5.1 *Bitis arietans*—courtesy R.D. Bartlett.

Plate 5.2 *Bitis atropos*—John H. Tashjian, courtesy Fort Worth Zoo.

Plate 5.3 *Bitis caudalis* — courtesy R.D. Bartlett.

Plate 5.4 *Bitis caudalis* — John H. Tashjian, courtesy California Academy of Sciences.

Plate 5.5 *Bitis cornuta*—John H. Tashjian, courtesy Dallas Zoo.

Plate 5.6 *Bitis cornuta*—John H. Tashjian, courtesy Dallas Zoo.

Plate 5.7 *Bitis peringueyi*—John H. Tashjian, courtesy Western Zoological Supply.

Plate 5.8 *Bitis schneideri*—John H. Tashjian, courtesy Dallas Zoo.

Plate 5.9 *Bitis xeropaga* — John H. Tashjian, courtesy Scott Bazemore.

Plate 5.10 *Bitis gabonica* — courtesy R.D. Bartlett.

Plate 5.11 *Bitis gabonica rhinoceros*—John H. Tashjian, courtesy California Academy of Sciences.

Plate 5.12 *Bitis nasicornis*—courtesy R.D. Bartlett.

Plate 5.13 *Bitis nasicornis*—courtesy R.D. Bartlett.

Plate 5.14 *Bitis worthingtoni*—John H. Tashjian, courtesy San Antonio Zoo.

Plate 5.15 *Bitis worthingtoni*—John H. Tashjian, courtesy San Antonio Zoo.

Plate 6.1 *Cerastes cerastes*—John H. Tashjian, courtesy San Antonio Zoo.

Plate 6.2 *Cerastes cerastes*—John H. Tashjian, courtesy San Diego Zoo.

Plate 6.3 *Cerastes vipera*—courtesy R.D. Bartlett.

Plate 6.4 *Cerastes vipera*—courtesy R.D. Bartlett.

Plate 7.1 *Daboia palaestinae*—John H. Tashjian, courtesy Houston Zoo.

Plate 7.2 *Daboia russelii*—courtesy R.D. Bartlett.

Plate 8.1 *Echis carinatus sochureki*—John H. Tashjian, courtesy Laboratoire des Toxines Animales.

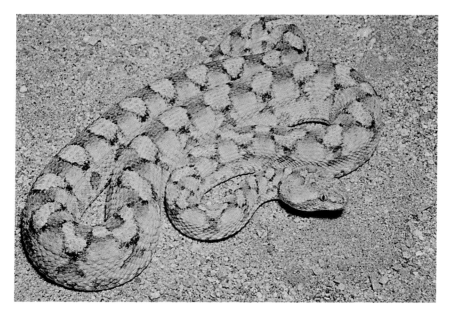

Plate 8.2 *Echis coloratus*—John H. Tashjian, courtesy Western Zoological Supply.

Plate 8.3 *Echis multisquamatus*—Göran Nilson.

Plate 9.1 *Eristocophis macmahonii* — courtesy R.D. Bartlett.

Plate 10.1 *Macrovipera lebetina cernovi* — Göran Nilson.

Plate 10.2 *Macrovipera lebetina turanica*—Göran Nilson.

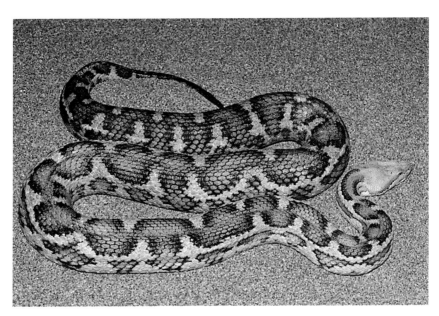

Plate 10.3 *Macrovipera mauritanica*—courtesy Claes Andrén.

Plate 10.4 *Macrovipera schweizeri*—Göran Nilson.

Plate 12.1 *Proatheris superciliaris*—John H. Tashjian, courtesy Dallas Zoo.

Plate 13.1 *Pseudocerastes persicus*—courtesy R.D. Bartlett.

Plate 13.2 *Pseudocerastes persicus*—John H. Tashjian, courtesy San Antonio Zoo.

Plate 14.1 *Vipera ammodytes* — John H. Tashjian, courtesy San Antonio Zoo.

Plate 14.2 *Vipera ammodytes* — courtesy R.D. Bartlett.

Plate 14.3 *Vipera aspis hugyi*—Claes Andrén.

Plate 14.4 *Vipera latastei gaditana*—Claes Andrén.

Plate 14.5 *Vipera transcaucasiana*—Claes Andrén.

Plate 14.6 *Vipera berus*—Göran Nilson.

Plate 14.7 *Vipera darevskii*—Göran Nilson.

Plate 14.8 *Vipera dinniki*—Göran Nilson.

Plate 14.9 *Vipera kaznakovi*—Göran Nilson.

Plate 14.10 *Vipera nikolskii*—Göran Nilson.

Plate 14.11 *Vipera pontica*—Göran Nilson.

Plate 14.12 *Vipera seoanei seoanei*—Göran Nilson.

Plate 14.13 *Vipera seoanei cantabrica*—Göran Nilson.

Plate 14.14 *Vipera eriwanensis*—Göran Nilson.

Plate 14.15 *Vipera lotievi*—Claes Andrén.

Plate 14.16 *Vipera renardi*—Göran Nilson/Claes Andrén.

Plate 14.17 *Vipera ursinii macrops*—Göran Nilson.

Plate 14.18 *Vipera ursinii rakosiensis*—Göran Nilson.

Plate 14.19 *Vipera albicornuta*—Göran Nilson.

Plate 14.20 *Vipera albizona*—Göran Nilson.

Plate 14.21 *Vipera bornmuelleri*—Göran Nilson.

Plate 14.22 *Vipera bulgardaghica*—Claes Andrén.

Plate 14.23 *Vipera latifii*—Göran Nilson/Claes Andrén.

Plate 14.24 *Vipera raddei*—courtesy R.D. Bartlett.

Plate 14.25 *Vipera wagneri* — Göran Nilson.

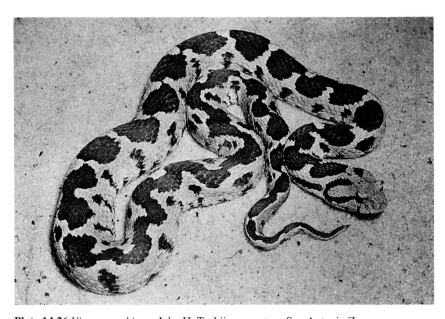

Plate 14.26 *Vipera xanthina* — John H. Tashjian, courtesy San Antonio Zoo.

is roughly linear from weeks 5 through 74 after rapid size increases from weeks 1 to 5. Males are slightly longer than females at each stage. In nature, the growth curve likely has seasonal fluctuations with abundant food in wet season and sparse food in dry season.

There are allometric changes in the male's tail as sexual maturity is attained. The ratio of total tail length with total length rises to a maximum of 7.55 cm at a body size of 26 cm, then decreases as total length increases, indicating maturation of hemipenal structures. Onset of sexual maturity in males is revealed by hemipenis structure (swelling, differentiation of spines and calyces) and begins at 30 weeks. Sexual maturity is reached at 18 months when mating occurs for the first time (Stevens 1973).

Food and Feeding

There is considerable difference of opinion in the literature regarding the diet, both in captivity and in the wild. Similarly, there are very divergent views regarding the ease with which individuals can be induced to feed in captivity and their relative hardiness under captive conditions. Sweeney (1961) felt that wild *P. superciliaris* preyed on cold-blooded animals. Morgan (1988) reported freshly captured specimens refused tadpoles but readily consumed week-old mice. Marais (1992) found captive individuals eat amphibians exclusively, especially reed frogs (*Hyperolius*). Weldon et al. (1992) reported that individuals could only be induced to feed on washed mice after scenting with treefrog (*Hyla*) skin. Cott (1936) and Broadley (1960) found it reluctant to feed in captivity, and therefore difficult to maintain. In contrast, Stevens (1973) called them easy to feed in captivity and reported the biggest problem was obtaining sufficient food to satisfy the large appetites of captive individuals. Young are initially difficult to feed, but most do so with persistent training.

At capture sites, individuals had eaten frogs almost exclusively during the rainy season, including *Hyperolius, Rana, Bufo, Ptychadena,* and *Leptopelis*. The spiny reed frog, *Afrixalus fornasini,* is not eaten unless all other potential prey items are scarce. In the dry season, other frogs and rodents are eaten. Shrews are apparently never eaten and the spiny mouse (*Acomys*) is taken only occasionally. A chameleon placed in a cage with several *P. superciliaris* was eaten. Young individuals less than 20 cm are cannibalistic, and losses can be heavy in captive broods or shipments of young (Stevens 1973). Loveridge (1933) reported that a rat and a frog (*Phrynobatrachus acridoides*) were recovered from stomachs of wild-caught individuals outside the rainy season.

Feeding behavior varies with size of prey relative to individual snake. Adults eat small frogs directly without using fangs. Larger frogs and rodents are struck, fangs are retracted and reset with a series of yawns. Prey is pursued after 30 seconds to 2 minutes. Frogs sometimes required a second strike, rodents are apparently more susceptible to envenomation and very rarely require additional bites (Stevens 1973).

In captivity, small prey items are consumed on a daily basis. Larger prey require several days for digestion, and prey items which overwhelm the digestive system are regurgitated within 12 hours of ingestion. They drink only rarely. Only one incident of drinking was observed in 1.5 years in a captive collection of many individuals maintained with a constant supply of fresh water (Stevens 1973).

Bite and Venom

Nothing is known regarding venom chemistry and toxicity. Els (1988) reported one known case history. A 20-cm captive-born neonate bit a 24-year-old 64-kg male. One fang penetrated the left index finger while the snake was being force-fed. There was immediate intense burning pain and slight bleeding from the bite. Suction removed 0.3 ml of blood from the wound. After 15 minutes, the bite site was blue-black with a well defined blister. At hour 3, a crepe bandage was applied to relieve the intense pain. The blister at this time was 4×7 mm and black. At hour 12, the bandage was removed. Pain was still severe but the victim spent a "peaceful" night. On day 2, pain was reduced but the blister was present and the bite area was very sensitive. On day 8, discoloration was gone and a small scar had begun to form. Skin around the nail on the bitten finger sloughed completely. The victim experienced full recovery with no lingering or permanent symptoms. The relatively mild symptoms were likely due to the small size of the snake and single fang penetration. Previous reports of hemotoxicity of *Atheris* venom were not borne out in this case. Stevens (1973) reported there is some autoimmunity. When individuals are accidentally bitten by cage mates, swelling and blistering at bite developed quickly but subsided after a few days with no discernible permanent damage.

Remarks

Spitting cobras are abundant within the natural range and probably consume *P. superciliaris* regularly, as do birds of prey. Local herders kill vipers on sight.

The egg-eating snake, *Dasypeltis scabra,* mimics *P. superciliaris* in the wild, lying at burrow entrances in similar fashion. *Dasypeltis* coloration in range of *P. superciliaris* appears to differ from elsewhere. Gans (1961) identified this ability in *Dasypeltis* to mimic other vipers, but questioned whether *P. superciliaris* was a suitable model because of perceived extreme coloration difference

Chapter 13
Pseudocerastes

The genus *Pseudocerastes* contains snakes commonly known as false horned vipers, Fields horned viper, or Persian horned vipers. These snakes are regarded as having "false" horns because the supraorbital hornlike structure comprises numerous small scales, compared to "true" horned vipers, *Cerastes cerastes*, which has horns made up by only a single elongated scale. A single species is recognized, *Pseudocerastes persicus*.

Pseudocerastes persicus
(Duméril, Bibron and Duméril 1854)

Recognition
(Plates 13.1 and 13.2)

Head—Broad, flat, triangular; distinct from neck; snout short and broadly rounded; nostrils dorsolateral, with valves, located within large undivided nasal shield; rostral small and wide; eye small to moderate with vertically elliptical pupil; crown covered with small imbricate scales; interocular scales 15–20; circumorbital ring with 15–20 scales; supraorbital hornlike structure comprising small imbricate scales; horns present on adults and juveniles; suboculars separated from supralabials by 2–4 rows of small scales; supralabials 11–14; sublabials 13–17.

Body—Dorsal scales weak to strongly keeled which terminate before posterior edge of scale forming a swollen knob; many are punctate; 21–25 nonoblique rows at midbody; lateral scales large, smooth, and nonserrated with a nodular prominence at the posterior tip; ventrals 134–163; subcaudals paired, 35–50; tail covered with 10 transverse rows of rough scales.

Size—Body stout with short tail; average total lengths range from 40 to 70 cm with 108 cm being the maximum reported length (Latifi 1991). Females are usually larger than males. *Pseudocerastes* also can attain considerable weight for its size. Gravid females and well-fed specimens from either sex can weigh in excess of 500 g (Mendelssohn 1965).

Pattern—A light brown band extending from each eye terminating at lower labials; body color palebrown, yellowish, gray, bluish, or khaki; dorsolaterally with 2 rows of light brown rhomboid blotches, often surrounded by a lighter zone; opposite blotches often fuse forming approximately 30 transverse crossbands; many dorsal scales, especially those bordering crossbands, with dark spots; laterally with a row of smaller brown blotches which become faded in larger individuals; venter usually pure white or straw colored; hatchlings unpatterned, acquiring it when approximately 25 cm; juveniles with 2 series of lateral dark spots, the upper alternating with dorsal bands; tails in adults and juveniles of both sexes conspicuously black-tipped (Flower 1930; Khalaf 1959; Mendelssohn 1965; Khan 1983; Mahendra 1984; Phelps 1989; U.S. Navy 1991; Latifi 1991).

Taxonomy and Distribution

The genus was established by Boulenger (1896) from the type species *Cerastes persicus* collected from Iran. Marx and Rabb (1965) suggested this genus be considered synonymous with *Vipera*. However, Minton (1966) and Price (1982, 1987) provided sufficient morphological and microdermatoglyphic differences for maintaining *Pseudocerastes* separately from *Vipera*.

Currently two subspecies are recognized:

Pseudocerastes persicus persicus Duméril, Bibron and Duméril 1854, and

Pseudocerastes persicus fieldi Schmidt 1930. These subspecies differ primarily by scale counts as shown in Table 4 (Joger 1984; Spawls and Branch 1995):

False horned vipers are found through the Middle East to Pakistan. They are not found on mainland Africa. The two subspecies are allopatric: *P. p. persicus* is found in Iran, southern Afghanistan, Pakistan, and the mountains of Oman; *P. p. fieldi* is found in the Sinai, Jordan, southern Israel, southwestern Iraq, and extreme northern Saudi Arabia (Flower 1933; Haas 1957; Joger 1984; Gasperetti 1988; U.S. Navy 1991; Leviton et al. 1992; Spawls and Branch 1995). Occasional reports from southern Turkey have never been verified.

Habitat

Favored habitat is semidesert with more or less level sandy areas. Soil may be light or heavy, interspersed with stretches of rocky or other hard ground (Minton 1966; Spawls and Branch 1995). Preferable habitat also bears a certain amount of vegetation, mostly shrubs. *Pseudocerastes* are not found on soft shifting sands or on rocky slopes. They often inhabit crevices among the rocks or under stones and pre-

Table 4. Comparison of scale counts.

	Pseudocerastes p. persicus	*Pseudocerastes p. fieldi*
Dorsals at midbody	23–25	21–23
Ventrals	144–158	127–142
Subcaudals	38–48	34–46
Number of scales between nasal and rostral	2	1

fer spacious hiding places. Mendelssohn (1965) reported a preference for inhabiting shrubs, which can be correlated with their feeding preference for birds. Rodent burrows are often used. In Oman they live in rocky country up to 2200 m (Joger 1984). Human habitations are avoided (Spawls and Branch 1995).

Food and Feeding

Despite the black-tipped tails caudal luring has not been reported. In captivity, adults are voracious feeders eating rodents, lizards, and small birds. False horned vipers readily accept dead animals as food, even animals not freshly killed. Their acceptance of carrion is so conspicuous that they could even be considered carrion feeders (Mendelssohn 1965). Newly captive specimens that initially refused to eat readily began feeding when offered dead birds. In the wild, the large amount of feathers in excreta suggests a definite preference for birds, and large numbers of migratory birds during spring and autumn provide two periods of relative food abundance for building fat reserves. Captive specimens may become quite fat, although specimens collected in the field are also fat which may be attributed to a large swallowing capacity. Among mammals, gerbelline mice are preferred and make up a large part of its natural diet. Frogs and toads are not eaten in captivity. Neonates feed only on small lizards, refusing insects and small rodents. At approximately 20 cm they begin to feed on young mice (Mendelssohn 1965).

Disgorgement of indigestible parts occasionally occurs in healthy specimens, which disgorge pellets consisting solely of hair. Observations of antiperistaltic movements of these pellets show they form in the stomach and intestine (Mendelssohn 1965).

Behavior

False horned vipers are generally slow-moving and use several types of locomotion. Sidewinding is used on level sand, hard ground such as roads or, when frightened, even on unsuitable surfaces. A combination of rectilinear, serpentine, or concertina motion is used when moving over uneven or rough surfaces (Gray 1946). Rectilinear rates of motion at 30°C in medium-sized (61 cm) snakes were 1.47 m in 1 minute while larger (79 cm) snakes covered 3.2 m in the same time. Sidewinding differences are less pronounced, with medium snakes covering 15 m per minute and larger ones covering 22 m per minute. It should be noted that rectilinear speed is fairly constant for each snake under similar conditions, while sidewinding is individually variable, depending on degree of motivation (Mendelssohn 1965).

Upon seeing another snake sidewinding, *Pseudocerastes* adopts a similar movement to pursue, overtake, and inspect the other snake. Conspecifics are inspected more thoroughly. Crawling snakes do not arouse the same interest (Mendelssohn 1965).

Almost exclusively nocturnal, they may be active during the day during cold seasons or twilight (U.S. Navy 1991; Spawls and Branch 1995). Mendelssohn (1965) reported it "frequently encountered" during the day under or in shrubs.

Specimens in shrubs rest mostly on the lower branches or on the ground underneath.

Pseudocerastes persicus is not a particularly aggressive snake, but it will hiss loudly when disturbed. C-shaped coils and strikes rarely occur (Mendelssohn 1965; U.S. Navy 1991).

False horned vipers use their heads to dig holes or enlarge preexisting holes for oviposition sites. One specimen dug its way through a sealed burrow of the gerbilline, *Meeriones crassus crassus,* which seals its burrow from the inside. *Pseudocerastes* are incapable of vertically sinking into the sand (Mendelssohn 1965).

Despite the snake's tendency to shy away from human habitations, Mendelssohn (1965) reported that during hot dry conditions individuals are often encountered at night in or near puddles formed under leaking taps or irrigation pipes. He also reported that a large male caught in an Israeli communal settlement dining hall was as fat as any captive-raised specimens, suggesting it thrived on house mice and rats.

Reproduction and Development

Pseudocerastes thrives in captivity and is relatively easy to breed. Cooling specimens during winter months results in greater breeding success (Mendelssohn 1965).

Mating primarily occurs in May and June but may occur as early as April, during unusually warm weather. Copulation is relatively brief, lasting 1–2 hours. Courting behavior has rarely been observed. Mendelssohn (1965) described precopulatory behavior in which a male and female, kept isolated prior to mating, underwent the following ritual: the female first inspected the male then crawled away with her tail raised and cloacal slit slightly agape. The male followed, and having reached the female he "proceeded to tongue her around the cloacal region." Next, he mounted her in a jerky fashion, rubbing his chin against her back, neck, and head. His tail coiled around hers to establish cloacal contact.

Pseudocerastes is oviparous, laying 11–21 eggs ranging from 8.4 to 10.1 g (mean 9.3 g). Eggs are laid in advanced developmental states, bearing embryos 8.5 cm long with supraorbital horns present. Consequently, egg development is relatively short lasting 30–32 days at 31°C (Mendelssohn 1965; Spawls and Branch 1995).

Mendelssohn (1965) reported on a female producing clutches during 3 consecutive years, with clutch size and individual egg weight decreasing from year to year. The female produced additional clutches of viable eggs after being separated from the male prior to its last oviposition, indicating sperm storage. However, unfertilized egg number increased in subsequent clutches.

Snout-vent length of hatchlings ranged from 14.0 to 16.2 cm (mean 15.4 cm) with weights ranging from 4.6 to 5.9 g (mean 5.4 g). First molts occurred 10–15 days posthatching. After 3 months, four specimens weighed almost 9.0 g, with body lengths over 18 cm. *Pseudocerastes* grows relatively fast, probably due to its voracious appetite and broad food preferences.

Bite and Venom

Epidemiology—No data are available regarding envenomation statistics.

Yield—Venom yield ranges from 11 to 100 mg (mean 44 mg) and differed between males and females, left and right fangs, and season (Latifi 1984). Mean yield for females was 77 mg and for males 68 mg. Males showed little difference between left and right fang yields (27 and 26 mg respectively), however, females varied with 26 and 35 mg respectively. Seasonally, the highest yields occurred in summer, followed by autumn, spring, and winter. Dried venom of *P. p. fieldi* appears as glittering white crystals and lacks typical yellow coloration. Venom from *Pseudocerastes* is fairly stable, remaining unchanged when heated at 80°C for 15 minutes and at an acidic pH (Grotto et al. 1967).

Toxicity—Latifi (1984) determined the following LD_{50}s from left and right fangs separately and together (total) for i.v.-injected male and female mice (values in µg). (See Table 5.)

Other reported LD_{50} values of 20 µg whole venom per 20 g mouse (Gitter et al. 1962), and Spawls and Branch (1995) reported an i.p. LD_{50} in mice as 1.0 mg/kg. Shabo-Shina and Bdolah (1987) found that the LD_{50} of the toxic fraction alone was approximately 10^3 times lower by intraventricular route as compared to a peripheral route.

Content—Although morphological, biological, and ecological differences between the two subspecies of *Pseudocerastes* are minor, substantial differences in their venom composition exists. These differences are so pronounced, they suggest that the two subspecies have been genetically isolated for a considerable time (Bdolah 1986).

Gel-isoelectric focusing revealed that venom from *P. p. persicus* had close to 30 protein bands spanning a wide pH range. *Pseudocerastes p. fieldi* had few bands and these were focused at extreme pH values. Phospholipase A_2 activity also differed, with venom from *P. p. fieldi* having about 2000 units/mg protein and venom from *P. p. persicus* having only 800 units. Venom from *P. p. persicus* is yellow, containing L-amino oxidase. It also contains hemorrhagic factors (Bdolah 1986).

Venom from *P. p. fieldi* contains proteolytic enzymes, protease, phospholipase, and hyaluronidase. A neurotoxin associated with the protease and phospholipase has been isolated by Gitter et al. (1962). The venom's main toxic fraction consists of two proteins: a lethal basic phospholipase A_2 and a nonlethal, nonenzymatic pro-

Table 5. LD_{50}s for i.v.-injected mice.

	Males			Females		
	Left fang	Right fang	Total	Left fang	Right fang	Total
$X =$	9.4–14.1 11.5	11.9–17.0 14.2	14.7–22.5 18.1	14.9–21.2 17.7	12.5–18.1 15.0	13.3–19.0 15.1

tein (Batzri-Izraeli and Bdolah 1982a, b). The nonlethal protein potentiates the neurotoxin's action on the central nervous system (Gitter and de Vries 1967; Batzri-Izraeli and Bdolah 1982a, b). Chromatographic analysis of venom from *P. p. fieldi* revealed two fractions, both neurotoxic and lethal (Gitter and de Vries 1967).

Symptoms and Physiological Effects—Venom from *P. p. fieldi* is unusual in causing a pronounced neurological disturbance. Both fractions isolated by Gitter and de Vries (1967) produced convulsions in mice followed by general paralysis. Fraction I had stronger convulsive effects, caused greater hepatic damage with severe congestion, and yielded large numbers of necrobiotic and necrotic foci. Fraction II had phospholipase A activity more than 100 times stronger than fraction I.

The major toxic fraction completely inhibits neuromuscular transmission (Tsai et al. 1983). Transmission inhibition is attributed to the binding of a phospholipase-independent toxin to a specific presynaptic site (Shabo-Shina and Bdolla 1985, 1987). Evidence for presynaptic interference comes from electron micrographs, which revealed markedly altered ultrastructure of presynaptic motor nerve terminals in envenomated mice. In addition, acetylcholine applied to postsynaptic receptors caused muscle contraction. Skeletal muscle ultrastructure in envenomated mice was unchanged (Tsai et al. 1983).

Pseudocerastes p. fieldi venom causes tissue damage, especially in the liver. However, damage occurs only after injection of low doses, giving venom sufficient time to act. High venom doses cause direct lytic activity but little cytopathic effect (Gitter and de Vries 1967).

Injections in mice of a single LD_{50} dose of venom from *P. p. fieldi* caused lung and liver congestion if death did not occur within 2–3 hours. Doses 1.5–2.5 LD_{50} produced microscopic signs of hemorrhage in the brain and lungs, with death occurring within 2.5 hours. Doses 2.5 LD_{50} caused weakness and immobility following 2–3 hours of jumpy movements, forced respiration, and prostration interrupted by heavy tonic and clonic convulsions. Paresis and paralysis of the hind legs occurred just prior to death. Doses 10 LD_{50} caused an earlier onset of neurological disturbance and mice died within 45 minutes with no abnormal hematological effects observed. Intraperitoneal injections of 0.2 mg caused death within 45 minutes without macroscopic or microscopic signs of hemorrhage (Gitter and de Vries 1967).

Hematological Effects—Hemorrhagic factors in *P. p. persicus* venom cause potent hemorrhagic activity typical of viperid snakes. Two micrograms of venom induced a hemorrhagic spot 0.5 cm in the skin of a mouse (Bdolah 1986). Venom from *P. p. fieldi* exhibits weak hemorrhagic activity. Intravenous injection did not cause bleeding and the blood remained coagulable. *In vivo* studies showed the presence of hemolysin but failed to show any coagulation acceleration or fibrinolytic activity.

Treatment—Spawls and Branch (1995) state that no existing antivenin protects against the venom. Latifi (1991) reported 1.0 ml of polyvalent antisera neutralized 0.3 mg of venom but had little neutralizing effect against the neurotoxic effects (Gitter et al. 1962).

Chapter 14
Vipera

The name *Vipera* is venerable, having first been used in 1753, and associated with a type species in 1768, when Laurenti described *Vipera francisci redi* (=*Vipera aspis francisciredi*). Linnaeus described *Vipera aspis* in 1758 (in his tenth edition of *Systema Naturae*) as *Coluber aspis* and did not provide a type species for the genus. His initial concept of this genus also included taxa now incorporated in *Cerastes*.

The genus *Vipera* s. str. has great ecological scope. Individuals in some populations feed opportunistically on a variety of prey while other species are specialists, consuming a narrow range of food items. Some species, particularly montane and Middle Eastern forms, inhabit dry landscapes with low vegetation and much bare and rocky ground. Others, including subpopulations of some of the more wide-ranging species, live in or near reedy wetlands or under the closed canopy of forested hillsides. In the short season and cold conditions of the Arctic and near-Arctic, *Vipera* populations are closely linked to their prey and to seasonal weather and regional climatic conditions. In more benign environments, vipers take advantage of seasonal shifts in prey distribution and of different ecological conditions in different areas of the landscape.

The Eurasian vipers (former *Vipera* s.l.) includes several evolutionary lineages that can be traced back to the Miocene, reflecting an old evolutionary history (Nilson and Andrén 1997; Szyndlar and Rage 1999). The oldest fossils are from early Miocene. Besides *Daboia* and *Macrovipera*, the subgenera *Vipera*, *Pelias*, and *Montivipera* are defined (Nilson and Andrén 1997; Nilson et al. 1999). The smaller Eurasian vipers (*Pelias* s.l.) is here divided into two subgenera: *Pelias* s.str. and *Acridophaga*.

As used here, the subgenus *Vipera* are mostly small to stout-bodied, viviparous, terrestrial forms, characterized by a raised snout or a pronounced snout-horn. They live in lower latitudes and at medium to low elevations, and prefer dryer and often rocky habitats.

The subgenus *Pelias* are mostly small and less stout-bodied, viviparous, terrestrial snakes, characterized by having larger plates on top of head. They live in more northern latitudes or at high elevations, and prefer moist and less rocky habitats.

The subgenus *Acridophaga* are mostly small, viviparous, terrestrial snakes, characterized by having larger plates on top of head. They live in lowland at lower latitudes, or in the mountains, and prefer dry meadow habitats with grassland vegetation.

The subgenus *Montivipera* are stout-bodied, viviparous, terrestrial forms without snout-horn and with all headscales, except supraoculars, strongly fragmented. They live in mainly mountainous habitats at lower latitudes but mostly at high elevations. They prefer rocky habitats.

Venom of most species of *Vipera* has both neuro- and hemotoxic components. In composition and effect they are, in a sense, "intermediate" between the crotalins and the elapids (Tu 1991). Bite severity varies widely. Some smaller northern forms inject little, only slightly toxic venom. Other species inject more venom with more devastating symptoms. Rarely, however, are bites of *Vipera* species as severe as those of the larger *Macrovipera* and *Daboia*.

Recognition

Head distinct from the neck and in many species covered with small scales. Some species have a few small plates on top of head. Body mostly small and stout. Dorsal scales distinctly keeled, subcaudals and anal divided.

Taxonomy and Distribution

A lot of literature concerning ecology, biology, and toxinology is based on older taxonomic concepts, and our presentation reflects this of necessity. Various arrangements of species and subspecies within the genus and its subgenera are discussed in detail in the species accounts. We present our detailed review of the literature for all species where information is available. Newly described taxa, or subspecies raised to species level, and with no or sparse information are as follows: *V. barani, V. darevskii, V. dinniki, V. kaznakovi, V. monticola, V. nikolskii, V. pontica, V. seoanei, V. wagneri, V. lotievi, V. anatolica, V. ebneri, V. eriwanensis, V. albicornuta, V. albizona, V. bulgardaghica,* and *V. transcaucasiana*. Recently, *V. orlovi* and *V. magnifica* have been described (Tuniyev and Ostrovskikh 2001).

The geographic range of this genus is enormous. It is found from the British Isles to Pacific Asia, and from north of the Arctic Circle to northern Africa. Across Europe, one or more species of *Vipera* may be found in most regions. In Africa and east Asia, the distribution is spottier and the number of species fewer. This pattern of distribution may reflect both the evolutionary origin of the group and the quality of ecosystems in place, with suitable habitat being less generally present beyond the relatively cool moist areas of Europe. In Asia, possible competition with pitvipers, like species of *Gloydius,* may have limited the dispersal eastwards.

Subgenus *Vipera* Laurenti 1768

Vipera aspis is the type species for the genus (*Vipera francisci redi,* Laurenti 1768) as well as for the nominate subgenus. The species in this subgenus are characterized by having a snout that is raised or by having a horn on the snout.

This subgenus is characterized by a reproductive pattern with both spring and fall mating periods, although spring mating is dominant. Male shedding of the skin has no impact on the start of mating activities. Most of the spermatogenesis (spermatocytogenesis and spermiogenesis) takes place during summer and fall, with ripe sperm in *vas deferens* during the fall and the following spring.

Vipera ammodytes (Linnaeus 1758)

Commonly called horned viper, long-nosed viper or sand viper, sometimes with qualification (for example, eastern sand viper).

Recognition
(Plates 14.1 and 14.2)

Head—Rostral wider than long, supporting a nasal "horn" of 9–17 scales arranged in 3 (rarely 2 or 4) transverse rows; 2 large supraoculars with posterior end extending beyond posterior margin of eye; remainder of head scalation with numerous small, smooth or only weakly keeled irregular plates; 7 interocular scale rows; frontal and parietals usually absent; large single (rarely divided) concave nasal contains nostril; nasal separated from rostral by single nasorostral scale; temporals smooth or weakly keeled; 10–13 small plates bordering eye; 2 rows of scales between eye and supralabials, which number 8–12 (usually 9–10), with fourth and fifth the largest (Nikolsky 1916; Steward 1971; Street 1979; Mehrtens 1987; U.S. Navy 1991).

Body—Dorsal scale rows 21 or 23 (rarely 25), strongly keeled, with row adjoining ventrals smooth or weakly keeled, 133–161 ventrals in males, 135–164 ventrals in females, 27–46 paired subcaudals in males, 24–38 paired subcaudals in females; anal single (Nikolsky 1916; Steward 1971; Street 1979; Mehrtens 1987; U.S. Navy 1991).

Pattern—Dorsum of head and nasal horn in males with irregular dark brown, dark gray, or black markings; back of head with characteristic dark blotch or V marking, often contiguous with dorsal zigzag (head blotch or V often lacking in females); thick black stripe from behind eye to behind angle of jaw; tongue usually black, iris of eye golden or coppery. Body of male dorsal ground color variable, light gray, ash gray, silver gray, pale gray, or grayish white, sometimes yellowish or pinkish gray or yellowish brown; thick dark gray or black zigzag dorsal stripe, sometimes finely bordered with thin darker edge; longitudinal row of indistinct dark (sometimes yellowish) spots along each side, sometimes joined in a wavy band. Females similarly patterned with less distinct and less contrasting colors shaded to browns and bronzes, dorsal ground color grayish brown, reddish brown, copper, "dirty cream," or brick red, dorsal zigzag shaded in browns. Venter in both sexes variable, grayish, yellowish brown or pinkish, "heavily clouded" with dark gray, gray-brown, or black spots, often ventral ground color black or dark bluish gray with white flecks and inclusions, often edged with white; chin lighter than and contrasting with belly; venter of tail tip yellow, orange, orange-red, red, or green. Juveniles similarly colored and patterned. Melanism is rare but is known. Fragmentation of the dorsal zigzag is relatively common (Steward 1971; Street 1979; Mehrtens 1987; U.S. Navy 1991)

Size—Males commonly less than 85 cm, maximum 95 cm; females smaller. Maximum length varies with population, northern forms are longest, southern race of the Balkans to only 60 cm (Nikolsky 1916; Steward 1971; Street 1979; Biella 1983; Mehrtens 1987; U.S. Navy 1991).

Taxonomy and Distribution

Originally described by Linnaeus in *Systema Naturae* in 1758. Subsequent work by Boulenger (1903, 1904, 1913a, b) provided the basic subspecific approach which is presently largely retained. Alternative taxonomies and distributional analyses are many. The division by Mertens and Wermuth (1960) has mostly been used during the last part of the 20th century: subspecies *V. a. ammodytes* from the former Yugoslavia, west through southern Austria to northern Italy, in southwestern Romania and adjacent Bulgaria; *V. a meridionalis* from Albania east through Greece, Cyclades Islands, and Turkey, *V. a. montandoni* from Bulgaria and adjacent Romania; and *V. a. transcaucausiana* from the "transcaucasian" section of the former U.S.S.R. and adjacent northern Turkey.

The nasal horn of subspecies *V. a. ammodytes* has a distinct and obliquely forward projection. Rostral is usually divided horizontally, the upper portion sometimes further divided into small plates. Tail tip venter red or yellow. The subspecies *V. a. meridionalis* and *V. a. montadoni* are generally darker in dorsal coloration with a more vivid distinct pattern. Head may be smaller and the horn is vertically erect. Rostral is not divided and the tail tip venter is green or yellowish green. There are 133–149 ventrals and 3–5 scales on the snout above the rostral (Steward 1971; Mehrtens 1987).

More elaborate taxonomic views include five subspecies (*ammodytes, gregorwallneri, meridionalis, montadoni,* and *transcaucasiana*) (Herprint International 1994) or six subspecies (*ammodytes, gregorwallneri, meridionalis, montadoni, ruffoi,* and *transcaucasiana*) (Tashjian pers. comm.) Subspecies *ruffoi* is included in a five-taxon scheme by Street (1979), and is also included in an approach that separates *meridionalis* and *montadoni* (Bruno 1968). The name *illyrica* is meanwhile used for the westernmost populations.

In recent times *transcaucasiana* has been given specific status, based on genetic distances larger than between other full species (e.g., between *V. aspis* and *V. latastei*) (Herrmann et al. 1987, 1992). Thus, an alternative concept of *V. ammodytes* includes three subspecies:

V. a. ammodytes, found in Austria (Styria and Carinthia), northern Italy, Slovenia, Croatia, Bosnia-Herzegovina, Macedonia, Albania, southwestern Romania, and northwestern Bulgaria;

V. a. montandoni, from Bulgaria, southern Romania, and

V.a. meridionalis from Greece (including Corfu and other islands), and Turkish Trace. The taxon *transcaucasiana* is here treated as a separate species.

Despite several literature citations, the species does not occur in Hungary or in the Czech and Slovakian republics, neither in Lebanon, Iran, or Syria.

Quantitative information regarding population densities for *V. ammodytes* is sparse. Locally, the species can be abundant in suitable rocky habitats. On one occasion, 25 specimens were observed in the Velebit mountains of Croatia (Nilson pers. obs.).

Habitat

Favors dry, brushy, rocky hillsides, preferring areas with only sparse vegetation, with scrubby relatively barren rocky slopes being the usual habitat. While it is not usually found in canopied woodlands, it does inhabit forest edges, clearings, and other disturbed areas where the ground is dry and vegetation scattered. It is sometimes common near human habitations, on railway embankments, and in agricultural areas, especially vineyard country, where rubble piles or stone walls are present. In the southern Tyrol, it may be restricted to limestone substrates. Reaches 2000 m of altitude and higher in southern areas, common to 900 m, occasionally to 1300 m in Austria (Steward 1971; Street 1979; Mehrtens 1987).

In some areas, there may be seasonal migratory movements along altitudinal gradients (Steward 1971). *Vipera ammodytes* may move downslope into woodlands during high temperature periods of midsummer and return to the preferred dry rocky scrub slopes at higher altitudes for the remainder of the year.

Food and Feeding

Primary prey items are small mammals and adult and nestling birds, with lizards reportedly favored by juveniles (Steward 1971; Street 1979; Mehrtens 1987). A captive juvenile in one case readily accepted small frogs (*Rana esculenta*) within 3 weeks of birth and continued to consume frogs until it was large enough to eat young mice (Holzberger 1980). Examples of gluttony are reported. A 17.3-cm *V. ammodytes* disgorged a 15.0-cm *Lacerta peloponnesiaca* on capture (Clark 1967). A 20-cm juvenile *V. ammodytes* consumed a 14-cm centipede. The snake died within an hour of capture and its death may have been due to this enormous meal (Clark 1967).

In captivity, rodents are accepted by adults (Clark 1967; Gulden 1988).

Feeding behavior varies with prey size. Larger prey are struck, released, tracked, and swallowed. Smaller prey such as nestling birds and mice are swallowed without invoking the venom apparatus (Steward 1971). Other snakes are occasionally eaten.

Behavior

This species has no great preference in daily activity period. In the south during summer it is primarily nocturnal. At higher altitudes, it may be active only during warm sunny periods. At lower altitudes, it is commonly found abroad both at night and by day, becoming primarily crepuscular or nocturnal when daytime temperatures become too high (Street 1979; Mehrtens 1987).

Vipera ammodytes often bask utilizing rocks, walls, and ground surface near shrubs and bushes, sometimes even climbing into bushes or trees (Street 1979). During hot weather in particular, "loafing" behavior is common, with individuals climbing into low branches, remaining motionless for long periods of time (Steward 1971).

In disposition it is generally sluggish and lethargic, not at all aggressive, tending not to bite without considerable provocation. When encountered in the wild it exhibits various behaviors. Street (1979) reported that it tends to remain motionless, hissing loudly if perturbed. Gulden (1988) reported that individuals hiss deeply and flee when disturbed. Immediate attempts at biting when provoked are also reported (Steward 1971; Street 1979; U.S. Navy 1991).

Hibernation varies with altitude and latitude. General hibernation dates are October to March or early April, with occasional individuals being seen abroad in winter on warm sunny days (Steward 1971; Street 1979; U.S. Navy 1991). In captivity, the effective artificial hibernation period is shorter (from 21 to 35 days), and is usually implemented later in the year (from November through January) than the natural inception of hibernation (Karstens 1986: Gulden 1988).

Reproduction and Development

Prior to mating, male combat, including competitive "dancing" similar to that occurring in adders, takes place (Street 1979). In one captive situation in which two males were kept in a single terrarium with a lone female, an active combat dance erupted between the males during the night on 4 April. On 14 April, the "strongest" male mated the female and 7 young were born 113 days after mating. Four developed young remained in the female's body, and she died several days after giving birth. The 7 young born alive ranged from 20 -23 cm in total length and weighed from 7 to 10 g (Holzberger 1980).

In another account of captive breeding, Gulden (1988) moved a male and female from hibernation boxes to a terrarium in early April. The female fed immediately. The male did not eat and showed no interest in the female. On 25 April mating occurred with a 1-hour copulation. The male began feeding on 9 May; the female, after eating a mouse on 30 April, ceased to feed. On 10 July the female began eating again. Four young were born on 3 August; 2 shed immediately, 2 after a day. The young were approximately 20 cm in total length and weighed between 6 and 8 grams. All 4 young snakes began eating nestling mice immediately and were raised successfully.

Hybrids with other viper species have been reported under captive conditions. In one case, a 4-year-old 75-cm male *V. ammodytes* copulated with an 18-month-old, 75-cm *V. aspis* female in a terrarium on 9, 12, and 15 March. Both animals were captive-bred. Eight live, 2 dead, and 2 unfertilized eggs were produced on 18 June (Karstens 1986). In another case, a second-generation litter born to a *V. ammodytes/V. aspis* cross yielded 1 healthy individual, 1 live deformed individual (no data on its fate), and 7 infertile eggs (Faoro 1987).

There is an account of newborns remaining near their mother and hiding themselves in her coils when disturbed (Gulden 1988).

Bite and Venom

Epidemiology—This species is reputed to be the most dangerous of European vipers due to both large size and high venom toxicity. Its bite is potentially more dangerous than any other European viper (Arnold and Burton 1978; Street 1979). The venom is characterized as dangerous and fang length up to 1.3 cm is a factor contributing to bite severity (U.S. Navy 1991).

Specific epidemiological data on *V. ammodytes* bites are not available. There are hundreds of cases of viper bites occurring annually over the area generally within the range of *V. ammodytes*. Although it is relatively rare and restricted in habitat, it may be assumed that *V. ammodytes* accounts for a portion of recorded bites (Gonzales 1991).

Yield—No data are available.

Toxicity—Toxicity varies over time in individual snakes and among populations (Mebs 1978; Detrait and Saint-Girons 1986). Intravenous mouse LD_{50} range from 8.9 to 16.4 μg for 20 g mice, or 0.44 to 0.82 mg/kg. Intraperitoneal mouse LD_{50} s for *V. ammodytes* range from 0.19 to 0.64 mg/kg (Novak et al. 1973). Subcutaneous LD_{50} in mice is 6.6 mg/kg (Minton 1974).

Humans respond rapidly to *V. ammodytes* venom. The rapid effect is reflected in a similarly quick and devastating effect on mammal prey. Mice usually die within 6 minutes, although under exceptional circumstances they may survive for 30 minutes. Small birds usually succumb within seconds, larger birds within 10 minutes. Lizards are susceptible, dying within 30 minutes, but amphibians are less susceptible and may survive a bite. European snakes of the genera *Natrix* and *Coronella* may be immune (Street 1979).

This species is commonly employed in polyvalent antivenin development because of the relative venom potency and the ease with which it may be maintained in captivity (Vogel 1964).

Content—Venom can be fractionated into several biologically active components, identified as "lethal," "proteolytic," "phosphodiesterase," "L-amino acid oxidase," and "phospholipase A_2." These fractions have various general effects in addition to proteolysis, hemorrhagic, and hypotensive effects. When partially purified, the lethal fraction has a toxicity of approximately 0.3 mg/kg and consists of three distinct molecular subunits (Orel et. al. 1983).

An alternative fractionation scheme yielded four components: "lethal," "hemorrhagic," "proteolytic," and "arterial vasodilation" (Gubensek et al. 1974). The latter is important in bite symptomology, is related to kinin-releasing activity (Mebs 1978), and causes a dramatic fall in blood pressure (Tu 1991). This precipitous decline in blood pressure is substantially reduced when various drugs which inhibit kinin-releasing enzymes are administered. Among the key venom components responsible for the hypotensive effect are phospholipases A_2 (Habermann 1961).

When administered intravenously, phospholipase A_2 (also identified as ammodytoxin A and "fraction K_2") is by far the most toxic component, having an LD_{50} of 0.21 mg/kg vs. a range of 0.36–3.60 for other fractions (Orel et al. 1983, Khole 1991). In general, the response of guinea pig tissue preparations to phospholipase A_2 are similar to those for crude venoms, but phospholipase A_2 induces a response at much lower concentrations. Chemicals that inhibit or eliminate hypotensive effects of other venoms do not do so when applied to systems dosed with *V. ammodytes* phospholipase A_2 (Sket and Gubensek 1976).

A neurotoxic complex has been isolated which retains phospholipase A_2 activity. The neurotoxic fraction comprises approximately 15% of the total venom. A most interesting component of *V. ammodytes* venom is nerve growth factor (NGF). NGFs regulate growth and differentiation of neurons of both the sympathetic and sensory nervous systems. Nerve growth factor was discovered serendipitously when crude venoms were applied in an attempt to eliminate nucleic acid contamination of preparations intended to quantify NGF effects in sarcoma transplants (Hogue-Angeletti and Bradshaw 1979). Active NGF can be purified from *V. ammodytes* venom (Cohen 1959; Banks et al. 1968; Bailey et al. 1975). Reasons for NGF in snake venoms are debated: NGF may accumulate as a general circulation artifact from such materials in tissues, or could possibly play a role in enhancing effectiveness of the venom as a toxin by stimulating bite victims' cells (Hogue-Angeletti and Bradshaw 1979).

Overall, amino acid composition of *V. ammodytes* venom is very dissimilar to other snake venoms (elapids, hydrophiids, and crotalids) (Gubensek et al. 1974).

Symptoms and Physiological Effects—Symptoms are pain, swelling, and discoloration, all of which may be immediate. Reports of dizziness and "tingling" exist (Street 1979; McMahon 1990).

Venom has notably low symptomatic neurotoxicity related to lack of activity affecting peripheral nerves (Sket and Gubensek 1976). However, its specific depressant effect on mammal respiration is due to nervous, not muscular, activity. Neurotoxicity is apparently induced by venom blockage of postsynaptic membranes without altering membrane electrical potential. The retention of phospholipase A_2 activity by the neurotoxic fraction may indicate both biochemical relationship and mechanisms of evolutionary development (Tchorbanov et al. 1978).

At the cellular level, *V. ammodytes* venom induces edema by causing vacuole numbers to increase and cytoplasm to swell (Tu 1991). It also obliterates heart muscle striations. When applied to isolated heart preparations, muscular contraction, not electrical activity, is halted and muscle mass stops in a contracted state. These effects are not fully reversible as contraction frequency and amplitude both fail to recover with time following administration of antivenin (Petkovic et al. 1979). Irreversibility is due to venom having direct degenerative impacts on myocardium. Excess calcium added to a perfusion testing system increased the degree of physical damage associated with the direct venom effects. This is likely due to increased penetration of the cell membrane by venom during active contraction (Petkovic et al. 1983).

Venom also inhibits proteolytic activities of trypsin and chymotrypsin (Takahashi et al. 1974a, b; Takahashi, Iwanaga, Kitagawa et al; Turk et al. 1979), and decreases blood lipid and lipoprotein content (Petkovic et al. 1979).

Hematological Effects—*V. ammodytes* venom has powerful blood coagulant properties, similar to and as powerful as crotalids (Copley et al. 1973). Other properties include anticoagulant effects (Meier and Stocker 1991), hemoconcentration, and hemorrhage (Petkovic et al. 1979).

Case Histories—Street (1979) was bitten on the right index finger, with penetration by one fang of what was presumably an adult individual. There was initially a slight tingling sensation but no pain. The finger swelled substantially within a few minutes but pain was still not present. After several hours, Street left his hospital bed, but became dizzy and collapsed, feeling "very weak and ill," possibly due to the characteristic hypotensive effect of the bite of this species (Mebs 1978). These symptoms decreased after he retired to bed. Antivenin was injected within 2 hours, by which time the finger was stiff, discolored, and burning with pain. The entire hand swelled overnight and through the next day, engaging the entire arm and armpit glands. By day 3, swelling began to recede, and discomfort was considerably reduced on day 4. On day 6, some finger movement of the bitten hand was possible, except for the bitten finger which remained stiff and numb. After minor surgery to remove a necrotic "bubble," Street left the hospital on day 7 postbite. The bitten finger remained partially numb for months and was noticeably tender for 18 months after the bite.

A second bite report (McMahon 1990) is also notable for the lack of systemic symptoms. The *V. ammodytes* was a 21-cm juvenile male. The victim was a 25-year-old male weighing 91 kg. He was bitten on the tip of the left index finger, with penetration by both fangs. He received no antivenin. The victim had a history of snakebite. Nine years prior, he had received three bites simultaneously from *Atractaspis bibronii*. The effect of this history on his responses to the *V. ammodytes* bite are unknown although the similarity of this report to that of Street suggests that it was not a factor. After being bitten, there was immediate throbbing pain involving the entire arm within 5 minutes. After 11 minutes, there was considerable discoloration at the bite site and skin in the area was shiny and taut. After three hours, the knuckle and finger were grossly swollen and finger movement was impossible. The hand and lower arm were swollen and tender, wrist flexion was very painful, and the upper arm was sensitive and painful to touch. On day 2, swelling subsided slightly and finger movement was possible. Discoloration spread from the bite site and movement of the entire arm was painful. The bitten hand was bluish purple and there was a large bruise on the upper inner arm. By day 3 pain was generally insignificant although there was still considerable discoloration at the bite site. The patient was unable to bend his fingers and the arm bruise was larger and more sensitive. On day 4, the vicinity of the puncture and upper arm was still discolored but pain was largely gone. Flexion of the finger was still difficult. On day 5, the bruise on the arm remained and on day 8, the patient was reported "normal." Long-term effects were not documented.

Treatment and Mortality—No information is available for treatment. Despite the reputation that *V. ammodytes* has for bite lethality and the obviously severe effects associated with envenomation, mortality statistics directly attributable to this species are not documented in the available literature.

Remarks

In captivity, this species is relatively easy to maintain and breed, requiring primarily dry living conditions and an annual hibernation period (Vogel 1964; Melichercik 1993). Given this stimulus, copulation can be expected within weeks of emergence and young may be born the same year (van de Beek 1991).

Certain populations of this species, particularly those in Romania, appear vulnerable and at risk (Honegger 1981). At least one reserve has been established within the range of this population. In Greece, *V. ammodytes* is not protected by national law and populations are vulnerable (Filio and Byron 1990). Under the Berne Convention, *V. ammodytes* is listed and therefore entitled to protection within the European Community and Finland. However, it does not appear on the IUCN "red list" (Corbett 1989).

Vipera aspis
(Linnaeus 1758)

Commonly known as the asp or asp viper.

Recognition
(Plate 14.3)

Head—Broad and triangular, well distinct from neck; snout tip slightly but distinctively upturned; snout dorsum flat with distinct and slightly raised sharp margins; nasal single (rarely divided) contains nostril, separated from rostral by a single nasorostral; head dorsum with numerous small irregular scales of various sizes which are generally smooth but may be slightly keeled; rostral generally higher than broad, touching 2 or 3 scales on upper side of snout; frontal and parietals not usually present, small and irregularly shaped when they are (in one clutch of 9 young born to a female who lacked frontals and parietals, all babies had a frontal and 7 had distinct parietals); when present, frontal separated from supraoculars by 2 scale rows; supraoculars large and distinct, separated by 4–7 scale rows; 10–12 (rarely 8–13) small scales surrounding the eye below the supraocular; eye separated from supralabials by 2 (rarely 3) scale rows; eye vertical diameter approxi-

Vipera

mately equal to distance between eye and mouth; supralabials 9–13; supralabials 4 and 5 (rarely 4–6 or 5 and 6) separated from eye by 2 (rarely 3) rows of small scales; sometimes a single scale between labial 4 and eye; temporals generally smooth, sometimes slightly keeled.

Body—Dorsals 21–23 (rarely 19 or 25) rows at midbody; dorsals strongly keeled except lowest row which varies and is sometimes smooth; ventrals 134–170 (subspecies *V. a. aspis* averages fewer, and subspecies *V. a. atra* averages more than 150); subcaudals 32–49 in males, 30–43 in females; anal single.

Size—Males to 85 cm, females rarely greater than 75 cm, generally 60–65 cm; tail very short: one-seventh to one-ninth of total in females, one-sixth to one-eight in males; males somewhat slimmer than females in overall body form.

Pattern—Head variable, ranging from unmarked to patterned; an inverted V or sometimes an X-shaped mark on back of head is relatively consistent, as are thick dark stripes along the sides of head from behind orbit to neck; often a dark blotch in the back of the V which may merge or even join the dorsal body pattern; sometimes the V is expanded into thick rounded blotches or it may be represented by finely etched, clear thin lines or teardrop shapes. A series of specimens from alpine Italy included individuals exhibiting elaborate linear or curvilinear closed figures on the back of the head, and others with trident, "lightning bolt," or S-shaped open patterns of various sizes and combinations (Bruno 1985). Upper lip shields whitish, cream, yellowish, or pinkish often with thin dark bars in sutures. Eye is small, usually golden or coppery red with vertical black pupil.

Body coloration is highly variable and the remarkable range of dorsal patterning is effectively illustrated by Bruno (1985). These illustrations schematically show a range from clear light to totally dark (noting the latter as exceptional), with a bewildering array of horizontal stripes, vertical stripes, crossbars, spots, saddles, and zigzags in between. In general, the dorsal ground color is cream, yellow, light gray, or gray-brown to reddish brown or brick red; some specimens are coppery red, orange, or straw yellow and some are totally black, especially in alpine populations; females are often darker than males, although dimorphism is not strongly expressed and either sex can be any and all colors; some specimens have a green tinge; basic dorsal pattern is a row of dark brown or black narrow crossbars or squarish blotches along entire body and tail; bars may be irregular or broken with halves diagonally above the other, often in the form of transverse triangular blotches; occasionally bars are connected by a thin dark longitudinal line; pattern only rarely configured in a clear zigzag pattern like *V. berus;* possibly with a single, thick middorsal stripe running the length of the body, with small spots or saddles adjoining or along the flanks; conversely, the pattern may be of finer thinner figures, with central transverse stripes or elongated ovals; venter often lighter than dorsum, although the range of shades includes white, yellowish, light to dark gray through dark blue or black; usually a whitish, yellowish, or reddish spotting or stippling; sometimes with a white or cream-colored spot near the outer edge of ventral plates, the prominence of this spot varies with subspecies, being inconspicuous if present at all in *V. a. aspis* and larger and more visible in *V. a. fransisciredi;* both the throat and underside of tail generally lighter than ventral ground color; throat often whitish, yellowish, or pale red, often stippled with darker shades, it may be uniformly dark, particularly in males; venter of tail tip usually yellow or orange, or it may be spotted or stippled with these colors.

Melanism is widely reported and has complex relationship to distribution. Melanistic individuals are rare in subspecies *V. a. aspis* and are always female. In subspecies *V. a. atra* melanism is common in alpine populations, members of which may be uniformly black with no trace of usual markings. In the Swiss Alps, populations studied by Naulleau (1973a, b, c) had two distinct color morphs present and intermingled: uniformly dark melanistic and very contrastingly patterned individuals with broad markings recognizably distinct from populations at lower altitudes. These morphs can mate and bear both color morphs in litters. Quantitative investigation of a melanistic population in western Switzerland yielded a melanism rate of 58% based on 36 snakes captured and marked in a 50-hectare area. (Nikolsky 1916; Steward 1971; Arnold and Burton 1978; Street 1979; Mattison 1982; Bruno 1985; Mehrtens 1987; Monney 1989).

Taxonomy and Distribution

Described as *Coluber aspis* by Linnaeus in 1758, redescribed and generalized in a later edition (1766) of *Systema Naturae*. The name Vipera was first used by Laurenti (1768) in his description of *Vipera francisci redi* (=*V. aspis fransisciredi*). Placed in *Vipera* by Daudin (1803) as a variety of *berus*. Type locality given as Poitou, France (Schwarz 1936). Despite its long taxonomic history and substantial interest from naturalists, it has generally remained in the genus *Vipera,* with the notable exception of an attempt to raise *aspis* to a genus by Fitzinger in 1826 (Herprint International 1994).

There is one wide-ranging and widely recognized subspecies,

V. aspis aspis, and a number of potential subspecies of more restricted range. *Vipera aspis aspis* is confined to and found over large areas of west Europe. It is found in most of France except areas bordering the English Channel. On the Atlantic coast it inhabits Île de Ré and Oléron, but it is not present on the coast south of the Gironde estuary. It is generally absent east of the Moselle River and from most of the Mediterranean region but is found near Montpellier and in Alpes-Maritimes. A discontinuous subpopulation exists in the Pyrenees, especially in Spain southwest of Bilbao. In Germany it is found in the southern Black Forest along the Swiss border but is rare in this region. It also occurs in northwestern Italy and western Switzerland, where it is common (Fretey 1975; Street 1979).

In Gascony, Andorra, and nearby Spain, *V. aspis* is represented by a subspecies described as

V. a. zinnikeri. This form has a distinctive dorsal pattern, with markings fused into a broad dark stripe, contrasting sharply with the lighter ground color and correspondingly dark vertical bars or blotches along the sides. In the south of Italy, subspecies

V. a. hugyi also is generally marked with a fused zigzag stripe and a recognizably raised snout. On the Mediterranean island of Monte Cristo, a relatively slender viper, meanwhile referred to as the subspecies *V. a. montecristi* has been described. It is similar to *V. a. hugyi* but with less tendency for the dorsal markings to fuse. In central Italy,

V. a. francisciredi is recognized by its head which is markedly swollen behind the eyes and upper lips, these features being distinctly visible from above in adults; a dark-colored morph described as

V. a. atra inhabits portions of Switzerland (Steward 1971; Street 1979; Mehrtens 1987). A poorly defined taxon is *V. aspis balcanica*.

Population differentiation in this species is very complex and it is presently unclear which described subspecies are valid and to what extent they are taxonomically distinct (Pozzi 1966; Kramer 1971; Bruno 1985; Zuffi and Bonnet 1999). In at least one published case, a specimen identified as *V. aspis balcanica* was actually *V. ammodytes* (Mertens 1950 referring to Buresch and Zonkov 1934).

Habitat

This species has relatively specific and clearly defined habitat requirements. The keys to the presence of this species in a particular area seem to be warm areas exposed to sun, with relatively dry soils and structured vegetation, (Luiselli and Rugiero 1990). In France and Italy, it is common in low mountains or hills, especially on limestone substrates, but it is sometimes found in lower plains. It prefers vegetated areas or habitats with at least some cover (Saint-Girons 1947). During the active season it is most abundant in wooded glades, scrublands, meadows, forest clearings, borders of woods, rubbish dumps, and stone quarries (Street 1979). In Italy, it is found in mesic chestnut/oak woodlands, often in the vicinity of streams. On slopes at 170–200 m altitude, it occupies canopied *Cytius* tree environs. In this region it prefers ecotones and edges, and generally favors heterogeneous landscapes (Luiselli and Rugiero 1990).

Food and Feeding

In general, diet consists of lizards and mammals. Mammals including mice (*Apodemys, Micromys, Mus*), voles (*Arvicola, Microtus*), shrews (*Sorex*), and moles (*Talpa*) are reported as food. Nestling birds have been recorded but not amphibians or fish. Neonates eat insects and worms, juveniles mainly wall lizards. Prey are killed by venom before ingestion and are only rarely swallowed alive (Lanceau and Lanceau 1974; Street 1979).

In southern France, prey is predominantly small mammals, with *Microtus arvalis* dominating gut contents (20 of 32 identifiable remains), followed by *Arvicola terrestris* (2), *Micromys minutus* (1), *Apodemus sylvaticus* (3), and *Sorex araneus* and *Crocidura russula* (total of 6). This distribution fits reasonably well with local abundance of these animals. Lizards were not recovered from these guts even though they are relatively abundant locally (Saint-Girons 1947).

In captivity, *V. aspis* eat mice, voles, and small quail, *Coturnix*. Crowding does not generally affect feeding except that cannibalism may occur when the excitement of food introduction peaks in confined groups. Captive females generally eat more than males. Newborn young can be induced to feed on small lizards or young mice. For neonates, lizards are preferred and mammals are only taken if sufficient feeding stimulation occurs.

Feeding individuals sometimes get so frenzied that they bite prey items many times before eating. In terraria in a feeding "frenzy," small snakes may be inadvertently eaten by larger ones (Smetsers 1990). This behavioral quirk was interpreted by Naulleau (1973a) to mean that "defense behavior is transformed into feeding behavior after biting the prey."

The preference of this species for warm, sunny habitats is reflected in its digestive physiology (Naulleau 1983a, b). *Vipera aspis* has higher thermal requirements for digestion than *V. berus*. In *V. aspis,* prey are regurgitated when environmental temperatures fall to 10°C. Regurgitation rate declines with increasing temperature to less than 10% at 20°C. Digestion rate shortens with increasing temperature and food acceptance rate increases 9% at 15°C, 40% at 20°C, 85% at 25°C, and 100% at 30°C degrees (Naulleau 1979).

In captivity, the average body temperature for individuals given thermal choices is 31°C–33°C. Digestion under these conditions requires 120–144 hours (Naulleau and Marques 1973). Optimum digestion rates are only obtained when thermal choice is constantly available (rather than only during the daylight hours).

Behavior

This is a generally sluggish species except when pursuing active prey or fleeing from a disturbance. It moves rapidly when necessary and inflates the body and hisses loudly when threatened (Saint-Girons 1947).

During the active season, March–October, individuals loyally inhabit a small, specific territory. This species is generally crespuscular, with behavior becoming nocturnal during warm weather and diurnal during cool. They bask frequently and openly, but are unable to flatten their backs while basking (as does *V. berus*) (Street 1979). In warm winters, individuals may begin basking in February and have been observed on the ground throughout winter (Fretey 1975). Under normal weather conditions, emergence and initial spring basking take place at the end of March or in April. During the course of the day, individuals exhibit a "plateau pattern" of daily body temperatures with three phases: a rapid warm-up in the morning, a steady temperature all day, and a long gradual overnight cool down (Seigel and Collins 1993). In captivity, it is largely diurnal with a small evening peak of activity (Saint-Girons 1947). In general, activity increases rapidly as temperature warms from 11°C–15°C degrees to 17°C–23°C, then falls rapidly at temperatures above 35°C.

Retirement to hibernacula takes place over several weeks from the end of October through the first half of November. Hibernation sites are underground cavities, crevices under tree trunks, or interstices in old stone walls. There are reports of individuals hibernating singly, but communal hibernation is more common. In

France, males retire to hibernacula earlier than females (Street 1979). During hibernation, there are structural bone changes and relatively rapid alterations of physiological state (Alcobendas et al. 1991).

Reproduction and Development

Mating takes place from mid-March to May and is preceded by ritualized combat among males (Street 1979). Prior to copulation males persistently follow females, tracking individuals by anal gland secretions. Copulation occurs by intercoiling, lasting 1–2 hours, and is often repeated 5–6 times. Pairs are monogamous during the mating season and may remain paired throughout the active season (Fretey 1975). In summer, gravid females congregate in favored gestation areas (Monney 1989).

Under ideal conditions it reproduces annually but the cycle often takes longer than a single year (Saint-Girons 1957; Duguy 1963). For breeding females, physiological changes like seasonal plasma cycles of phosphate and calcium levels reflect resorption of bone as a mineral reservoir to support vitellogenisis (Alcobendas et al. 1992). Pregnant females do not feed until after giving birth (Naulleau 1973a).

In lowland areas, mating may take place in autumn. Otherwise, mating occurs after hibernation in spring and young are born late in summer. In captivity, reproduction may be forced into an 8–10 month cycle, and one individual gave birth twice in 1 calendar year (Naulleau 1973b). Sexual cycles may be seasonally suppressed under captive conditions, and sexual behavior may occur throughout the year (Naulleau 1973a).

Under natural conditions, gestation varies from 97 to 133 days. In captivity without hibernation, gestation can be shortened. With 12 hours of light and thermal gradient of 18°C–40°C, average gestation is 67 days with a minimum of 54 days. In captivity, the reproductive thermal minimum is 23°C–26°C. Gestation period increases with decreasing temperature to approximately 60 days at 31°C and 92 days at 26°C. At 26°C, there is consistent backbone malformation in young. When continued embryonic development is impossible, females die (Naulleau 1983c, 1986).

Parturition generally requires from one-half to 1 day. Clutch size is from 2 to 9 (mean 5.6). In 19 broods the sex ratio at birth was 53:47 male:female. At birth, young are from 14.7 to 25.5 cm in total length and weigh 3–11 g (Phisalix 1940; Naulleau 1973a). In captivity, the percentage of babies feeding spontaneously is very low. There is a minimum 2-day fast from birth to first meal for lizards, and 4 days for mice. However, this initial fast can last for more than 2 months (Naulleau 1973a).

Young grow rapidly for 2 years, then plateau. The initial growth rate is from 47–67 cm over 2.5 years. In general, growth is at least 7–8 cm/year for the first 2 years. After maturity, growth occurs at about 2 cm per year to full size. Maturity is at 3 years for males and 4 for females.

In outdoor terraria, individuals copulated in April and May, and young were born in September. In one enclosure, a total of 25 young were produced from four females. In this situation neonates sloughed between 4 and 14 days. Newborns were force-fed pink mice, after which they fed actively on their own (Jooris 1986).

Vipera aspis is one of few snakes to exhibit sexual size dimorphism at birth (Siegel and Collins 1993). Newborn males average longer (21.6 cm vs. 20.5 cm) and heavier (7.1 vs. 6.7 g), and the maxima are even more divergent (the smallest female was 14.7 cm and weighed 3 g, the largest was a male at 25.5 cm and 10.5 g) (Naulleau 1970).

Shedding usually lasts 10–45 minutes. Most individuals stop feeding prior to shedding, but some have been observed to eat within 24 hours prior, and there are even observations of feeding during shedding. Before shedding, animals become less active and bask in warmer spots for longer times. Individuals sometimes shed twice without feeding between shedding events (Naulleau 1973a).

Hybrids of *V. aspis* with other species of *Vipera* are known. For example, a male *V. aspis* from Andorra mated with a female *V. ammodytes* which later produced offspring (Mertens 1964). Other accounts include a hybrid *V. ammodytes/V. aspis* in which a second-generation captive breeding yielded only dead and deformed young (Faoro 1986).

Bite and Venom

Epidemiology—*V. aspis* is responsible for 90% of reported venomous snakebites in Italy. In Tuscany over the period 1960–1970, 526 viper bites were recorded, nearly all attributed to this species, of which 7 were fatal (Lombardi and Bianco 1974, Street 1979). In Spain from 1965 to 1980, 80 *V. aspis* bites were treated. Of these, about half were severe cases requiring antivenin (Gonzalez 1991).

Yield—Yield is relatively low, on the order of 9–10 mg dry per extraction (Boquet 1964).

Toxicity—Toxicity varies among populations. Of several populations from Switzerland tested, snakes from Passwang District had distinctly more toxic venom (Stemmler 1971a). Toxicity is relatively high on a comparative basis. Reported LD_{50} values in mice are approximately 1.0 mg/kg with a minimum lethal dose of 1–2 mg/kg (Boquet 1967b; Mebs 1978). Intramuscular LD_{50} is higher, approximately 4.7 mg/kg (Tu et al. 1969).

Toxicity is reduced if venom is heated to 60°C, and at 80°C phospholipase and anticoagulant activity disappear (Boquet 1967a).

Content—Venom is reported similar to *Daboia russelii* except for electrophoretic fraction locations of the coagulants and anticoagulants (Devi 1968). Several proteins are reported including proteolytic enzymes, L-amino acid oxidase, phospholipase, hyaluronidase, hemorrhagic factors, and coagulation accelerators and inhibitors (Boquet 1967a). Although proteases are not present, necrosis occurs in envenomated test animals. However, venom from specimens collected in southern France lack L-amino oxidase and riboflavin, and fail to cause necrosis (Sarkar and Devi 1967).

Symptoms and Physiological Effects—In general, the bite is painful and destructive (Street 1979). Symptoms include rapidly spreading acute pain followed by edema and discoloration. Severe hemorrhagic necrosis may begin a few hours postbite (Boquet 1967b). Vision may be severely impaired, probably through degradation of blood and blood vessels in the eyes (Flichy 1978).

The venom has coagulant and anticoagulant effects. The mechanisms and balance between these are poorly understood. Anticoagulant activity is reportedly stronger in *V. aspis* venom than in Russell's viper (*Daboia russelii*) venom, and affects the conversion of prothrombin to thrombin (Devi 1968). Fibrinogenolytic activity is also present along with platelet aggregation inhibitors of three kinds: ADPase, fibrinolytic enzymes, and phospholipase A. In *in vitro* tests, diluted *V. aspis* venom inhibited migration of fibroblast cells on initial contact. After 24–48 hours, cells transferred to fresh medium lacking venom migrate normally (Lasfargues and Di Fine-Lasfargues 1951; Boffa et al. 1972a, b; Boffa and Boffa 1975)

The coagulant component of the venom is structurally linked to proteolytic components (Grasset et al. 1956), and coagulation occurs within the plasma (Rosenfeld et al. 1967).

Glomerular structure may be impacted, leading to death from renal failure (Hofstetter and Stolz 1982).

Vipera aspis venom does not affect neuromuscular contractions in *in vitro* preparations (Cheymol et al. 1973). Lack of this neurotoxic effect means that fatal cases involving the cardiovascular system result from direct muscle injury or reduced oxygen exchange (Tu et al. 1969). However, Gonzalez (1991) reported on two bite victims who developed neurotoxic symptoms including difficulties breathing and swallowing, and paralysis of the bitten limbs.

Venom introduced directly into a small blood vessel can cause rapid death (Boquet 1967b). Overall, about 4% of untreated bites are fatal, with rapid fatalities sometimes occurring in individuals bitten directly in a blood vessel (Stemmler 1971a).

Remarks

Until recently, *V. aspis* was considered a difficult species to maintain in captivity. Street (1979) reported that individuals would never become tame and that most refused food, characterizing it as an "unsuitable" captive. However, it is now known that with care, *V. aspis* can be easily maintained in captivity (Naulleau 1973a). It is best reared in large outdoor (6 m × 6 m) or indoor (7 to 14 m^2) terraria. Fluorescent light should be provided at 300 lux at ground level with a 700- or 1000-W infrared light over a basking site for warmth, with the surface temperature at the basking spot of 20°C–38°C and a maximum of 45°C. Humidity should be maintained at 52%–75%, with no watering of substrate. It can be kept in smaller terraria, and maintained without hibernation and allowed to shed and feed throughout the year (Mattison 1982). The best and simplest captive diet is small mice. When more than one individual snake is kept in a single terrarium, it is imperative to observe them during feeding to ensure that small snakes are not inadvertently swallowed by larger individuals (Smetsers 1990).

There are reports of *V. aspis* being eaten in the wild or in captivity by hedgehogs, cats, dogs, pigs, poultry, and birds of prey (Street 1979).

Populations of *V. aspis* are threatened by habitat pressure and overcollecting throughout Europe. In central Italy, all reptiles except *V. aspis* are protected by regional law and populations suffer correspondingly. In this region, it is an ecological associate of *Coluber viridiflavus, Coronella austriaca, Elaphe longissima, E. quatuorlineata, Natrix natrix,* plus seven species of lizards and one turtle, all occurring within a single 4-km^2 area (Luiselli and Rugiero 1990).

Vipera latastei Bosca 1878

Also called Lataste's viper or snub-nosed viper.

Recognition
(Plate 14.4)

Head—Rostral longer than wide, covering the front part of a nasal "horn"; 2 large supraoculars with posterior end extending beyond posterior margin of eye; remainder of head scalation with numerous small, smooth, or only weakly keeled, irregular plates; 7 interocular scale rows; frontal and parietals usually absent; large single (rarely divided) concave nasal contains nostril; nasal separated from rostral by single nasorostral scale; temporals smooth or weakly keeled; 10–12 small plates bordering eye; 2–3 rows of scales between eye and supralabials, which number 9–11, with fourth and fifth the largest.

Body—Dorsal scale rows 21, strongly keeled, with row adjoining ventrals smooth or weakly keeled, 122–147 ventrals, 29–47 paired subcaudals; anal single (Steward 1971; Saint Girons 1977).

Pattern—Dorsum of head occasionally with irregular dark markings, a dark streak runs from the eye to the angle of mouth, dorsal pattern a series of black-edged, dark brown rhombs that form a wavy or zigzag band. A series of dark spots along each side. Body dorsal ground color gray, brown, or reddish brown. Juveniles similarly colored and patterned (Nikolsky 1916; Biella 1983).

Size—Maximum length around 72 cm, but normally smaller.

Distribution and Taxonomy

Originally described by Bosca (1878) as *Vipera latasti*. Over the years it has been regarded as subspecies both to *Vipera aspis* and to *V. ammodytes* (Schwartz 1935). Today it is divided into two subspecies:

V. l. latastei in most of the Iberian peninsula south of the Pyrenees, and

V. l. gaditana in southern Spain and Portugal, as well as in North Africa (Saint Girons 1977).

Habitat

Favors dry, brushy, rocky hillsides, often in open pine forests, and in maquis and garigue vegetation. Occasionally found in sand dune areas (Nilson pers. obs.).

Food and Feeding

Food contents for wild specimens consisted of mammals, reptiles, arthropods, and birds. Younger specimens, especially, feed on orthopterans and other insects and reptiles (Bea and Brana 1988).

Behavior

The southern subspecies, *V. l. gaditana* is a good sidewinder in the sand dunes of Coto Donana in southern Spain.

Reproduction and Development

Breeding takes place during fall, with birth of 2–13 young almost a year later (Parellada 1995).

Bites and Venom

No information is available.

Vipera monticola
Saint-Girons 1954

Recognition

Head—Broad and triangular, well distinct from neck; snout tip distinctively upturned into a small horn; snout dorsum flat with distinct and slightly raised sharp margins; nasal single (rarely divided) contains nostril, separated from rostral by a single nasorostral; head dorsum with numerous small irregular scales of various sizes which are generally smooth but may be slightly keeled; rostral twice as high as broad, forming anterior part of the horn; frontal and parietals not present; supraoculars large and distinct, separated by 4–7 scale rows; 8–10 small scales surrounding the eye below the supraocular; eye separated from supralabials by 2 scale rows; supralabials 9 (rarely 10); temporals generally smooth.

Body—Dorsals 19 (rarely 20) rows at midbody; dorsals keeled except lowest row which varies and is sometimes smooth; ventrals 123–138; subcaudals 30–35 in females, 35–43 in males; anal single.

Size—A very small viper, with maximum length less than 40 cm.

Pattern—Head variable, ranging from unmarked to patterned with small dark markings; temporal region with a dark streak from eye to neck.

Upper lipshields whitish, cream, yellowish, or pinkish often with thin dark bars in sutures. Eye is small, usually golden or coppery red with vertical black pupil.

Body ground color is usually gray with a dorsal rather narrow undulating or zigzag stripe along the back. This stripe may be fragmented into cross-bars or rhombic blotches. Laterally on each side there is a series of dark spots.

Taxonomy and Distribution

Described as *Vipera latastei montana* by Saint-Girons (1953), and changed to *Vipera l. monticola* the year after (Saint-Girons 1954) (The name *montana* was preoccupied by *Vipera berus montana* Méhely 1894). Found in mountain valleys in central and western Haut Atlas, Morocco.

Later, Beerli et al. (1986) proved its status as a full species.

Habitat

This species occurs at elevations between 2100 and 3900 m. Trees and big bushes are absent in these valleys, and the main vegetation is thorny small bushes and big grass tussocks on stony and gravel ground. This dense low vegetation is important for protection and shelter.

Food and Feeding

The main food items are lizards and insects, including carabid beetles. Rodents are absent in the habitat (Schleich et al. 1996).

Behavior

The cold winters and long snow cover in the high Atlas mountains make hibernation necessary for a long period. It is a diurnal species as nights are cold at these altitudes throughout the summer. Much time is spent in thorny bushes and grass tussocks (Nilson pers. obs.)

Reproduction and Development

Clutch size is small and covers 2 to 4 young (Schleich et al. 1996). Mean total length of newborns is 14.9 cm and mean weight is 2.99 g (Saint-Girons and Naulleau 1981).

Bite and Venom

Epidemiology—No cases of envenomation are known.

Vipera transcaucasiana
Boulenger 1913

Known as the Transcaucasian sand viper, and occasionally Armenian sand viper (albeit wrong as the species does not occur in Armenia).

Recognition
(Plate 14.5)

Head—Rostral wider than long, supporting a nasal "horn" of 9–17 scales arranged in 3 (rarely 2 or 4) transverse rows; 2 large supraoculars with posterior end extending beyond posterior margin of eye; remainder of head scalation with numerous small, smooth or only weakly keeled, irregular plates; 7 interocular scale rows; frontal and parietals usually absent; large single (rarely divided) concave nasal contains nostril; nasal separated from rostral by single nasorostral scale; temporals smooth or weakly keeled; 11–12 small plates bordering eye; two rows of scales between eye and supralabials, which number 9–10, with fourth and fifth the largest (Nikolsky 1916; Biella 1983).

Body—Dorsal scale rows 21, strongly keeled, with row adjoining ventrals smooth or weakly keeled, 148–160 ventrals, 32–40 paired subcaudals; anal single (Nikolsky 1916; Biella 1983).

Pattern—Dorsum of head and nasal horn without irregular dark markings, except a weak V marking on back of head, dorsal pattern of narrow transverse bands, iris of eye golden or coppery. Body dorsal ground color light gray, ash gray, silver-gray, pale gray, or grayish white. Juveniles similarly colored and patterned (Nikolsky 1916; Biella 1983).

Size—Maximum length around 75 cm, but normally smaller.

Taxonomy and Distribution

Originally described by Boulenger (1913a, b) as *Vipera ammodytes transcaucasiana*. In recent times *transcaucasiana* has been given specific status, based on genetic distances larger than between other full species (e.g. *V. aspis* and *V. latastei*) (Herrmann et al. 1987, 1992).

Quantitative information regarding population densities is not available for *V. transcaucasiana*. There is general agreement among collectors and field biologists that individuals are relatively rare, even within suitable habitat (Bozhanskii and Kudryavtsev 1986). Low population densities possibly result from localized habitat conditions favored by *V. transcaucasiana*. In dry, sparsely vegetated areas, resources (food, water, and shelter) are diffuse, being scattered across the landscape. Under such circumstances, dense snake populations cannot be supported, and the reported "rarity" of individuals may simply represent natural correspondence between *V. transcaucasiana* and its ecological system.

The distribution is restricted to parts of Georgia and northern Turkish Anatolia (Nilson et al 1988). It does not occur in Armenia, Azerbaijan, or Iran, as stated in literature.

Habitat

Favors dry, brushy, rocky hillsides, preferring areas with only sparse vegetation, and scrubby, relatively barren rocky slopes being usual habitat. Occasionally found in open coniferous woodlands on rocky ground (Nilson pers. obs.).

Food and Feeding

Food contents for wild specimens in Georgia were sparrows, rodents, and lizards, and especially young agamid lizards (Biella 1983).

At the Moscow Zoo, where this species was kept successfully, the adult diet was 66% birds, and smaller snakes were fed lizards and insects. Feeding rates in captivity during the active season range from once per week to once every 2 weeks (Kudryavtsev et al. 1993).

In captivity, the effective artificial hibernation period is shorter (from 21–35 days), and is usually implemented later in the year (from November through January) than the natural inception of hibernation (Kudryatsev et al. 1993).

Behavior

No information is available.

Reproduction and Development

Mating may occur in March, April, and May, but the main mating period in Georgia is the last half of May (Biella 1983). The young are born in late summer or early autumn, usually August or September. Two to 15 young are common, with an average litter of 6 individuals. Length at birth ranges from 15 to over 20 cm. Birth weight is approximately 6 to 8 grams, although in captivity birth weights up to 10 grams are reported (Kudryatsev et al. 1993).

Kudryatsev et al. (1993) reported on a long-term breeding and maintenance program. Terraria used varied from 800–1,000 cm^2, depending on size of individual animals. A high thermal gradient was maintained betwen the relatively warm cage floor (30°C) and cool overlying air (20°C) to simulate the mountain environment.

Bite and Venom

Much of what is said about *Vipera ammodytes* also applies to *V. transcaucasiana*.

Subgenus *Pelias*
Merrem 1820

The name *Pelias* was given for *Pelias berus* (=*Vipera berus*) by Merrem in 1820. *Pelias* represents the smaller Eurasian vipers that are characterized by having larger and normally irregularly fragmented head plates. Members of *Pelias* normally have two apical plates in contact with rostral, and upper preocular separated from the nasal plate by loreals. Also, members of *Pelias* have a reproductive pattern characterized by a spring mating period, and which is triggered by the spring molt of the males in the local population. Spermatocytogenesis takes place during fall and spermiogenesis during the following spring (Nilson and Andrén 1997).

Vipera barani
Böhme and Joger 1984
Known as Baran's adder.

Recognition

Head—Moderately large and fairly distinct; sides flat and almost vertical; edge of snout commonly raised into a low ridge; eye fairly large; eye about equal nasal shield; rostral barely or not visible from above; 1 or 2 apical shields in contact with rostral; plates on top of head normally fragmented, a small frontal can be discerned in some specimens, parietals fragmented, 2 big supraoculars; large nasal contains nostril in shallow depression; 8–14 (mean number 10.8) small scales bordering eye below supraoculars; temporals smooth (rarely weakly keeled); 7–10 supralabials, fourth and fifth the largest, numbers 4 and 5 separated from eye by single row of small scales; 9–12 sublabials.

Body—Dorsals increase in width with proximity to ventrals, yielding outer rows twice as broad as those along midline; dorsals 21 (rarely 23), strongly keeled except weak or smooth keels on last row adjoining venter on each side; ventrals 138–144 in males, 139–146 in females; anal single; paired subcaudals 35–36 in males, 26–37 in females.

Size—A short thick-bodied snake; male generally shorter than females; adult females reach 60 cm and males 54.5 cm total length.

Pattern—Melanism is common and total except on supralabials, which are white.
In non-melanistic specimens, head with a weak V on the back; with dark streak from eye to corner of mouth; upper labials whitish with brown or black sutures; iris of eye red; black pupil is vertical in bright light.

Body color pattern is grayish brown with a dark, blackish zigzag band, or dorsal crossbars along the entire body and tail; venter is black with dense white spots; tail tip is yellowish red (Böhme and Joger 1983; Franzen and Heckes 2000; Baran et al. 2001).

Taxonomy and Distribution

A rare species that was originally described as *Vipera barani* by Böhme and Joger (1983) from 60 km north of Adapazari in Turkey.

The species is distributed in northern Turkish Anatolia. Albeit very few specimens have been found, genetic investigations indicate that the same taxon is distributed all the way from Adapazari in the northwest to between Ardesen and Camlihemsin, Province Rize, in the northeast. Earlier, specimens of *Vipera barani* from northeastern Turkey (Camlihemsin) have incorrectly been determined as *Vipera pontica* (see Baran et al. 1997; Baran and Atatür 1998; Baran et al. 2001). Nevertheless, it seems to be parapatric with *Vipera pontica* in the Artvin region.

Habitat

It inhabits rocky shrub, edges of cultivated fields with tea plantations, dry open oak woodlands, and rocky slopes at lower elevations (700 m) (Franzen and Heckes 2000) to the upper treeline in the alpine zone. At this altitude, 1950 m, the habitat consists of open, stony fields with a sparse vegetation cover (Baran et al. 2001).

Food and Feeding

Rodents are probably an important prey item for adults. The holotype had a *Microtus majori* in the stomach (Böhme and Joger 1983). Lizards are abundant in the habitat.

Behavior

No information is available.

Reproduction and Development

No information is available.

Bite and Venom

No information is available.

Vipera berus (Linnaeus 1758)

Known as the common European adder.

Recognition
(Plate 14.6)

Head—Moderately large and fairly distinct; sides flat and almost vertical; edge of snout commonly raised into a low ridge; eye fairly large, often smaller in female; eye equal or slightly smaller than nasal shield; rostral barely or not visible from above; 2 (rarely 1) small shields just behind rostral; normally 5 large plates on dorsum: squarish frontal (longer than broad and rectangular in some specimens), 2 parietals (sometimes with a tiny scale between frontal and parietals), 2 long narrow supraoculars, each usually separated from frontals by 1–4 small scales (in subspecies *bosniensis* plates on head, especially parietals, reduced or fragmented); large nasal contains nostril in shallow depression; 6–13 (usually 8–10) small scales bordering eye below supraoculars; supraoculars usually large and distinct; temporals smooth (rarely weakly keeled); 6–10 (usually 8–9) supralabials, third and fourth the largest, numbers 4 and 5 (rarely 3 and 4) separated from eye by single row of small scales (double row occasionally in alpine specimens); 10–12 sublabials.

Body—Dorsals seem loosely attached to skin, and increase in width with proximity to ventrals, yielding outer rows twice as broad as those along midline; in subspecies *V. berus berus,* dorsals 21 (rarely 19, 20, 22, or 23), strongly keeled except weak or smooth keels on last row adjoining venter on each side; ventrals 132–150 in males, 132–158 in females; anal single; paired subcaudals 32–46 in males, 23–38 in females; in subspecies *V. berus bosniensis,* ventrals 140–155 with 26–37; paired subcaudals.

Size—A short thick-bodied snake; male generally shorter and slimmer than female in all seasons; in subspecies *V. berus berus,* adults reach 60 cm and average 55 cm; individuals of *V. berus bosniensis* are larger, with adults commonly to 70 cm; exceptional individuals reported at larger than 100 cm. Twice, specimens of 104 cm have been observed in Sweden (e.g. Bruno 1985), maximum size varies by region, with largest individuals (over 90 cm) from Scandinavia, and maximum lengths of 80–87 cm in France and Britain; tail one-sixth to one-seventh of total length in males, one-eight to one-ninth of total length in females; tail terminates in distinct horny point.

Pattern—Head with a dark V or X on the back (the most common, consistent and distinctive characteristic); with dark streak from eye to neck, continuing as a longitudinal row of roundish or oval spots along flanks; commonly the V or X on the neck dissolves into dark coloration on top of head; upper labials yellowish or whitish with brown or black sutures; chin and throat white or yellowish, often tinged with black, brown, orange, or red; in males, chin and throat scales often spotted or edged with black; iris of eye coppery red or reddish brown; black pupil is vertical in bright light; tongue black, sometimes with reddish coloration.

Body patterns commonly range from very light-colored individuals with small, incomplete dark dorsal crossbars to entirely dark, largely lacking discernible dorsal pattern; consistently with some kind of dark zigzag dorsal pattern along the entire body and tail (in some individuals, dorsal zigzag markings may be indistinct or absent, rarely replaced by continuous stripe); dorsal pattern may be fragmented into row of oval spots; a row of spots in parallel concavities in saddles or bars of dorsal zigzag; sides darker than back; in subspecies *berus berus,* individuals may usually be sexed by color: males are whitish, silvery white, pale gray, yellowish, or rarely pale greenish; markings are vivid uniform black; during mating season, contrast between pattern and ground color increases; gray males with brown markings occasionally occur; females have ground yellowish, whitish brown, brown, reddish brown or coppery with dark brown or dark red markings; venter brownish with black spots especially in females, or gray, light blue, navy, or black, speckled at edges with white, ivory, pinkish white, or yellow; tail tip bright yellow to pale orange, darker in males.

Melanism is common and is sometimes total. Some individuals in Britain are colored deep blue-black with markings difficult to discern. Young are born normal but may develop melanistic pattern by 3–4 years. In populations with melanism, about 50% of the individuals are melanistic (Andrén and Nilson 1981). Juveniles in general are patterned similar to normal adults. Albinism has been reported (Sochurek 1953; Marian 1956; Hopkins 1957; Frommhold 1969; Steward 1971; Street 1979; Schiemenz 1985; Bruno 1985).

Taxonomy and Distribution

Described as *Coluber berus* from a specimen from Sweden (type locality restricted to Uppsala, Sweden). It was placed in the genus *Vipera* as *Vipera berus* in 1803 by Daudin. The subspecies *bosniensis* was described relatively early (1889) from the Balkan region. The subspecies *sachalinensis* was erected in the Russian literature in 1917. *Vipera seoanei,* formerly incorporated as a subspecies of *berus*, has received specific recognition since the early to mid-1980s.

Vipera berus has an enormous range, occurring from Britain and northern Scandinavia to northern Greece and to the Sakhalin islands on the Pacific coast. The range may be approximately defined by cartographic boundaries: above the Arctic Circle in the north from 5° to 145° longitude and south to 40° latitude (Bruno 1985). Within that range, distribution is not continuous and there are both large areas covered by single populations and disjunct local populations. Insufficient knowledge exists to support erection of more than three presently recognized subspecies:

V. berus berus,
V. berus bosniensis,
V. berus sachalinensis.

However, both *bosniensis* and *sachalinensis* have been considered as full species in recent literature (Joger et al. 1997; Ananjeva et al. 1998). In addition, *Vipera nikolskii* is genetically similar to *V. berus berus,* as is *V. berus bosniensis* to *V. barani* (Joger et al. 1997). All these taxa could be looked upon as a subspecies of *V. berus* or as full species. They all differ from each other in morphology (Franzen and Heckes 2000). Future studies may reveal additional taxa within the huge range.

Distributional details are of some interest because of the remarkable lack of taxonomic variation in most of the range in this wide-ranging species. The adder is the most common snake in Britain, being found in all Welsh counties to northern Scotland but absent from Ireland and the Isle of Man. On the mainland, it reaches northern Norway, Russia, and the whole of Sweden. In France, it is found north of the Loire, in the Massif Central, and on Mont Ventoux in Vaucluse and the Alps. It does not occur in Alsace and Lorraine. Most of Holland and Belgium and much of Germany are included in the range, but the actual occurrence in these countries is fragmented. It is very common in parts of Denmark, especially western Jutland and occurs on many large islands. It is found in southern Poland and in the Czech and Slovakian republics and in the Baltic states. It is extinct in much of Switzerland but is found in the north. In Italy, it occurs only in the north, in parts of Piedmont, Lombardy, Trentino-Alto Adige, Veneto, and Emilia to Ferrara. On much of the Balkan Peninsula *V. berus berus* is replaced by *V. berus bosniensis*. Distribution is limited in Austria and Hungary and it is found in isolated high altitudes in Slovenia, Croatia, Bosnia, Herzegovina, Serbia, and Macedonia, and western half of Bulgaria. The range extends across asiatic Russia to Kazakhstan, southern Siberia and northern Mongolia (Bodenheimer 1944). The subspecies *V. berus sachalinensis* occurs in Russian Far East (Amur) China (Jilin) and on the Sakhalin Island as well as in North Korea (Nilson et al. 1994).

Population densities do not vary greatly when normalized to a "suitable habitat" basis. Populations as high as six individuals per hectare of suitable habitat were found following a year of particularly high rodent population density on a Swedish island (Andrén 1982a). This density was unusually high, and there was a substantial decrease the following year after rodent populations fell. Continental population densities range from one individual per suitable hectare in Poland (Pomianowska-Pilipiuk 1974) to as high as 2.7 individuals per suitable hectare in European Russia (Boshansky and Pishchelev 1978).

Habitat

Crucial requirements for its presence seem to be sufficient habitat complexity to support hibernating, basking, and foraging areas, and some protective covering from predators and human harassment (Bruno 1985). Habitats are varied: moors, sandy heaths, meadows, chalky downs, rocky hillsides, rough commons, edges of woods, sunny glades and clearings, bushy slopes and hedgerows, stone quarries, dumps, and coastal dunes. *Vipera berus* does not avoid wetlands if dry ground is available within the local landscape. Thus, it inhabits stream, lake, and pond banks. In Europe it inhabits rocky shrub areas along shores, cultivated fields with hedgerows, dry open woodlands, and rocky slopes to the treeline in alpine regions. It is not found in clay soil areas, and in southern Europe it is found either in low marshy land or at high altitudes reaching 3000 m in the Swiss Alps (Appleby 1971). It is absent from the open steppes of Hungary and Russia, which are favored by *Vipera ursinii* and *V. renardi*. It is present in forested steppelands in Russia, preferring damp and cool woods on sandy soil near marshes ranging from 130–160 m altitude.

Melanistic forms (black adder) live in moorlands and mountains and are common in alps to 2000 m.

In any particular location, individuals commonly inhabit deserted rodent burrows, rock piles, wood, haystacks, and grass clipping heaps.

In Hungary, *V. b. bosniensis* dispersal is clearly differentiated by habitat (Fritsche and Obst 1966). Snakes bask in dry oakwood borders in spring upon emerging from hibernation. In summer, individuals migrate to lower wetter areas, and in dry years they go deep into alder forests. In relatively wet years, populations inhabit drier poplar plantations within the alder landscape (Prestt 1971).

Food and Feeding

Diet varies with locality and prey availability. This is a very adaptable species and is not specialized on particular foods (Andrén 1982a; Bruno 1985,).

Diet consists mainly of small mammals including voles (*Arvicola, Clethrionomys, Microtus*), shrews (*Sorex, Neomys, Crocidura*), mice (*Apodemus*), and lizards (Street 1979). They also eat frogs (*Rana temporaria*), newts, and salamanders, but not toads. Nestling birds and eggs are taken, and individuals will climb to obtain them (Street 1979). Adders on Luneburg Heath in Germany eat slow worms, baby weasels, and moles (Frommhold 1969). Young eat nestling mammals, small lizards, frogs, insects, worms, and spiders. When juveniles reach 30 cm (approximately their second year) their diet is similar to that of adults. In continental environments lizards are primary prey of young and subadults, and lizards and amphibians are locally important for adults (Mertens 1947; Pielowski 1962; Prestt 1971; Sebela 1978; Saint-Girons 1980). Several locally common prey such as toads, sand lizards (*Lacerta*), and water voles (*Arvicola*) have not been recorded in gut contents (Prestt 1971). Active avoidance of these potential food items may be surmised. All individuals drink water "relatively frequently" (Prestt 1971).

Prestt (1971) recovered 97 prey items from 54 wild-caught snakes. In 43% of these snakes food consisted of more than one item, mostly differing in digestive stage. However, when prey consisted of nestling birds or suckling mammals, all were at the same stage of digestion, indicating that they had been taken at the same time. *Vipera berus* stalks prey slowly and strikes from an S posture. Prey are released and tracked after variable waiting time. Individuals have been observed to strike and track an adult rodent, and consume both the tracked adult and litter of young. Prey is invariably swallowed headfirst (Street 1979). Frogs and newts are eaten directly and alive without invoking the venom apparatus (Smith 1969).

After emerging in spring, individuals do not generally feed until after mating. Adult males and nonbreeding females feed from June to August, but breeding females, having fat from the previous year, eat little until after giving birth. During the late summer peak in rodent population (a period of about 1 month), all individual adders hunt until hibernation (Prestt 1971). Feeding occurs in a separate summer habitat, up to 2 km away from hibernacula (Viitanen 1967). *Vipera berus* has a well developed homing instinct (Frommhold 1969).

Feeding rates and trophic dynamics vary with latitude, presumably based on temperature, with thermal tolerance range for digestion much lower in *V. berus* than in *V. aspis* (Naulleau 1983a, b, 1986).

In Sweden among the most northerly populations of adders, rodents are dominant prey for individuals of all ages (Andrén 1982a). Rodent populations on Swedish islands fluctuate widely, more so than on the mainland. Adder populations were monitored following high rodent populations in 1973 and low rodent populations in 1974 on a single island. Total adder population of this island based on recaptures was approximately 200 individuals, or 4 individuals per hectare. This yields 6 individuals per hectare of habitat suitable for producing rodents (grass meadows and moorlands). Adder population decreased 50% from 1974 to 1975 (Andrén and Nilson 1981). An index of food resources based on rodent density fell enormously with a drastic reduction in rodent populations between 1973 and 1974. That rodent population density was crucial to adder nutritional status was evidenced by dramatic individual mass decrease in snakes coinciding with rodent population decline (Andrén 1982a). In general, island snakes maintain a higher weight status and length than mainland individuals of the same sex. From 1974 to 1975 in this system, male mass decreased from an average of 99–75 g (spring) and to 61 g in autumn. Mean length increased over the same period indicating greater survivorship of larger males. Males losing 55%–65% body mass survived 2 years of low vole populations, no males losing more than 45% mass were recaptured. These individuals probably starved to death. There were similar mass losses for females. Survivorship was much higher in nonreproductive females than in reproductive females. Lack of food in late summer impacts survival of reproductive females, which need energy to recover from physiological stress of breeding. In general, the late summer rodent population peak yields a cycle perfectly timed to accommodate this.

Energy content of rodents consumed from island populations was 1.5 kcal per g body mass. An energy budget for rodent-feeding island populations averaged an annual adult consumption of 350 kcal per capita. Extrapolating to population densities yields 9 voles per adult per year (Andrén 1982a). Total number of voles on the Swedish island investigated in spring 1975 was estimated directly at fewer than 400. This would provide the annual ration for fewer than 50 adult vipers. Thus, adder predation in 1974 and 1975 was probably crucial in controlling rodent populations. Similarly, reduction in rodent populations yielded a concomitant reduction in the local adder population (Pomianowska 1972; Pomianowska-Pilipiuk 1974).

Behavior

In the north adders are largely diurnal, and in the south they are nocturnal or crepuscular (Prestt 1971; Street 1979).

There are three active phases to the overall annual cycle: basking and mating; dispersal and feeding; and return to hibernacula and birth of young (Prestt 1971).

Throughout the nonbreeding season, individuals bask but routinely avoid locations subject to full sun. Basking periods are dependent on local warming conditions. Daily emergence is at approximately 1000 hrs during the spring in England

and individuals have been found on the surface as early as 0840 hrs in Hungary in April.

Basking is a complex behavior, tightly linked to thermoregulation. During cooler periods individuals flatten their backs and orient toward the sun to obtain the largest warming area and maximize thermal return. Adders generally avoid windy conditions and tend to sit coiled in cool or damp weather (van der Rijst 1990). Basking occurs mainly on the litter layer in spring, under the litter layer in summer, and always under some vegetation canopy. Basking sites are increasingly in competition with humans due to activities such as picnicking and sunbathing (van der Rijst 1990).

In September foraging declines with temperature and autumn basking is replaced by hibernation when daily maximum air temperatures fall to 13°C (Viitanen 1967). In general, hibernacula face south and occupy high dry ground in areas of thick vegetation. Adders hibernate in mammal burrows, tunnels, root cavities, and crevices in vegetated banks and other sheltered well drained situations. Roots of "gorse" are favored for shelter (Leighton 1901). Some individuals hibernate singly, but in Scandinavia and elsewhere in the north most hibernate in groups, in rare instances as many as 300 to 800. Winter quarters are up to 2 m deep. Adders often share hibernacula with other snakes, slow worms, common lizards, and common toads (*Bufo bufo*). No segregation occurs by age or sex and individuals do not always return to the same hibernaculum in successive years even if that is the norm. There are up to 70 days between the first and last individuals entering hibernation and hibernation date varies with sex. Although there is overlap, females generally remain on the surface 2 weeks longer than males (Prestt 1971). This may be in response to longer feeding periods needed to accommodate stresses of parturition (Andrén 1982b).

Adders are cold-adapted and emerge on mild winter days. They may bask where snow is melted and will readily travel across snow. Males in southern Dorset in England hibernate for 150 days, females for 180. In northern Scandinavia, hibernation lasts 8–9 months. On much of the continent it is one of the first spring reptiles, sometimes emerging by February, and usually by March or April. Approximately 15% of adults and 30%–40% of young regularly die during hibernation (Viitanen 1967).

Across the wide geographic range of the adder, spring emergence is temperature dependent (Viitanen 1967; Andrén 1982a). In spring, males emerge 2–5 weeks before females and young when maximum daily air temperature is 8°C, followed by females in mid-April when maximum temperature is approximately 12°C, then immatures. In exceptionally cold years, all may not emerge until May. At a monitored study site the earliest male emerged on 20 February, followed by the earliest female on 2 March. In spring, there is only local dispersal followed by intense foraging. Fat reserves are replenished during the last half of May, June, and early July. In Scandinavia, females accumulate only a small quantity of fat and do not reaccumulate full reserves until the second season when they do not breed (Nilson 1981). In late summer, 43-cm nonbreeding females weighed 78 grams with 1.37 g ovaries and 7.6 g fat. Breeding individuals weighed 35 grams, with 0.4 g ovaries and 0 g fat (Street 1979).

Vipera berus is not aggressive and bites only when cornered or alarmed. Accidents are most common from relatively lethargic individuals being stepped on. Defense responses include a loud sustained hiss, especially evident in gravid females. It retreats from disturbance by preference then returns to the same basking spot after the irritant passes (Street 1979).

Climbing is mostly restricted to occasional forays into low bushes, but occasionally it can be found in bird nests several meters above the ground. The adder is a good swimmer, sometimes found in water. It swims on the surface, and although it can submerge, it rarely does so (Street 1979).

Reproduction and Development

Spring sloughing in males is synchronized and the first shed of the year begins the mating period. Females shed later, and the period is much more prolonged (Frommhold 1969). In Hungarian populations mating occurs the last week of April in the south and the second week of May in the north. Smith (1969) observed a late mating on 19 June and a male combat in early October. Sometimes a late autumn mating occurs. It is not known if autumn matings produce young. Males apparently follow female scent trails, sometimes traveling hundreds of meters a day. If the female is encountered and responds and flees, the male pursues. After mating the pair remain together for 1–2 days. Active courtship involves tongue flicks along the back and excited tail lashing. Biting does not occur during courtship. Males chase away rival males. Both male combat and mating involve side-by-side parallel "flowing" behavior, a combat stance yielding the dramatic "adder dance." The biggest male will always win the combat (Andrén and Nilson 1981; Andrén 1986). Challengers seldom or never win these combats. After the mating period, males leave the area to forage. Females hunt only after giving birth (Marian 1963).

In Sweden, vitellogenesis and preovulatory follicle formation start in autumn of nonreproductive years and continue to May of reproductive years. Ovulation occurs in early June. Copulation precedes ovulation, with sperm present in the female uteri for a month. Juveniles are born in early August. Ovary, fat body, and liver reach peak weights in spring of the reproductive year, then shrink. Fat body and liver reach their lowest weights at the end of the gestation period (Nilson 1981). These are clearly adaptive responses to short seasons and severe climate in Scandinavia. Females breeding every third year are known in Switzerland (Saint-Girons and Kramer 1963) and Finland (Viitanen 1967).

Clutch size is a linear function of body size. Young number 3–20 per brood, and are born from August to September or even October, rarely in July (Marian 1963). In Denmark, the first half of August is typical. In particularly cold summers birth may be delayed until the following spring (Smith 1969).

Young range in size from 14 to 23 cm (mean 17 cm) and shed the first or second day. The sex ratio is nearly 1:1 at birth and this equability is generally maintained. Young have been found with mothers after several days but no maternal care is apparent (Smith 1969). Neonates grow at rate of 10 cm/yr for 5 years, then growth slows. Maturity is reached in males in 3–4 years, females in the third summer (2 years old).

Molting occurs frequently during the active season, approximately every 4–6 weeks. The first slough is often in April, the last in August.

Average mortality based on recapture data and therefore likely overestimated is 24% per year and suggests 45% cohort survival in year 1, 20% in year 2 and 10% in year 3. They may reach 15–20 years of age (Frommhold 1969).

Occasionally, females can mate with multiple males, but multiple matings with different partners are rare (Stille et al. 1986, 1987; Andrén et al 1997). Females might copulate more than once with a single male. The copulatory plug of the contracted sphincter blocks the uteri in response to a renal sex secretion produced following insemination. This plug is believed to prevent multiple fertilization (Nilson and Andrén 1982). DNA fingerprinting of 12 litters of adders showed that in 83% of the cases only one male had sired the offspring (Andrén et al. 1997). In addition, the order of mating is important in reproductive success as first male will sire most number of young. There is a "first male advantage" in male combats (Höggren 1995). Long-term sperm storage, over one or several reproductive cycles, does not occur (Höggren and Tegelström 1996).

Bite and Venom

Epidemiology—Bites are relatively common because of the rapid rate of human expansion over much of this species range. Dogs, cattle, sheep, goats, and horses are frequent victims. In Great Britain, most bites occur from March through October (Walker 1945). In Sweden, *V. berus* bites are estimated at 1300 per year with about 12% of those requiring hospitalization (Gitter and de Vries 1967).

Yield—10–18 mg from individuals 48–62 cm (Minton 1974).

Toxicity—Venom toxicity is relatively low. LD_{50}s for mice were reported as 0.55 mg i.v., 0.80 mg i.p., and 6.45 mg/kg s.c. (Minton 1974). A minimum lethal dose for laboratory guinea pigs is approximately 40–67 mg/kg compared with *Daboia russelii* venom for the same test organism at 1.7 mg/kg. Common lizards die 30 seconds after bite (Smith 1969), however, Street (1979) observed a young adder bite a young lizard which took 4 hours to die.

Content—Active proteins are not well-known but are presumed similar to *V. aspis* (Boquet 1967a). Proteolytic enzymes are present at relatively high concentrations (Samel et al. 1987). L-amino acid oxidase activity is present but at low levels (Iwanaga and Suzuki 1979). Whole venom contains several proteins with phospholipase activity. Two anticoagulant protien complexes have been isolated that interfere with prothrombinase formation by blocking phospholipid active sites. These complexes are easily dissociated by antivenins, allowing normal clot formation (Boffa et al. 1972b, 1976, 1978). Other proteins identified are hyaluronidase, hemorrhagic factors, and coagulant accelerators and decelerators (Boquet 1967a).

Symptoms and Physiological Effects—In general, to humans bites from *V. berus* are not severe. Bite victims experience intense severe pain, swelling, and perspiration. Petechiae may form and there are large areas of discoloration, often remote from the bite site. Lymph nodes become swollen and blisters, usually filled with blood, appear after 48 hours. Despite the presence of anticoagulants, enhanced blood coagulation often occurs and may be dose-dependent. Coagulant activity occurs on fibrinogen, plasma, and prothrombin. Hemoconcentration (hemoglobin 113%, red blood cells 5.2 million) also has been reported. In severe cases, persistent vomiting and cardiovascular failure may occur (Gitter and de Vries 1967; Rosenfeld et al. 1967; Gonzalez 1991).

Vipera berus bite may initiate myocardial infarction from three possible sources: hyperhistaminemia, activation of bradykinogen, or abolition of the Jarisch-Bezold reflex (Brown and Dewar 1965). The effect is postulated to result from arterial thrombosis caused by severe hypotension. Recovery is variable and may take as long as a year.

Case Histories—A 22-year-old man was admitted to the hospital 30 minutes after a bite. There was severe abdominal pain, with vomiting and watery diarrhea. Local edema, lymphangitis with axillary lymphadenopathies, and ecchymosis were pronounced. At 30 hours, arterial pressure fell despite increasing doses of norepinephrine. Sinus tachycardia occurred followed hours later by pulmonary edema. The patient was given intravenous digoxin, chlorothiazide, norepinephrine, and hydrocortisone and began to improve. This critical period lasted almost 2 hours. An ECG taken during this time showed T-wave inversion with elevated ST in I and aV_L. Alternate-day ECGs taken over the next 2 weeks disclosed increasingly deep T-wave inversion with reduction in ST elevation.

A similarly abnormal ECG, lasting 2–4 weeks, was reported for a envenomated 14-year-old boy (Chadha et al. 1968).

A 69-year-old man with a 10-year prior history of myocardial infarction was bitten (Moore 1968). There was extreme abdominal pain and reflex vomiting. Symptoms were local necrosis, systemic epigastric tenderness, and a heart rate of 60 beats/min. Within a few minutes, diarrhea began. Electrocardiogram showed intermittent 2:1 second-degree heart block, which was reversed 6 hours later to a normal sinus rhythm, when gastrointestinal symptoms resolved. Electrocardiogram and cardiac enzymes were normal after that.

Treatment and Mortality—The Royal Society of Tropical Medicine and Hygiene consider bites by *V. berus* less dangerous than the antivenin used. A 30% incidence of sensitization using specific antivenin is reported. Of eight bites between 1950 and 1958, the only fatality was due to anaphylactic response to antivenin (Chapman 1967).

In England and Wales, only seven deaths were documented in 50 years up to 1945. From 1940 to 1948, two fatalities were reported (Walker 1945). In Scotland, two deaths were reported in 1941 and none from 1950 to 1958 (Morton 1960). In Sweden from 1915 to 1944, only 15 of 4736 (0.3%) patients died. In Denmark, seven fatalities were reported in a 50-year period (Gitter and de Vries 1967). No fatalities occurred in a number of bites to dogs, cats, cows, and goats (Prestt 1971).

Remarks

Adder populations throughout Europe are at risk from habitat destruction, and locally from hunting and collecting for captivity (Honegger 1978b). In Poland where nearly all reptiles and amphibians are rigidly protected, *V. berus* is explicitly not protected and, as a result, is at risk (Corbett 1989).

Adders are eaten by hedgehogs, polecats, wild pigs, foxes, badgers, and birds. In one case, a frog (*Rana esculenta*) regurgitated a baby adder (Frommhold 1969). Adders have been found in stomachs of pike and eels (Smith 1969).

Adders are sometimes difficult captives. Holzberger (1981) successfully maintained them in large (60 cm × 120 cm) terraria with branches for climbing and a high relative humidity maintained by a heat source under a water dish. Adults ate mice and juveniles frogs. On approximately 1 September, feeding was stopped and, after the animals voided, they were placed in hibernation terraria. They were gradually warmed in the spring and, when fully hydrated, they began feeding. Adults commonly refuse food for up to 4 weeks after hibernation.

The grass snake and smooth snake are immune to venom. Hedgehog immunity is claimed but is erroneous as hedgehogs are not usually fully struck and injected (Steward 1971).

Vipera darevskii
Vedmederja, Orlov and Tuniyev 1986

Known as Darevsky's viper.

Recognition
(Plate 14.7)

Head—Fairly distinct, sides flat and almost vertical; edge of snout raised into a low ridge, which is a little rounded anteriorly; eye fairly large; eye equal or slightly smaller than nasal shield; rostral barely or not visible from above; 1 or 2 apical shields in contact with rostral; normally 5 large plates on dorsum: squarish frontal, 2 parietals, 2 supraoculars, each usually separated from frontals by 1–4 small scales; large nasal contains nostril in shallow depression; 8–9 small scales bordering eye below supraoculars; supraoculars usually large and distinct; temporals smooth; 9–10 supralabials, numbers 4 and 5 (rarely 3 and 4) separated from eye by single row of small scales; 9–10 sublabials.

Body—Dorsals increase in width with proximity to ventrals, yielding outer rows twice as broad as those along midline; dorsals mostly 21 (occasionally 19), strongly keeled except weak or smooth keels on last row adjoining venter on each side; ventrals 134–140; anal single; paired subcaudals 29–35 in males, 25–30 in females.

Size — Adults reach up to 42.1 cm, but normally somewhat smaller; of 3 examined Armenian males the biggest was 25.8 cm, of 5 examined females the biggest was 42.1 cm (Orlov and Tuniyev 1990); tail terminates in distinct horny point.

Pattern — Head with light yellowish spots along the edges of the frontal, parietals and lower oculars, with yellowish temporals; dark streak from eye to neck, the head pattern is separated from the dark dorsal band along the back.

Body color is yellowish gray. A brown dorsal zigzag band runs along the back. Occasionally the band is fragmented into transverse spots. On each side is a row of inconspicuous spots merging into a light brown stripe. Belly is black, but marked with lighter contours of ventrals (Orlov and Tuniyev 1990; Nilson et al. 1995).

Taxonomy and Distribution

Described as *Vipera darevskii* by Vedmederja et al. (1986) from Armenia. Originally, this isolated Armenian population was descovered by Darevsky, who considered it to be an isolated part of the range of *Vipera kaznakowi dinniki* (=*V. dinniki*) (Darevsky 1956).

The total known range for this viper is southeastern Dzhavakhet Mountains (Mt. Legli) in Armenia, and possible adjacent mountain areas in Georgia (McDiarmid et al. 1999).

Habitat

Typical habitats are on elevations between 2500 m and 3000 m, and can be characterized as alpine meadows with local outcrops of black volcanic rocks and small moraines (Orlov and Tuniyev 1990; Nilson pers. obs.).

Food and Feeding

Nothing is known of foraging ecology in the field, but orthopterans and lizards (*Lacerta valentinii*) occur in the habitat (Nilson pers. obs.).

Behavior

It has the habit of staying under flat stones to thermoregulate. The microclimate under such stones is favorable also at the very high elevations where it occurs. The open ground is mostly very chilly and seems to be avoided by this viper (Nilson pers. obs.). Also a potential predator, *Coronella austriaca,* was observed under such stones.

Reproduction and Development

The mating period is from late April to May in the field, and birth occurs in late August to September. Clutch size is from 3 to 5 young in most cases. The newborn vipers have a mean length of 131 mm and a mean weight of 3.1 g. They reach sexual maturity at the age of 3 years, and the female has a biennial (every 2 years) reproductive cycle (Orlov and Tuniyev 1990)

Bite and Venom

No information is available.

Vipera dinniki
Nikolsky 1913

Known as Dinnik's viper, or the Caucasus subalpine viper.

Recognition
(Plate 14.8)

Head—Large and fairly distinct, somewhat triangular; sides flat and almost vertical; edge of snout commonly raised into a low ridge; cheeks are swollen and posterior part of head massive; eye fairly large; eye equal or slightly smaller than nasal shield; rostral barely or not visible from above; 1 or 2 apical shields in contact with rostral; normally 5 large plates on dorsum: squarish frontal, 2 parietals, 2 supraoculars, each usually separated from frontals by 1–4 small scales; large nasal contains nostril in shallow depression; 9–12 small scales bordering eye below supraoculars; supraoculars usually large and distinct; temporals smooth; 8–11 supralabials, numbers 4 and 5 (rarely 3 and 4) separated from eye by single row of small scales; 8–12 sublabials.

Body—Dorsals increase in width with proximity to ventrals, yielding outer rows twice as broad as those along midline; dorsals mostly 21 (occasionally 23), strongly keeled except weak or smooth keels on last row adjoining venter on each side; ventrals 126–141; anal single; paired subcaudals 31–37 in males, 18–30 in females.

Size—Adults reach up to 50 cm, but normally somewhat smaller; of 29 examined Russian males the biggest was 41.2 cm; of 20 examined females the biggest was 48.6 cm (Orlov and Tuniyev 1990); tail one-sixth to one-seventh of total length in males, one-eight to one-ninth of total length in females; tail terminates in distinct horny point.

Pattern—A very polymorphic species. Head dark; with dark streak from eye to neck, continuing as a longitudinal series of dark spots along flanks; the black head pattern is normally continuing into the dark dorsal band along the back, but mostly with a light triangular spot on neck, inside the black or dark pattern, occasionally the dark dorsal pattern is totally separated from the head pattern, or with some fragmented connection; upper labials light, occasionally with dark sutures; chin and throat blackish, often tinged with orange or red; iris of eye coppery red or reddish brown; black pupil is vertical in bright light.

Body color is grayish, silver grayish, or green grayish, occasionally with orange or yellowish colors; normally a longitudinal black wavy zigzag band along the back and a series of black dots along the lateral sides. Occasionally, specimens can be similar to *V. kaznakovi* in pattern with more yellowish colors, occasionally similar to *V. berus* or *V. lotievi* in color pattern. Some specimens have a series of narrow transverse bands along the back ("tigrina pattern"), or have a uniform grayish to brownish or blackish ground color covering all parts of the body except the head, sometimes unicolored with a darker narrow or broad vertebral stripe. Melanism is common in some populations. Belly is black (Orlov and Tuniyev 1990; Nilson et al. 1995).

Taxonomy and Distribution

Described as *Vipera berus dinniki* by Nikolsky (1913) from specimens from the upper parts of the small Laba River, north Caucasus, Russia, and Svanetia, Georgia. For a longer period of time these mountainous populations above the coniferous belt was included in *Vipera kaznakovi* as *V. k. dinniki*. In 1986 this taxon was separated as a full species (Vedmederja et al. 1986: Orlov and Tuniyev 1986). All the populations of *Vipera kaznakovi* and *V. dinniki* in Great Caucasus are to some extent different from each other, indicating a great isolation of local populations for longer periods of time (Nilson et al. 1994, 1994).

The distribution for *Vipera dinniki* is Great Caucasus in Russia and Georgia (high mountain basin of the Inguri River) and eastward to Azerbaijan.

Habitat

Typical habitats are on elevations between 1500 m and 3000 m, and can be characterized as upper forest belt, subalpine and alpine meadows, rocky outcroppings and montane moraines, often near water (Orlov and Tuniyev 1990; Nilson pers. obs.).

Food and Feeding

Diet varies with locality and prey availability and consists mainly of small mammals including voles (*Apodemys silvaticus, Microtus majori, Sicista caucasica,* fledgings of ground nesting birds, such as *Anthus spinoletta,* and lizards (*Darevskia alpina*) (Orlov and Tuniyev 1990). Young vipers eat orthopterans.

Behavior

No information is available.

Reproduction and Development

The mating period is from late April to May in the field, and birth occurs in late August to September. Clutch size is from 3 to 5 young in most cases. The newborn vipers have a mean length of 131 mm and a mean weight of 3.1 g. They reach sexual maturity at the age of 3 years, and the female has a biennial reproductive cycle (Orlov and Tuniyev 1990).

Bite and Venom

No information is available.

Vipera kaznakovi Nikolsky 1909

Known as the Caucasus viper, or Kaznakow's viper.

Recognition
(Plate 14.9)

Head—Large and fairly distinct, somewhat triangular; sides flat and almost vertical; edge of snout commonly raised into a low ridge; cheeks are swollen and posterior part of head very massive; eye fairly large; eye equal or slightly smaller than nasal shield; rostral barely or not visible from above; 1 or 2 apical shields in contact with rostral; normally 5 large plates on dorsum: squarish frontal, 2 parietals, 2 long supraoculars, each usually separated from frontals by 1–4 small scales; large nasal contains nostril in shallow depression; 7–12 small scales bordering eye below supraoculars; supraoculars usually large and distinct; temporals smooth (rarely weakly keeled); 8–10 supralabials, third and fourth the largest, numbers 4 and 5 (rarely 3 and 4) separated from eye by single row of small scales; 8–12 sublabials.

Body—Dorsals increase in width with proximity to ventrals, yielding outer rows twice as broad as those along midline; dorsals mostly 20 to 21 (ranging from 18 to 21), strongly keeled except weak or smooth keels on last row adjoining venter on each side; ventrals 124–143; anal single; paired subcaudals 31–40 in males, 22–32 in females.

Size—A thick-bodied snake; male generally shorter and slimmer than female; adults reach 65 to 70 cm, but normally smaller; of 23 examined Russian males the biggest was 47.5 cm, of 16 examined females the biggest was 60 cm (Orlov and Tuniyev 1990); tail one-sixth to one-seventh of total length in males, one-eight to one-ninth of total length in females; tail terminates in distinct horny point.

Pattern—Head dark; with dark streak from eye to neck, continuing as a longitudinal black line along flanks; the black head pattern continues directly into the dark dorsal band along the back; upper labials reddish or yellowish; chin and throat blackish, often tinged with orange or red; iris of eye coppery red or reddish brown; black pupil is vertical in bright light.

Body color is reddish, orange, or yellowish; pattern is normally a longitudinal black band along the back and black bands along the lateral sides, separated by the yellow to red color, which in turn appears as two longitudinal, dorsolateral bright-colored bands. The dorsal black band can meanwhile have a zigzag shape. Belly is black.

Melanism is common in the northern part of the range (Russia) and is often total (Nilson et al. 1995).

Taxonomy and Distribution

Described as *Vipera kaznakovi* by Nikolsky (1909) from a specimen from Tsebelda, Sukhumi, Abkhasia, Georgia. For a long period of time the mountainous populations above the coniferous belt originally described as *Vipera berus dinniki*, were included in *Vipera kaznakovi* as *V. k. dinniki*. However, in 1986 this taxon was separated as a full species (Vedmederja et al. 1986: Orlov and Tuniyev 1986), with the remaining populations, below the coniferous belt in Great Caucasus, as *V. kaznakovi*. All the populations of *Vipera kaznakovi* and *V. dinniki* in Great Caucasus are to some extent different from each other, indicating a great isolation of local populations for longer periods of time (Nilson et al. 1994, 1995).

Two new taxa have recently been described from this complex:

Vipera orlovi from Papai Mountain to the peak of Mount Bol'shoi Pseushkho in the northwestern part of the Greater Caucasus, and

Vipera magnifica from the Rocky (Skalistyi) range at the boundaries of Afonka Mountain and Malyi Bambak ridge at the northern slopes of Greater Caucasus (Tuniyev 2001). Additional research will demonstrate whether these taxa are full species or incipient species (subspecies).

The distribution for *Vipera kaznakovi* is from northeastern Turkey, through Georgia into Russia along the Black Sea coast.

Habitat

Typical habitats in the Russian section of the range are small meadows in the Colchian forests with subtropical climate. Suitable sites are located near standing water with rocks suitable for hibernation nearby. It often occurs in mountain wooded slopes, in the bottom of humid canyons and meadows adjacent to forests (Orlov and Tuniyev 1990).

In the Turkish section of its range it prefers fern meadows, inside or in close connection to subtropical deciduous forests. It is often basking on thick layers of dead ferns, with lots of hiding places (Nilson pers. obs.).

Food and Feeding

Diet varies with locality and prey availability and consists mainly of small mammals including voles (*Apodemys silvaticus, A. agrarius, Microtus majori, M. gud,* but it also takes shrews (*Sorex raddei*) and lizards (*Darevskia saxicola, D. dejurgini, D. praticola,* and *Lacerta agilis*) (Orlov and Tuniyev 1990). Young vipers eat orthopterans.

Behavior

No information is available.

Reproduction and Development

The mating period is from late March to April in the field, and birth occurs in late August. Clutch size is from 3 to 5 young in most cases. The newborn vipers have a mean length of 145 mm and a mean weight of 4.1 g. They reach sexual maturity at the age of 3 years, and the female has an annual reproductive cycle (Orlov and Tuniyev 1990).

Bite and Venom

No information is available.

Vipera nikolskii
Vedmederja, Grubant and Rudayeva 1986

Known as Nikolsky's adder, or forest-steppe viper.

Recognition
(Plate 14.10)

Head—Moderately large and distinct; sides flat and almost vertical; rostral slightly longer than wide; canthal edge of snout raised into a low ridge; eye large, equal to nasal shield in size; rostral barely visible from above; 2 apical shields in contact with rostral; often 5 large plates on dorsum: squarish frontal (longer than broad), 2 parietals (sometimes with a tiny scale between frontal and parietals), 2 long narrow supraoculars, each usually separated from frontals by 1–4 small scales; meanwhile plates on head, especially parietals, reduced or fragmented; large nasal contains nostril in shallow depression; 7–12 small scales bordering eye below supraoculars; supraoculars usually large and distinct; temporals smooth; 8–11 supralabials, fourth and fifth the largest and separated from eye by 1 or 2 rows of small scales; 10 sublabials.

Body—Dorsals increase in width with proximity to ventrals, yielding outer rows twice as broad as those along midline; dorsals 21 (rarely 20 or 23), strongly keeled except weak or smooth keels on last row adjoining venter on each side; ventrals

142–157 in males, 146–159 in females; anal single; 33–45 paired subcaudals on average 42.2 in males, 33.7 in females;

Size—A short thick-bodied snake; male generally shorter than female; adults reach 680 mm;

Pattern—Always melanistic as adult. Supralabials may have white spots.
Juveniles are born with dark zigzag pattern along the back, but overproduction of melanin covers the juvenile pattern during early growth.

Taxonomy and Distribution

A species described by Vedmederja, Grubant and Rudayeva (1986) from the vicinity of Uda River, Charkov, Ukraine. The type series consists of an adult female and her clutch of 16 juveniles. Due to its special color, ecology, and behavior *Vipera nikolskii* has been considered as a good species (e.g. Nilson and Andrén 1997b; Ananjeva et al. 1998). But it is genetically similar to *V. berus berus* (Joger et al. 1997), although differing from it in morphology (Franzen and Heckes 2000). *Vipera nikolskii* could either be looked upon as a subspecies of *V. berus,* or as full species.

The distribution is concentrated to the forest-steppe zone in the Charkov region of Ukraine.

Habitat

It is basically a lowland form of the Ukrainian forest steppe (Nilson and Andrén 1997b). The forest steppe has been described as a macromosaic of forests and meadows, with dense vegetation cover and lots of water. Large areas are submerged during periods of the year.

Food and Feeding

No information is available.

Behavior

No information is available.

Reproduction and Development

Studies in outdoor enclosures (Stettler 1993) gave the information that mating was early, in last half of March. Pregnancy period is 130 to 133 days and clutch size is 14 to 15 young. Together with the type series of 16 young, it indicates a rather big clutch size in this species. Mean individual size per cutch varied between 216 to 224 mm, and 4 to 4.5 g. The young become melanistic after the fourth or fifth slough.

Bite and Venom

No information is available.

Vipera pontica
Billing, Nilson and Sattler 1990

Known as the pontic adder.

Recognition
(Plate 14.11)

Head—Moderately large and distinct; sides flat and almost vertical; snout tip slightly but distinctively upturned; rostral longer than wide, covering the front part of the nasal "horn"; canthal edge of snout raised into a low ridge; eye large, equal to nasal shield in size; rostral not visible from above; 2 apical shields in contact with rostral; plates on top of head fragmented, a small frontal and divided parietals can be imagined, 2 big supraoculars; large nasal contains nostril in shallow depression; 9–12 small scales bordering eye below supraoculars; temporals smooth (rarely weakly keeled); 8–9 supralabials; 10–12 sublabials.

Body—Dorsals increase in width with proximity to ventrals, yielding outer rows twice as broad as those along midline; dorsals 21–23, strongly keeled except weak or smooth keels on last row adjoining venter on each side; ventrals 142–147 in males, anal single; paired subcaudals 32–36 in males.

Size—A small snake, probably much less than 60 cm. The largest type specimen measures 277 mm total length.

Pattern—Head with 2 dark brown oblique bands running from posterior end of parietal area to join the outer margin of the first element of the dorsal pattern; wide black bordered band from eye to corner of mouth; upper labials whitish with a few spots; black pupil is vertical in bright light.

Body color pattern is beige, lateral areas with a reddish tinge, a dark, brown and black bordered zigzag band with a total of 66–69 lateral windings on total body and tail; venter is red brown marbled with bigger black and white spots along posterior margins of ventrals; tail tip is yellowish green (Billing et al. 1990).

Taxonomy and Distribution

A rare species described by Billing et al. (1990) from Coruh Valley, Province Artvin, northeastern Turkey. The type series consists of two specimens, and up to now very few additional specimens have been recorded. In a recent study (Baran et al. 2001) an additional specimen of *V. pontica* from northwest of Artvin, near the Turkish-Georgian border was discussed.

The known distribution is restricted to northeastern corner of Turkish Anatolia and adjacent parts of Georgia. Baran et al. (2001) is giving Camlihemsin/Rize further west as locality for snakes similar to *V. pontica*. However, according to photos published and characters given these Camlihemsin specimens are all *V. barani* (also verified by a referred unpublished genetic study in Baran et al. 2001) (Baran et al. 1997; Baran and Atatür 1998; Franzen and Heckes 2000; Baran et al. 2001).

Habitat

It inhabits steep wooded mountain slopes with small rocky outcrops and stone piles (Billing et al. 1990).

Food and Feeding

No information is available.

Behavior

No information is available.

Reproduction and Development

No information is available.

Bite and Venom

No information is available.

Vipera seoanei Lataste 1879

Known as the Baskian viper.

Recognition
(Plates 14.12 and 14.13)

Head—Large and fairly distinct, somewhat triangular; sides flat and almost vertical; edge of snout commonly raised into a low ridge; cheeks are swollen and posterior part of head very massive; eye fairly large; eye equal or slightly smaller than nasal shield; rostral barely or not visible from above; 1 or 2 apical shields in contact with rostral; pronounced variation in head scalation, from 5 large plates on dorsum (frontal, 2 parietals, 2 supraoculars) to total fragmentation of all plates; large nasal contains nostril in shallow depression; 6–12 (normally 9–10) small scales bordering eye below supraoculars; supraoculars large and distinct; temporals smooth (rarely weakly keeled); 8–10 supralabials, numbers 4 and 5 separated from eye by single row of small scales (occasionally 2 rows); 9–13 (mostly 10–12) sublabials.

Body—Dorsals increase in width with proximity to ventrals, yielding outer rows twice as broad as those along midline; dorsals mostly 21, strongly keeled except weak or smooth keels on last row adjoining venter on each side; ventrals 129–148 in males; 132–150 in females; anal single; paired subcaudals 32–42 in males, 24–35 in females.

Size—Adults reach 75 cm, but normally smaller.

Pattern—*Vipera seoanei* is a highly polymorphic species, and four main types of color patterns can be distinguished.

A: A classic type of pattern resembilng *V. berus* by having a well-developed zigzag band along the back; the ground color is beige or light gray, the dorsal zigzag band is brown.

B: A bilineata kind of pattern, with the ground color expressed as two narrow, straight, dorsolateral longitudinal light lines along the body. To some extent this morph resembles *V. kaznakovi*.

C: A uniform brownish morph without pattern.

D: A morph with fragmented zigzag pattern, that can apear as transverse bands along the back. This morph is approaching a pattern meanwhile seen in *Vipera aspis* (=*V. s. cantabrica*).

In addition, melanistic morphs occur at higher altitudes in the range of *V. s. cantabrica* and unicolored *V. s. seoanei* (Cantabrian Mountains).

In some specimens, head with weak dark streak from eye to neck, continuing as a longitudinal black line along flanks; the head pattern is always separated from the dark dorsal band along the back; black pupil is vertical in bright light (Bea et al. 1984).

Taxonomy and Distribution

Described as *Vipera berus seoanei* by Lataste in 1879. Type locality is restricted to Cabanas, province Coruna, Spain. Two subspecies are recognized:

V. s. seoanei, which is very polymorphic including unicolored, bilineate, melanistic, and classic color morphs (pattern types A, B, C from above). The different pattern types are irregularly distributed within the range. Some populations contain only bilineate specimens and other only unicolored ones while still others are mixed (Nilson pers. obs.). Most populations are of the "classical" morph.

V. s. cantabrica, from the south central parts of the Cantabrian mountains. It is characterized by the type D pattern described above.

The distribution for *Vipera seoanei* is extreme southwestern France and northern regions of Spain and Portugal.

Habitat

The species is distributed from sea level to 1500 m altitude and prefers open glades in moist deciduous forests, forest edges, hedges, rocky slopes, stonewalls (Nilson pers. obs.).

Food and Feeding

Diet varies with locality and prey availability and consists mainly of small mammals including voles (*Microtus lusitanicus, M. arvalis, Apodemys sylvaticus*) but it also takes lizards (*Lacerta monticola, Podarcis muralis*). According to Brana et al. (1988) mammals constitute of 72.0% of the prey items, birds 5.1%, amphibians 9.3% and reptiles 13.6%.

Behavior

No information is available.

Reproduction and Development

Sperm production (spermiogenesis) takes place during fall (during spring in *V. berus*) and the mating season is from the end of March to early May. No fall mating activities have been observed. The first slough of the season is at the end of May, and after the mating period (in *V. berus*, it triggers the mating activities). The female reproductive cycle is running over two years (biennial). Clutch size is from 3 to 10, and the young are born in August or September (Saint-Girons and Duguy 1976).

Bite and Venom

The venom of *V. seoanei* is completely neutralized by different anti-venomous immune sera against European vipers. Its toxicity, lower than *V. berus berus* and *V. aspis zinnikeri* venoms, is close to *V. a. aspis* venom (Detrait et al. 1982).

Subgenus *Acridophaga* Reuss 1927

The name *Acridophaga* was originally set by Reuss (1927) for the small meadow vipers that had the characteristic of feeding on orthopterans. The members of the group are characterized by having a single apical plate in contact with rostral, upper preocular in contact with nasal, a more or less developed ocellated parietal spot, feeding mainly on insects and inhabiting dry meadows. Like *Pelias* they are characterized by larger and normally irregularly fragmented head plates, and a spring mating period.

Two main lineages can be defined within the smaller Eurasian vipers, based on molecular studies (Herrmann and Joger 1992, 1997): the *ursinii* group (=subgenus *Acridophaga*), and the remainig *Pelias*. *Acridophaga* consists of *V. anatolica*, *V. ebneri*, *V. eriwanensis*, *V. lotievi*, *V. renardi* and *V. ursinii*. Immunological studies based on albumins indicate that the *Vipera ursinii* complex was separated, according to the Biological Clock hypothesis, from the base of the *Pelias* lineage during the periods of grassland dispersals at the upper Miocene, about 10 million years ago (Nilson and Andrén 2001; Nilson 2002).

Immunological studies of the different taxa within the *Vipera ursinii* complex show differences in the albumin serum profile that are similar to differences between full viper species. The altogether 12 different species and subspecies within the complex have branched off during the Pliocene and Pleistocene and correspond to series of sibling species rather than subspecies (Nilson and Andrén 2001).

The range for the complex is spotty and discontinuous in Europe and western Asia, and more integral over portions of provincial China. Reported from France, Italy, Austria, Hungary, Slovenia, Croatia, Bosnia-Herzegovina, Serbia, Montenegro, Macedonia, Albania, Romania, Bulgaria, Greece, Turkey, Iran, Armenia, Ukraine, Russia, Azerbaijan, Kazakhstan, Kirgiziya, and China (Nilson and Andrén 2001).

The *Vipera ursinii* complex (*Acridophaga*) has a complex taxonomic history. Originally, Boulenger (1913b) defined a "typical" form from France, Italy, and Hungary, and a Balkan form identified as *V. u. macrops*.

Hellmich (1962) listed three subspecies: *V. u. macrops*, with distinctively large eyes (from what was then Yugoslavia, northern Albania, Romania, Macedonia, and parts of Bulgaria), *V. u. renardi* with a wavy band on ash-gray, yellow-brown, or olive gray backgrounds and a tendency for the pattern to break into squares or diamonds (from the Danube delta, Ukraine, Caucasus, and "Bessarabia"), and the typical form *V. u. ursinii* from France, Italy, and parts of eastern Europe.

A *Vipera ursinii* complex description prior to the current concept was a five subspecies approach provided by Steward (1971). In this scheme, *V. u. ursinii* was assigned to mountainous populations of Italy, based on the type specimen collected in the Abruzzi. This subspecies has a relatively long narrow head with small eyes, white throat, and white lower labials sometimes edged with black. There are 19 scale rows at midbody, and the dorsum has a gray to yellowish white ground color with a dark brown stripe. This tends to separate into rounded spots anteriorly, each with a narrow white border. Dark "flanks" are reduced to short streaks of color between scale rows 5 and 6. Ventrals lightly "powdered" with black central spots and edges have 4–7 large spots that coalesce into lines on the posterior third of the body. Tail tip venter is pale yellow. The second subspecies in Steward's list, *V. u. macrops*, is from western and southern areas of the former Yugoslavia, Bosnia, Macedonia, Krk Island near Istria, and northern Albania. This form has a short head, blunt snout, large eyes, and narrow scales between eyes and supralabials. Throat and sublabials edged with black. Nineteen dorsal scale rows, and dorsal stripe is variable and irregular, frequently broken into blotches or saddlelike crossbars. The venter may be dark or light in ground color. A third subspecies, *rakosiensis*, is larger than those described above and occurs in the lowland Danube valley from Vienna to the Black Sea, and on the plains of Hungary, Slavonia (Croatia), southern Romania, and northern Bulgaria. This population has 19 or (rarely) 21 dorsal scale rows, 120–142 ventrals, and 20–37 paired subcaudals. The ground color is light gray or light brown, darker on the sides than on top, particularly posteriorly. Exceptional males are described as "golden yellow or greenish gray." The dorsal stripe is always dark-edged with distinctive smaller side spots. The throat is white, the tail tip dark, and the undersides are marbled dark gray and white. A fourth subspecies, *renardi*, is the largest of all and the brightest colored. This form occurs from the Danube delta to southern Russia, along the Black and Caspian Seas to Turkmenistan. There are 21 scale rows, up to 152 ventrals, and 15–38 subcaudals. The ground color is gray, yellow-gray, or yellow-brown and the "flanks" are suffused with black. The dorsal stripe is wavy, blotches (if present) are angular and often joined by a thin black line. The final subspecies in Steward's scheme was *wettsteini* from the Basses-Alps of southern France. This form has large eyes with a light dorsal ground color and little darkening of the lower dorsum. Dark spots are reduced to obscure streaks of color and the venter is light in adults with a distinctive rosy tinge on the posterior.

A four-subspecies concept was provided by Welch (1983), who listed *V. u. ursinii*, *V. u. ebneri*, *V. u. rakosiensis*, and *V. u. renardi*. Street (1979) included one additional form, listing *V. u. wettsteini* (France), *V. u. ursinii* (Italy), *V.u. macrops* (the former Yugoslavia), *V. u. renardi* (Romania, Ukraine to central Russia, Greece, northeastern Turkey, northwestern Iran, China and Mongolia), and *V. u. anatolica* (southwestern Turkey).

Joger (1984) discussed three complicated subspecies. One, *V. u. anatolica*, was confined to isolated populations in southwestern Turkey. Distinguishing characteristics include a white-tipped tail, 19 scale rows at midbody, and fewer than 120 ventrals. Coloration is reddish brown with a brown stripe ending at the neck and 2 brown bands from the top of the head to above the angle of the mouth. A second subspecies is *V. u. eriwanensis*, described from Yerevan, Armenia. This taxon has meanwhile appeared under the name *V. u. ebneri*, described from northern Iran (Knoepffler and Sochurek 1955; Saint-Girons 1978). The taxon *eriwanensis* (in the sense of Joger) is fragmented into three isolated populations which Joger (1984) thought distinguishable in their own right: one from the Transcaucasus and adjacent Turkey, one from the Talysh and Alborz Mountains, and one from eastern Uzbekistan and the Naryn River area. This form has 21 scale rows, 120–142 ventrals, and light overall coloration. The third subspecies was *V. u. renardi*, which Joger (1984) thought might be a distinct species incorporating *V. u. eriwanensis* as subspecies. This form has 21 scale rows, a high ventral count (142–152), and is pale gray with a dark brown stripe and rhomboidal patches. The venter is grayish with black spots. This subspecies occurs in the northern Black Sea region and eastwards.

One source (Gruber 1989) listed nine subspecies. These included *V. u. ursinii* (central Italy), *V. u. anatolica* (southwestern Turkey), *V. u. ebneri* (Transcaucasus to central Asia), *V. u. eriwanensis* (Armenia and northeastern Turkey), *V. u. graeca* (central Greece, Nilson and Andrén 1988a), *V. u. macrops* (the former Yugoslavia and Albania), *V. u. rakosiensis* (Hungary, Austria, Bulgaria), *V. u. renardi* (the Don River region, Kazhakstan, and Kirgiziya), and *V. u. wettsteini* (southern France).

Golay et al. (1993) catalogs four subspecies in the complex: *V. u. ursinii* (incorporating *macrops*)—southeastern France, central Italy, Slovenia, Bosnia-Herzegovina,

Serbia-Montenegro, Macedonia, northeastern Albania, northwestern Bulgaria, northern Greece, and southwestern Turkey; *V. u. eriwanensis* (incorporating some populations formerly grouped with *renardi*)—northwestern Iran, northeastern Turkey, and Armenia; *V. u. rakosiensis*—eastern Austria, central and northwestern Hungary, western Romania, and northern Bulgaria; and *V. u. renardi*—northeastern Romania, Ukraine, southern Russia, Kazakhstan, Kirgiziya, and China. Chinese populations are found in western Xinjiang in relatively continuous distribution.

In a series of recent publications additional taxa within this complex have been described (*Vipera u. moldavica, Vipera lotievi*) or raised from subspecies to species status (*Vipera anatolica, Vipera ebneri, Vipera eriwanensis*) in Eastern Europe and western Asia (Joger et al. 1992; Nilson et al. 1993, 1994, 1995; Höggren et al. 1993). The most recent revision, by Nilson and Andrén (2001), includes a systematic reorganization and the description of new taxa as subspecies of *Vipera renardi* in the eastern section of the range. These taxa are *Vipera r. parursinii* and *V. r. tienshanica*.

Vipera anatolica Eiselt and Baran 1970

Anatolian mountain steppe viper.

Recognition

Head—Oval, not clearly distinct from the neck; snout concave on dorsal side in half of the specimens; canthus raised; crown with 3–5 large scales: frontal, 2 parietals, 2 supraoculars; other head scales small; single central scale above rostral (vs. 2 *Pelias*); rostral deeper than broad, visible viewed from above; 8–10 circumorbital scales; a single scale row between eye and supralabials, which number 7–9 (average 8); fourth supralabials aligned directly below eye; 3–4 sublabials in direct contact with single pair of chin shields; nasal single, usually in contact with upper preocular (Nilson and Andrén 2001).

Body—Body moderately slender. Dorsal scales 19 at midbody, highly keeled, with exception of outer row on either side; scales "wavy" in cross-section, ventrals 114–124, with no sexual dimorphism. Subcaudals in females 19–23; anal entire.

Size—A small size viper with maximum length 43 cm; one-tenth total length is the tail (one-seventh to one-eighth for males, one-nineth to one-twelfth for females); females generally larger than males (Nilson and Andrén 2001).

Pattern—Dark labial sutures mostly absent; occipital and postorbital stripes on dorsal and lateral sides of head. Overall body is ash-gray or olive gray, with a dark zigzag dorsal stripe with rounded corners of windings; lateral sides of body with blotches or spots. Belly is white.

Taxonomy and Distribution

Described by Eiselt and Baran (1970) as *Vipera ursinii anatolica*. The type locality is between Ciglikara Ormani, Kuhu Dagh Mountains, province Antalya, south Turkey.

Based on the genetic distances to all other members of the *ursinii* complex it was considered to be a good species by Joger et al. (1992).

The main range is the Kuhu Dagh Mountains, south Turkey (Nilson and Andrén 2001).

Habitat

The habitat is small grass meadows on limestone ground (dolines). These dolines are situated in a mixed Cedrus-Juniperus forest (Eiselt and Baran 1970; Nilson pers. obs.; Nilson and Andrén 2001).

Food and Feeding

No information is available.

Behavior

No information is available.

Reproduction and Development

No information is available.

Bite and Venom

Treatment and Mortality—Antivenin for *V. ursinii* (=*V. ebneri*) is neither available nor necessary. Nonspecific treatment is effective for the relatively mild bite of this species. No human fatalities are reported associated with bites of this species.

Vipera ebneri
Knoppfler and Sochurek 1955
Iranian mountain steppe viper.

Recognition

Head—Oval, not clearly distinct from the neck; snout concave on dorsal side in half of the specimens; canthus raised; crown with 3–5 large scales: frontal, 2 parietals, 2 supraoculars; other head scales small; single central scale above rostral (vs. 2 *Pelias*);

rostral deeper than broad, visible viewed from above; 8–10 circumorbital scales; a single scale row between eye and supralabials, which number 8–10 (average 9); fourth supralabials aligned directly below eye; 3–4 sublabials in direct contact with single pair of chin shields; nasal single, usually in contact with upper preocular (Nilson and Andrén 2001).

Body—Body moderately slender. Dorsal scales 21 at midbody, highly keeled, with exception of outer row on either side; scales "wavy" in cross-section, ventrals 123–134, with no sexual dimorphism. Dimorphism distinctive in subcaudals, males 23–34, females 19–25; anal entire.

Size—A small size viper with maximum length 44 cm; one-tenth total length is the tail (one-seventh to one-eighth for males, one-nineth to one-twelfth for females); females generally larger than males (Nilson and Andrén 2001).

Pattern—No dark labial sutures; occipital and postorbital stripes on dorsal and lateral sides of head. Overall body is ash-gray, or olive gray, with a dark zigzag dorsal stripe with rounded corners of windings; lateral sides of body light with weakly developed blotches or spots. There is very little sexual dimorphism. Belly is white.

Taxonomy and Distribution

Described by Knoppfler and Sochurek (1955) as *Vipera ursinii ebneri*. The type locality is between Rhema and Demarvand, Elburz Mountains, North Iran. For a period it was treated as *Vipera ursinii eriwanensis,* when all transcaucasian and Iranian populations were treated as a single taxon. The name *eriwanensis* has priority over *ebneri*. However, all investigated transcaucasian taxa are genetically separated from *Vipera ursinii* and *Vipera renardi* (Nilson et al. 1994), and recently the population in Iran, south of the Aras river, has been considered as a separate species carrying the name *Vipera ebneri* (Nilson and Andrén 2001).

The main range is the Elburz Mountains in Iran, but it also occurs in the Talyish Mountains in northern Iran and southern Azerbaijan (Nilson and Andrén 2001).

Habitat

The habitat is rural plains, alpine mountains, grasslands, and rock beds (Latifi 1991). In the Lar Valley the habitat can be characterized as alpine steppe (pers. obs; Nilson and Andrén 2001). It seems to prefer altitudes around 2700 m (type locality altitude, as well as own observations in northern Iran).

Food and Feeding

The diet of *V. ursinii* (=*V. ebneri*) is largely comprised of insects, lizards, and small mammals (Latifi 1991). *Darevskia raddei* and *D. defilippi* are abundant at known sites.

Behavior

No information is available.

Reproduction and Development

No information is available.

Bite and Venom

Yield—The yield is relatively low (Latifi 1991).

Toxicity—The LD_{50} for *V. ursinii* (=*V. ebneri*) venom is 21.7 "gammagrams" (Latifi 1991).

Treatment and Mortality—Antivenin for *V. ursinii* (=*V. ebneri*) is neither available nor necessary. Nonspecific treatment is effective for the relatively mild bite of this species (Latifi 1991).

No human fatalities are reported associated with bites of this species.

Vipera eriwanensis (Reuss 1933)

Called the Armenian mountain steppe viper.

Recognition
(Plate 14.14)

Head—Oval, not clearly distinct from the neck; snout concave on dorsal side in half of the specimens; canthus raised; crown with 3–5 large scales: frontal, 2 parietals, 2 supraoculars; other head scales small; single central scale above rostral (vs. 2 *Pelias*); rostral deeper than broad, visible viewed from above; 8–11 circumorbital scales; a single scale row between eye and supralabials, which number 8–10 (average 9); fourth supralabials aligned directly below eye; 3–4 sublabials in direct contact with single pair of chin shields; nasal single, usually not in contact with upper preocular (contact only in 22% of the snakes) (Nilson and Andrén 2001).

Body—Body moderately slender. Dorsal scales 21 at midbody, highly keeled, with exception of outer row on either side; scales "wavy" in cross-section, ventrals 133–143, with no sexual dimorphism. Dimorphism distinctive in subcaudals, males 32–39, females 23–30; anal entire.

Size—A medium-sized viper with maximum length 50 cm; one-tenth total length is the tail (one-seventh to one-eighth for males, one-nineth to one-twelfth for females); females generally larger than males (Nilson and Andrén 2001).

Pattern—Dark labial sutures in about half of the specimens; occipital and postorbital stripes on dorsal and lateral sides of head. Overall body is ash-gray, or olive gray, with a dark zigzag dorsal stripe with rounded corners of windings; lateral sides of body light with weakly developed blotches or spots. There is very little sexual dimorphism. Belly is white.

Taxonomy and Distribution

Described by Reuss (1933) as *Acridophaga renardi eriwanensis*. The type locality is Erivan (at 2000 m altitude), Armenia. For some time it was treated as *Vipera ursinii ebneri,* and later as *Vipera ursinii eriwanensis.* Investigated Armenian and Turkish vipers are genetically separated from *Vipera ursinii* and *Vipera renardi* (Nilson et al. 1994), and the populations in eastern Turkey, Armenia, and adjacent Azerbaijan, north of the Aras River, has been considered as a separate species carrying the name *Vipera eriwanensis* (Höggren et al. 1993).

The range covers Lesser Caucasus in Armenia and western Azerbaijan, as well as eastern Turkey (Nilson and Andrén 2001).

Habitat

In Armenia the habitat includes dry high mountain meadows. In Turkey it occurs in hilly grassland, occasionally with some rocky outcrops. It always prefers microhabitat with dry grass tussocks (Nilson pers. obs; Nilson and Andrén 2001).

In the Lesser Caucasus, according to Bozhanskii and Kudryavtsev (1986) five species of vipers occur in a relatively restricted area. Within this area (=Armenia), *V. ursinii* (=*V. eriwanensis*) is rigidly restricted to a very narrow altitudinal band of mossy, dense oak thickets 3–5 m high, with well-developed coppiced shoots, and a surface litter layer 20–30 cm deep. *Vipera eriwanensis* was never seen beyond this belt where other viper species are commonly observed. Since members of the *V. ursinii* complex have relatively catholic taste in habitats elsewhere in its great geographic range, habitat partitioning may be inferred for the Caucasus viper community.

However, in the investigated area of Armenia, *V. eriwanensis* is sympatric with *Macrovipera lebetina obtusa* and *Vipera r. raddei* only. Personal observations (Nilson) at the same locality also showed specimens of *V. eriwanensis* on dry grassy-rocky habitats in the juniperus belt. *Vipera raddei* prefers rocky habitat and *M. l. obtusa* is found at lower elevations. There is no competition between these species. *Vipera eriwanensis* forages on insects, *V. raddei* on small rodents, and *M. l. obtusa* on large rodents.

Food and Feeding

The diet of *V. eriwanensis* largely comprises insects, lizards, and small mammals. In eastern Turkey *Darevskia valentinii* was seen preyed upon (Nilson pers. obs.).

Behavior

No information is available.

Reproduction and Development

No information is available.

Bite and Venom

No information is available.

Vipera lotievi
Nilson, Tuniyev, Orlov, Höggren and Andrén 1995

Called the Caucasian meadow viper.

Recognition
(Plate 14.15)

Head—Oval, not clearly distinct from the neck; snout concave on dorsal side; canthus raised; crown with 3–5 large scales: frontal, 2 parietals, 2 supraoculars (parietals fragmented in some subspecies); other head scales small; single central scale (canthal) above rostral (vs. 2 in *Pelias*); rostral as deep as broad, visible viewed from above; 7–11 (usually 9–10 in males, 8–9 in females) circumorbital scales; a single scale row between eye and supralabials, which number 8–9; fourth supralabials aligned directly below eye; 3–4 sublabials in direct contact with single pair of chin shields; nasal single, usually in contact with upper preocular (Nilson and Andrén 2001).

Body—Body moderately slender. Dorsal scales 21 at midbody, highly keeled, with exception of outer row on either side; scales "wavy" in cross-section, ventrals 137–146, with no sexual dimorphism. Dimorphism distinctive in subcaudals, males 33–38, females 23–27; anal entire.

Size—A medium-sized viper with maximum length 60 cm; one-tenth total length is the tail (one-seventh to one-eighth for males, one-nineth to one-twelfth for females); females generally larger than males (Nilson and Andrén 2001).

Pattern—No labial sutures; occipital and weak postorbital stripes on dorsal and lateral sides of head. Overall body is light brown, with a dark zigzag dorsal stripe with rounded corners of windings or pronounced bilineate type of pattern; some specimens are unicolored bronze; lateral sides of body light with pronounced blotches or spots (except in unicolored specimens). There is very little sexual dimorphism. Belly is white.

Taxonomy and Distribution

Investigated Caucasian vipers are genetically separated from *Vipera ursinii* and *Vipera renardi* (Nilson et al. 1994), and the population in Caucasus has been described as a separate species carrying the name *Vipera lotiev* (Nilson et al. 1994,

1995). The type locality is Armkhi, Chechnya, below Mt. Stolovaya in northern Caucasus, at 2000 m altitude, Russia.

The range covers the northern slopes and main range of Greater Caucasus. Altitudinal span from 1200 m to 2700 m. (Nilson et al. 1995; Nilson and Andrén 2001).

Habitat

Habitat range includes oreoxerophytic landscapes with semiarid forests with shrubs and brushwood (shibliak), and thorny and summer deciduous vegetation (phrygana), which are similar to east Mediterranean types of vegetation. At higher elevations *V. lotievi* reaches the subalpine mountain belt.

Food and Feeding

The diet of *V. lotievi* largely comprises insects, lizards, and small mammals.

Behavior

No information is available.

Reproduction and Development

No information is available.

Bite and Venom

No information is available.

Vipera renardi (Christoph 1861)

Commonly called the steppe viper.

Recognition

(Plate 14.16)

Head—Oval, not clearly distinct from the neck; snout concave on dorsal side; canthus raised; crown with 3–5 large scales: frontal, 2 parietals, 2 supraoculars (parietals fragmented in some subspecies); other head scales small; single central scale above rostral (vs. 2 in *V. berus*); rostral as deep as broad, visible viewed from above; 8–11 (usually 9–10) circumorbital scales; a single scale row between eye and supralabials, which number 8–10 (average 9); fourth supralabials aligned directly below eye; 3–4 sublabials in direct contact with single pair of chin shields; nasal single, usually in contact with upper preocular (Nilson and Andrén 2001).

Body—Body moderately slender. Dorsal scales 21 at midbody, highly keeled, with exception of outer row on either side; scales "wavy" in cross-section, ventrals 129–151, with some discernible sexual dimorphism (males 129–148, females 134–151). Dimorphism distinctive in subcaudals, males 28–38, females 23–30; anal entire.

Size—A medium-sized viper with maximum length 70 cm; one-tenth total length is the tail (one-seventh to one-eighth for males, one-nineth to one-twelfth for females); females generally larger than males (Nilson and Andrén 2001).

Pattern—Labial sutures often with dark broad pattern; broad occipital and postorbital stripes on lateral and dorsal sides of head. Overall body is ash-gray, yellow-brown, or olive gray, with a dark zigzag dorsal stripe with rounded corners of windings; stripe often broken into spots or saddles which may be oval, elliptic, or rhomboidal; dark pattern areas often edged with a 1–2-scale thick black line; lateral sides of body light with pronounced blotches or spots. There is very little sexual dimorphism.

Melanistic forms have been found in the northern Caucasus (Ostrovskikh 1997).

Taxonomy and Distribution

Described by Christoph (1835) as *Pelias renardii*. The type locality is Sarepta, Lower Volga, Russia. For a long period of time it was treated as a subspecies of *Vipera ursinii,* but is today considered as a good species with a number of subspecies (Nilson and Andrén 2001).

The range covers eastern Europe and western Asia, and includes parts of northwestern China (Xinjiang). Reported from Ukraine, Russia, Azerbaijan, Kazakhstan, Kyrgyzstan, and China (Nilson and Andrén 2001).

Recent analyses of larger series of specimens have resulted in descriptions of subspecies in the eastern range of the distribution (Nilson and Andrén 2001):

Vipera r. renardi occurs in a western, Ukrainian-Russian morph and an eastern Kazakh morph.

Vipera r. tienshanica occurs in mountain areas of Kirgyzstan and adjacent areas of Kazakhstan and Uzbekistan, and

Vipera r. parursinii is distributed in Xinjiang in China. This last subspecies is characterized by having only 19 dorsal scalerows on midbody. All other *V. renardi* populations have 21 dorsal scale rows.

Habitat

Habitat range includes dry plains, flatlands with few trees or bushes, moist grassland, and open areas near dry clay or loamy soil (U.S. Navy 1991; Nilson and Andrén 2001). It occurs in hilly grassland, occasionally with some rocky outcrops and near water. However, it always seems to prefer microhabitat with dry grass tussocks (Nilson pers. obs.).

Food and Feeding

The diet of *V. renardi* largely comprises insects, lizards, and small mammals. However, there may be strong seasonal shifts in diet among vertebrate and invertebrate prey. In the Ukraine, invertebrates dominated the diet by mass and number (Kotenko 1989).

Lizards are eaten in proportion to availability in local environments. Rodents are eaten relative to snake size. In many cases, nestling rodents are consumed but adults are not a major dietary component.

The narrow feeding niche breadth and seasonal dietary shifts are clearly related to prey availability and the need to maximize energy resources over a short active season. Similar phenomena occur in Black Sea island populations, which feed largely on nestling waterfowl and sandpipers when these are available early in the year, shifting to lizards and small mammals later in summer. Immatures actively feed only during the last half of the warm season. Females feed during gestation. Feeding continues even during molts but at a slower rate.

Behavior

No information is available.

Reproduction and Development

No information is available.

Bite and Venom

No information is available.

Vipera ursinii
(Bonaparte 1835)

Commonly called the meadow viper or Ursini's viper.

Recognition
(Plates 14.17 and 14.18)

Head—Oval, not clearly distinct from the neck; snout flat or varying by subspecies; canthus slightly raised; eye with vertical pupil, in most cases smaller than nasal shield, the horizontal diameter of which is less than or equal to distance from eye to posterior of nostril; vertical diameter of eye less than or equal to minimum distance from eye to mouth; crown with 5 large scales: frontal, 2 parietals, 2 supraoculars; frontals elongated, their length usually greater than parietals; other head scales small; single central apical scale above rostral (vs. 2 in *V. berus*); rostral as deep as broad, visible viewed from above; 8–9 circumorbital scales; a single scale row between eye and supralabials, which number 6–8; third or third and fourth supralabials aligned directly below eye; 3–4 sublabials in direct contact with single pair of chin shields; nasal single, usually in contact with upper preocular.

Body—Body moderately slender (Steward 1971). Dorsal scales mostly 19 at midbody, highly keeled, with exception of outer row on either side; scales "wavy" in cross-section, relatively short, so dark underlying skin shows through (these characteristics give a "rough" surface appearance, a distinctive field mark (Arnold and Burton 1978), and the last one is especially pronounced in the subspecies *V. u. rakosiensis,* ventrals 120–145, with no sexual dimorphism (males 121–144, females 120–145) (Nilson and Andrén 2001). Dimorphism distinctive in subcaudals, males 27–41, females 20–32 (exception is *V. u. graeca,* where males have 20–27 subcaudals, females 18–21); anal entire.

Size—Smallest European viper, length 63–80 cm (the latter figure from Gruber 1989); adults average 40–50 cm; one-tenth total length is the tail (one-seventh to one-eighth for males, one-nineth to one-twelfth for females); females generally larger than males (Bruno 1985).

Pattern—General body patterning, head ornamentation, and venter appearance varies among populations. However, the relative uniformity of overall coloration and degree of variability within populations make collection location a good help as indicator of population affinity (Kramer 1961b; Hellmich 1962; Steward 1971; Bruno 1985).

As a group, *V. ursinii* pattern and colors are subdued and generally less variable than other wide-ranging viper species (Arnold and Burton 1978). This uniformity within the entire complex (*Acridophaga*) is surprising in light of the enormous geographic range (from western Europe to central Asia).

Subspecies and local populations vary in pattern with some consistency. Head ornamentation is variable, ranging from paired oblique slashes behind eyes to elaborate figures or masses lighter or darker than ground color; head sometimes with dark spots haphazardly distributed over and around basic pattern; often with oblique band from eye to angle of mouth, which may be confluent with dorsal "slash" marks; rostral and labial shields uniformly yellowish or whitish, sometimes with small dark spots or brown borders; chin and throat often yellowish white or matching venter in darker specimens.

Overall body is gray, pale brown, or yellowish with dark zigzag dorsal stripes; stripes often broken into spots or saddles which may be oval, elliptic, or rhomboidal; dark pattern areas often edged with a 1–2-scale thick black line. Flanks often dark gray or brown; sometimes 2–3 longitudinal series of brown or black spots running along the sides, with the lower row following the outer row of scales; venter highly variable, may be blackish, whitish, dark gray, or suffused with a rosy tint; with or without dark spots; sometimes with white spots on dark specimens, or checkered or clouded patterns on ventrals; tail tip may be dark with or without yellow markings and subcaudals may be darker than rest of underside.

Early authors (Boulenger 1913b) reported little or no sexual dimorphism in coloration. However, more recent work resulting from accumulated specimens across the geographic range suggests some consistent, although not great, distinction between the sexes. Females' tails sometimes tipped with yellow, males generally without. Males in some subspecies may have darker dorsal coloration than females. With these exceptions, variation among individuals and populations seems to override sexual differences.

Melanistic individuals are extremely rare, although occasionally reported in literature (e.g. Bruno 1985; Arnold and Burton 1978). One specimen from Bosnia had an unmarked venter and uniformly brown dorsum (Mehely 1911). However, individuals with unusual color patterns are found throughout the range (Janisch 1993).

Taxonomy and Distribution

Described by Bonaparte (1835) as *Pelias ursinii* on an unnumbered page in Folio 12 of his inventory of Italian fauna. The type locality was a mountainous area in the Abruzzi, near Ascoli. Modern taxonomic history began when it was renamed *V. ursinii* by Boulenger (1893). Bonaparte described the taxon on a single specimen which is the type (McDiarmid et al. 1999).

As ongoing Viperid taxonomic revisions are published and understanding improves, it is expected the *V. ursinii* complex will change. For example, one cladistic analysis—which suggests a polyphyletic nature of the Viperidae—also suggests that *V. ursinii* is taxonomically linked to some congeners (*V. aspis, V. ammodytes,* and *V. berus*) (Ashe and Marx 1988). However, more recent biochemical investigations have demonstrated that *Vipera ursinii* complex (*Acridophaga*) is related to the *berus* group (*Pelias*) only, at a basal position (Herrmann and Joger 1992, 1997).

Vipera ursinii is divided in five subspecies (Nilson and Andrén 2001) where:

V. u. ursinii occurs in Italy and France;

V. u. rakosiensis in Hungary (nowadays extinct in Austria and Romania);

V. u. macrops in Bosnia-Herzegovina, Croatia, Montenegro, Macedonia, and northern Albania;

V. u. moldavica in Romania, and perhaps in northern Bulgaria and Moldavia;

V. u. graeca in Greece.

Habitat

Vipera ursinii has narrow ecological scope. The principle habitat is dry grasslands. Habitat range includes dry plains, flatlands with few trees or bushes, montane grasslands, and open areas near dry clay or loamy soil (Nilson and Andrén 2001). In much of southern Europe, it is a montane snake, occurring at altitudes from 2000 to 3000 m on dry, well drained, rocky hillsides with plenty of grass tussocks, or in dry mountain meadows (Street 1979; Nilson and Andrén 2001). It is particularly fond of rocky limestone slopes and subalpine steppe, always with grassy vegetation, and on grassy ledges and hillsides overlooking mountain lake shores. In Austria, Hungary, and Romania, it lives in lowland meadows. In central Europe, the subspecies *V. u. rakosiensis* inhabits wide plains and low hills up to a few hundred m of altitude, dry hillocks in marshes, and well drained broad meadows. Populations living in drier meadows and pastures may be limited by cattle grazing.

In isolated mountains above 1000 m, populations are relatively sparse across the range of habitats. This is particularly true of *V. u. ursiniii* in Italy and southern France, where this snake historically has always been rare. Lowland populations in Austria and in meadow habitats in the Danube basin were once very high. In the late 1800s, the superintendent of the Imperial Castle near Luxemburg, Austria, offered a bounty for vipers, and paid out for 1000 specimens a year. Population densities have fallen from historical levels and is today extinct in Austria, but still frequently occurs in the Danube delta area (*V. u. moldavica*) where the habitat is dry, sandy islands (Boulenger 1913a; Honegger 1978a, 1981; Nilson and Andrén 2001).

Food and Feeding

The diet of *V. ursinii* largely comprises insects, lizards, and small mammals. However, there may be strong seasonal shifts in diet among vertebrate and invertebrate prey. Orthopteran insects are scarce during June, and early season feeding is dominated by vertebrates. Later in summer, as grasshopper populations increase, feeding shifts and vertebrates nearly disappear from the diet. The insectivorous nature of meadow vipers has long been known (Boulenger 1913b). At times, these snakes consistently refuse mice and lizards, eating only grasshoppers. In many cases, stomach contents contained only grasshoppers. One wild-caught specimen had a single mass in its gut containing over 100 grasshoppers (Mehely 1911). In France, excrement analysis revealed the ubiquity of an orthopteran diet (Dreux and Saint-Girons 1951). Dietary insects may also include many beetle species (Steward 1971). In France, invertebrates dominated the diet by mass and number. (Baron 1980, 1989). In feeding on insects, the venom apparatus is apparently not used (Steward 1971).

In marshes of Austria, the diet included lacertid lizards and mice (Boulenger 1913b). Lizards are often swallowed without waiting for venom to take effect, but larger lizards may be held in the mouth until struggling ceases (Steward 1971). Street (1979) pulled a lizard from a hole in Hungary to find that a meadow viper had engulfed its anterior end. The latter refused to release the lizard until both were placed in a collecting bag where the lizard exhibited clear signs of envenomation. The snake later consumed the lizard.

Lizards are eaten in proportion to availability in local environments. Rodents are eaten relative to snake size. In many cases, nestling rodents are consumed but adults are not a major dietary component.

Feeding behavior of *V. ursinii* is distinctive. Small prey may simply be swallowed, with or without being envenomated. Larger prey are envenomated, partially swallowed, and held until dead when swallowing proceeds. The venom, lacking potency and volume to be effective defensively, may be well suited to immobilizing small vertebrate prey items. Venom may also contribute to the rapid digestion of vertebrate prey as noted by the condition of individual gut content items (Agrimi and Luiselli 1992).

Recent field investigations have elucidated feeding in this species. Grasshoppers of all available species are consumed and the individual size of insects eaten is somewhat proportional to snake size. In larger snakes (longer than 28 cm), orthopteran insects and other invertebrates (including beetles and opilionids) dominate gut contents by frequency, comprising 84% of recorded prey items and 80% of biomass. Larger vipers feed only on grasshoppers greater than 15 mm in length. Lizards, small mammals, and ground-breeding nestling birds (Motacillidae) were eaten much less frequently, less than 20%, but represented a substantial biomass, nearly 60%, because of the greater mass per individual. Smaller snakes (less than 28 cm) fed nearly exclusively on invertebrates and small insects, with only 1 of 14 identifiable prey items representing a lizard (Agrimi and Luiselli 1992).

Feeding period for meadow vipers in mountain habitats is short, lasting from June through September. Few individuals feed early in the season but nearly 100% feed in July–September. Feeding in June and early July is irregular but individuals feed almost daily throughout the active season. Females feed at rates similar to those of males throughout the active season, with no apparent halt in feeding during gestation period. Males do not feed until the mating season ends in June (Agrimi and Luiselli 1992).

The narrow feeding niche breadth and seasonal dietary shifts in montane populations are clearly related to prey availability and the need to maximize energy resources over a very short active season. Immatures actively feed only during the last half of the warm season. Females feed during gestation. Feeding continues even during molts but at a slower rate. It is likely that the tendency to feed through physiologically stressful conditions (molting, gestation) is an adaptation to the very short (3–4 months) activity period available to montane populations (Baron 1989).

Meals are taken frequently, averaging every 2–4 days. Each meal is generally small, about 3%–7% of live weight (Baron 1989).

Behavior

Across its range, this species is normally diurnal but it may become nocturnal for relatively brief periods during hot weather (Steward 1971; U.S. Navy 1991).

As expected in a species exhibiting great geographical and altitudinal range, the annual activity cycle is variable and flexible. In general, individuals enter hibernation in October or November and emerge in early spring (March or April) depending on regional and local conditions. Males emerge first, as early as snow melt will allow (Steward 1971; Corbett 1989). In alpine regions of France, males emerge in mid-April coincident with snow melt, and females emerge the first half of May (Baron 1980).

Neither males nor females feed heavily the first few weeks after emergence, although whether this is due to low availability of insect prey and the necessity to feed on scarce vertebrate resources or to inherent aspects of the biological cycle is presently unclear (Baron 1980; Agrimi and Luiselli 1992).

Immatures become active only late in the year in montane populations (Baron 1980). Immatures emerge when the summer feeding period begins in June, months after the adult males, and feed heavily until the return to hibernacula in autumn.

In temperament, *V. ursinii* is consistently reported as placid and unaggressive, rarely biting unless seriously molested (Steward 1971; Arnold and Burton 1978). It is characterized as a gentle animal, even when handled (U.S. Navy 1991). According to Mehrtens (1987), "... they are commonly caught and carried about by children as 'pets'." Even early literature remarked on the relatively gentle nature of this species. Boulenger (1913b) reported it to be handled without biting and (echoed by Mehrtens) to be safely "carried about" by village boys. Boulenger (1913b) was impressed that no bites were known from Laxenburg despite the very high density of the population of this snake in the area.

In comparison to other European *Vipera*, *V. ursinii* is relatively active and agile (Steward 1971; U.S. Navy 1991).

Reproduction and Development

Little information is available on reproduction of this species because captive breeding is rarely reported (Steward 1971; Mattison 1982). Mating period is prolonged, running from emergence of females as early as April until well into June. Young are born July or August. In some populations births may occur as early as June or as late as September. Generally litter size is small, usually averaging 5–8 young, but sometimes 10–18 for larger and older females. Some individuals have reportedly produced 17–22 but this is exceptional. Young are approximately 140 cm total length at birth (Boulenger 1913b; Steward 1971).

Bite and Venom

Epidemiology—It is generally agreed the combination of small size, rarity, population isolation, gentle temperament, and weak venom make this species little threat to man. The U.S. Navy (1991) placed *V. ursinii* in the lowest threat category due to infrequent contact, small venom yield, and low toxicity.

Yield—The yield is relatively low.

Toxicity—No information is available.

Content—Proteolytic compounds have been isolated. The swelling of rat liver mitochondria from disruption of the electron transport chain and degraded mitochondrial structure suggests phospholipase is present. Phospholipase presence is also indicated because of its role in hemolytic activity by disrupting red blood cell membranes (McKay et al. 1970).

Vipera ursinii venom is distinctly yellow colored due to the presence of L-amino acid oxidases. This constituent is associated with tissue necrosis but is apparently present at low concentrations based on mildness of symptoms.

Other active components of *V. ursinii* venom include proteases, arginine ester hydrolases (associated with coagulant properties), and kinin-releasing compounds. Bradykinin is released as a result of autopharmacological action. Bradykinin released from tissues causes hypotension because of its powerful properties as a vasodilator and its role in increasing capillary wall permeability. However, this effect in *V. ursinii* is among the weakest in venoms demonstrating such activity (McKay et al. 1970).

Symptoms and Physiological Effects—Symptoms are generally mild. A major effect of *V. ursinii* envenomation is hemolytic disturbance caused by destruction of red blood cell membranes (McKay et al. 1970). Local pain and swelling occur in the bite area. There may be dizziness and nausea but recovery is rapid and complete (Steward 1971; U.S. Navy 1991).

Treatment and Mortality—Antivenin for *V. ursinii* is neither available nor necessary.

No human fatalities are reported associated with bites from this species.

Remarks

The degree of evolutionary adaptation represented by isolated montane populations of *V. ursinii* has been the subject of some debate in the literature. According to Nilson and Andrén (1987, 2001), montane subspecies of *V. ursinii* are the most derived populations within the *ursinii* complex (*Acridophaga*).

Subgenus *Montivipera*
Nilson, Tuniyev, Andrén, Orlov, Joger and Herrmann 1999

Subgenus *Montivipera* consists of the *Vipera xanthina* complex, and where the members are characterized by having the head covered by small scales with the exception of big supraocular plates. All species in this group are mountain-dwellers and prefer rocky habitats. Only *Vipera xanthina* is found both at sea level and in high mountains. Subgenus *Montivipera* has a reproductive pattern that is characterized by a spring mating, which is triggered by the spring molt of the males in the local population.

Vipera albicornuta
Nilson and Andrén 1985

Suitable names are the Iranian mountain viper, or zigzag mountain viper.

Recognition
(Plate 14.19)

Head—Relatively small and elongate, fairly distinct from neck; supraoculars raised and separated from eye by row of small scales; nostril centered in a large nasal shield which is partially fused with prenasals; loreal present between upper preocular and nasal; supraoculars separated by 7 scales at their shortest distance; supraoculars separated by total 24–28 scales; supraoculars separated from supranasals by 2 canthals;

2 apicals bordering rostral; rostral wider than high; 12–16 intercanthals; total scales on head dorsum 39–40; circumorbital scales 13–15 scales, with an incomplete outer ring of 13–17 scales; supralabials 9, separated from eye by 1–2 scale rows; 11–12 sublabials; 2 large anterior chin shields; 4 posterior chin shields; 2–3 preventrals.

Body—23 midbody dorsal scale rows; 165–171 ventrals; 35–38 subcaudals (in males); anal single.

Size—Maximum for adult males is 66 cm.

Pattern—Head with 1 dark line from eye to corner of mouth; row of dark blotches along each side of dorsum; venter dark and mottled with lighter shade; throat whitish with dark mottling; back of head with distinct teardrop-shaped deep black spots; similar deep black bands run from posterior border of eye to corner of mouth, and from lower border of eye down to mouth; supraoculars noticeably pale. Body ground color grayish brown with darker brown zigzag pattern, consisting of about 44 to 52 windings and edged with black (Nilson and Andrén 1986; Latifi 1991; Leviton et al. 1992).

Taxonomy and Distribution

Originally described from three specimens taken in the Zanjan valley, northwest Iran, by Nilson and Andrén (1985). The distribution is restricted to parts of the Elburz, Talysh, and Zanjan mountains.

Habitat

All forms included in this complex are mountain snakes living in dry, sparsely vegetated habitats. The Zanjan valley is characterized by dry rocky slopes with large boulders and little vegetation (Nilson pers. obs.). The habitat may also be dry areas with sandy substrates. Vegetation at these sites includes grasses and shrubs (U.S. Navy 1991), and there is heavy grazing pressure from goat herds (Nilson and Andrén 1986).

Food and Feeding

Feeds on insects (grasshoppers), lizards, and mice (Latifi 1991). Adults switch to small mammals, primarily voles (*Microtus*) while young feed on insects and small lizards (*Darevskia raddei*).

Behavior

Nothing has been published, but is probably similar to *Vipera raddei* in many aspects.

Reproduction and Development

No information is available.

Bite and Venom

In older Iranian literature on venoms and envenomation, many Iranian species of vipers were named *Vipera xanthina,* or *Vipera xanthina* ssp. (e.g. Latifi 1973, 1984; Latifi et al. 1973). Before taxonomic studies were performed on Iranian material many studies on venoms and envenomation, based on *Vipera albicornuta, Vipera latifii* and *Vipera raddei,* have been published under with the name *Vipera xanthina.*

Epidemiology—Data not available.

Yield—On a dry-weight basis, venom yield for nearly 4500 animals milked at the Razi Serum Institute in Tehran, Iran, is approximately 7–18 mg per snake (Latifi 1984). There is little seasonal variation, although yields appear somewhat lower (about 7 mg/snake) in fall vs. summer and spring (10–12 mg/snake). Individual variation in yields is low. Standard deviation of average venom yield is around 10%–17%. Similarly, little variability occurs between left and right fang yield. For 10 similar sized snakes milked the same time, males 9 mg (left fang) and 8 mg (right fang), and females 5 mg for each fang. Sexual differences in venom yield is interesting assuming that animals were truly matched for size across sexes.

Toxicity—Lethality is consistent among venoms milked from right vs. left fangs (Latifi 1984).

Content—Data not available.

Symptoms and Physiological Effects—Data not available.

Treatment and Mortality—Polyvalent antivenins prepared for either Near Eastern, Middle Eastern, or European vipers in general provide appropriate treatment for *V. albicornuta* envenomation (Latifi 1984, Al-Joufi et al. 1991).

Vipera albizona Nilson, Andrén and Flärdh 1990
Central Turkish mountain viper

Recognition
(Plate 14.20)

Head—Relatively large and distinct from neck; snout rounded, covered with small keeled scales; large supraoculars in broad contact with eye; 7–10 supralabials; 2 scale rows between eye and supralabials; sublabials usually 10–13; 9–13 circumorbital scales; nostril within a single scale; temporal scales keeled; 2 or 3 apical scales in contact with rostral; usually 1 canthal on each side of head.

Body—Usually 23 keeled midbody dorsal scale rows; 149–155 ventrals; 2 preventrals; 23–30 paired subcaudals; anal single.

Size—Maximum length little less than 78 cm (male), but normally smaller.

Pattern—Dorsum of head usually with two large black oblique spots; there is commonly a dark stripe running from corner of eye to angle of mouth or beyond. Body dorsal ground color grayish; along midline from back of head to tail is a series of about 30 transversed and pronounced white- and black-edged narrow bands separated by a brick-red brown zone 3–4 scales long and 9–12 scales wide. Lateral spots may be small and in a double series. Venter grayish, finely speckled with darker spots (Nilson et al. 1990; Bettex 1993; Mulder 1994).

Taxonomy and Distribution

Described by Nilson et al. (1990) as *Vipera albizona* from mountain areas of central Turkey. It is parapatric with *Vipera xanthina,* which occurs nearby, on the mountain Erciyas Dagh (Nilson et al. 1990).

Habitat

The habitat is dry and very rocky mountain slopes and fields.

Food and Feeding

In captivity, *V. albizona* feeds readily on mice, and adults can be maintained in breeding condition on a ration of one freshly killed laboratory mouse per week.

Behavior

No information is available.

Reproduction and Development

No information is available.

Bite and Venom

No information is available.

Vipera bornmuelleri
Werner 1898

Known as Bornmuellers viper.

Recognition

(Plate 14.21)

Head—Relatively large and roundish, distinct from neck; snout rounded, covered with small keeled scales; large supraoculars in broad contact with eye; interoculars 6–9; 9–10 supralabials; 2 scale rows between eye and supralabials; sublabials usually 8–9; 11–15 circumorbital scales; nostril within a single scale; temporal scales keeled; 2 or 3 apical scales in contact with rostral; usually 1 canthal on each side of head, sometimes one and one-half.

Body—Usually 23 (rarely 21) keeled midbody dorsal scale rows; 142–153 ventrals; 2 preventrals; 23–31 paired subcaudals; anal single.

Size—Maximum length approximately 75 cm; usual range much less. In some populations, males tend to be larger than females. In females from Mt. Liban, Lebanon, maximum length was 47.3 cm and males from the same locality had a maximum of 53.8 cm (Nilson and Andrén 1986). Tail length approximately 7%–10% of total length.

Pattern—Dorsum of head usually with two larger but weakly developed spots; there is commonly a dark stripe running from corner of eye to angle of mouth or beyond. Body dorsal ground color grayish to brownish; along midline from back of head to tail is a dorsal pattern consisting of a series of thin, irregular, dark crossbands, 47 to 64 in number; lateral blotches ofter reduced. Venter grayish, finally dotted without dark blotches (Joger 1984; Nilson and Andrén 1986).

Taxonomy and Distribution

Vipera bornmuelleri was described by Werner (1898) based on material from Lebanon and from southern Turkey. The Lebanon specimen was considered as lectotype (Werner 1922) while the Turkish specimens later were moved to a separate species, *Vipera bulgardaghica* (Nilson and Andrén 1985). Both *Vipera bornmuelleri* and *V. bulgardaghica* have by various authors been considered as full species or as subspecies of *V. xanthina*. We regard them as good species.

Vipera bornmuelleri is confined to southern Lebanon, western Syria, and northern Israel and Jordan (Joger 1984; Nilson and Andrén 1986; Nilson et al. 1990).

Habitat

Although inhabiting dry regions, most populations require some moisture and vegetation. The specific habitats identified for this species include densely vegetated rocky areas, rocky scrublands, stream valleys and other moist vegetated places, talus slopes and rock fields, cedar forests. The altitudal distribution is between 1450 m and 2000 m (Nilson and Andrén 1986). In southern Lebanon and in the border vicinities of Lebanon, Syria, Jordan, and Israel, *V. bornmuelleri* is clearly a montane snake, inhabiting high mountain areas, some with relatively dense vegetation and some subject to snow cover (U.S. Navy 1991). On Mount Hermon at the Lebanon-Syria border, it is confined to altitudes greater than 1800 m (Werner and Avital 1980).

Food and Feeding

Nothing is reported from wild populations. In captivity it eats mice readily (Nilson pers. obs.).

Behavior

No information is available.

Reproduction and Development

Individuals from populations of Mt. Hermon in the Bekáa Valley area are bred at Tel Aviv University. Clutch sizes here range from 2 to 18 and young may be born as late as early September (Nilson and Andrén 1986).

Bite and Venom

Epidemiology—No information is available.

Yield—No information is available.

Toxicity—A *Vipera bornmuelleri* subpopulation evaluated had an i.v. toxicity of about 0.6 mg/kg in laboratory mice (or about 12 μg/mouse assuming an individual weight of 20 g per mouse). Intraperitoneal and s.c. toxicities are 1.9 mg/kg (38 μg/mouse) and 6.3 mg/kg (126 μg/mouse) respectively (Weinstein and Minton 1984).

Content—Several proteins separable by gel electrophoresis are present in whole venom of *V. bornmuelleri* populations, the effects of which can be tested individually (Bernadsky et al. 1986). In general, heavier fractions have little or no proteolytic property although other studies demonstrate some proteolysis. Of six protein fractions separated from whole venom, only one, a relatively high molecular weight fraction, has lethality in itself equivalent to that of whole venom. However, toxicity of this venom results from an important synergistic effect. When one lighter fraction is injected in conjunction with several heavier fractions, lethality is substantially enhanced. This synergistic interaction may be characteristic of snake venoms in general and indicate a property of common evolutionary origin. For example, the protein filtration pattern for *V. bornmuelleri* closely follows *V. xanthina* in general but does not clearly track *D. palaestinae* (Bernadsky et al. 1986).

Symptoms and Physiological Effects—Data not available.

Treatment and Mortality—Data not available.

Vipera bulgardaghica
Nilson and Andrén 1985

Known as the Bulgardagh viper.

Recognition
(Plate 14.22)

Head—Relatively large and roundish, distinct from neck; snout rounded, covered with small keeled scales; large supraoculars in broad contact with eye; interoculars 6; 9 supralabials; 1 (sometimes 2) scale rows between eye and supralabials; sublabials usually 11–13; 9–13 circumorbital scales; nostril within a single scale; temporal scales keeled; 2 or 3 apical scales in contact with rostral; usually 1 canthal on each side of head.

Body—Can vary between 23 and 25 (sometimes 21) keeled midbody dorsal scale rows; 145–156 ventrals; 2 preventrals; 24–33 paired subcaudals; anal single.

Size—Maximum length approximately 78 cm.

Pattern—Dorsum of head usually with two black oblique drop-shaped spots; there is commonly a dark stripe running from corner of eye to angle of mouth or beyond; there may be a dark spot below the eye. Often there is a series of small blackish spots or a narrow band across the snout in front of the eyes.

Body dorsal ground color brownish gray. Dorsal longitudinal band pronouncedly right-angled or partly consisting of more or less round or squarish dark blotches; the color of the dark blotches and sinuous band is somewhat reduced in intensity in central parts. Venter grayish, finely dotted with only very weakly developed dark spots. (Nilson and Andrén 1985b, 1986; U.S. Navy 1991; Bettex 1993; Mulder 1994).

Taxonomy and Distribution

Described by Nilson and Andrén (1985b) as *Vipera bulgardaghica* from specimens originating in the Bolkar Daghlari mountain range in south-central Turkey (Joger 1984; Nilson and Andrén 1986; Nilson et al. 1990; U.S. Navy 1991; Bettex 1993).

Habitat

Vipera bulgardaghica is found in mountain regions of southern Turkey. The original locality was at Bulgar Dagh, in Cilician Taurus, and since then additional specimens have been taken in the area. The habitat is open thorn-shrub vegetation and grasslands, with rocky hillsides and talus slopes throughout. The altitudinal range of *V. bulgardaghica* in this region is from 1450 m to altitudes greater than 2000 m (U.S. Navy 1991; Bettex 1993).

Food and Feeding

Captive specimens feed readily on rodents.

Behavior

In high mountain areas in cooler weather, particularly where there are seasonal snows, it becomes diurnal in order to bask (U.S. Navy 1991).

Reproduction and Development

A captive female gave birth to 13 young on September 11 (Bettex 1993).

Bite and Venom

No information is available.

Vipera latifii
Mertens, Darevsky and Klemmer 1967

Commonly called Latifi's viper.

Recognition
(Plate 14.23)

Head—Relatively small and elongate, fairly distinct from neck; supraoculars raised and separated from eye by row of small scales; nostril centered in a large nasal shield which is partially fused with prenasals; loreal present between upper preocular and nasal; supraoculars separated by 7–9 scales at their shortest distance; supraoculars separated by 21–35 scales; supraoculars separated from supranasals by 1–3 canthals; 2–3 apicals bordering rostral; rostral wider than high; 10–20 intercanthals; total scales on head dorsum 35–48; circumorbital scales 10–15 scales, with an incomplete outer ring of 12–16 scales; supralabials 8–12, separated from eye by 2 scale rows; 9–13 sublabials; 2 large anterior chin shields; 4–7 posterior chin shields; 1–3 preventrals.

Body—21–25 midbody dorsal scale rows; 161–169 ventrals; 26–39 subcaudals; anal single.

Size—Maximum for adult males 78 cm, maximum female length 70 cm; male tail length 6.2% to 11.1% of total length.

Pattern—Head with a vague dark line from eye to corner of mouth. The species has several relatively distinct color morphs including a brownish gray ground with central zigzag pattern, yellowish gray ground with darker blotches, silverish, grayish, or brownish ground with narrow dark central stripe or indistinct blotches, and unpatterned grayish or brownish neck with 2 oblique darker blotches; venter pale yellow mottled with blackish and whitish speckles.

Taxonomy and Distribution

Originally described by Mertens et al. (1967) from several specimens from the upper Lar Valley in northern Iran.

Population distribution is north central Iran in the Upper Lar Valley of the Elburz Mountains.

Habitat

All forms included in this complex are mountain snakes living in dry sparsely vegetated habitats.

Lar Valley populations of *V. latifii* are found along steep valley walls in a relatively narrow altitudinal range (2200–2900 m). These snakes are found among weathered rock fragments and boulders at the borders of valley walls and floor (Nilson and Andrén 1986). Habitat here is very dry with little vegetation.

Food and Feeding

Feeds on insects (grasshoppers), lizards, and mice (Latifi 1991).

Behavior

In the Upper Lar Valley in Iran, *V. latifii* populations may be subterranean except during the mating season (Andrén and Nilson 1979). This habit could be a response to great diurnal temperature fluctuations that occur in this dry and sparsely vegetated area. In their very rocky habitat the vipers can hide for longer periods of unsuitable temperatures. At this altitude vipers are normally not nocturnal as the nights are too cold throughout much of the summer.

Reproduction and Development

In the wild in the Lar Valley in Iran, mating takes place in June and young are born late September. Young per litter ranged from 5 to 10 (Nilson and Andrén 1986).

Bite and Venom

In older Iranian literature on venoms and envenomation many Iranian species of vipers were named *Vipera xanthina* or *Vipera xanthina* ssp. Before taxonomic studies were performed on Iranian material many studies on venoms and envenomation based on *Vipera latifii, Vipera albicornuta,* and *Vipera raddei* have been published together with the name *Vipera xanthina*.

Epidemiology—Few bites are reported. This may be due partially to the steep and rugged habitat of these snakes, and partially to the care taken by local visitors to the remote Upper Lar Valley where the species occurs.

Yield—On a dry-weight basis, venom yield for nearly 8750 animals milked at the Razi Serum Institute in Tehran, Iran, is approximately 2–11 mg per snake (Latifi 1984). There is little seasonal variation, although yields appear lower (about 2 mg/snake) in winter vs. the rest of the year (6 mg/snake). Similarly, little variability occurs between left and right fang yield. For 10 similar sized snakes milked the same time, males 8 mg (left fang) and 7 mg (right fang), and females 4 mg (left fang) and 5 mg (right fang). Also in this species sexual differences in venom yield is interesting assuming that animals were truly matched for size across sexes.

Toxicity—A comparative study of intravenous toxicity reported an LD_{50} of 4.8 µg/g in laboratory mice (Latifi et al. 1973). This dose is substantially higher than comparable lethal doses reported for *V. aspis, V. berus, D. palaestinae, V. xanthina,* and *D. russelii* (Tu 1991).

In a detailed investigation (Latifi 1984), a range of intravenous LD_{50}s was reported for *V. latifii*. Dose ranged from 3.2 to 10.5 µg/mouse. No seasonal pattern of toxicity was evident, and there was little difference between toxicity from left and right fangs of individual snakes or between venoms extracted from male and females. Of seven species tested in this study, (*Naja oxiana, Echis carinatus, Pseudocerastes persicus, Macrovipera lebetina, Vipera latifi,* and *Gloydius intermedius,* only *E. carinatus* had a more lethal venom (Latifi 1984). Similar intravenous toxicity (0.348 µg/kg) is reported for laboratory mice by Weinstein and Minton (1984). Simple experimental exposure of mice to unreported quantities of venom yielded rapid mortality. Lizards are quickly paralyzed and die after some minutes. Experimentally administered venom is fatal to both rabbits and dogs. Lethality is consistent among venoms milked from right vs. left fangs (Latifi 1984).

Content—Little is known regarding venom composition. Gel permeation patterns of Middle Eastern and European species may provide some clues to taxonomic relationships. *Vipera raddei* and *V. latifii* had very similar filtration patterns and these patterns were closely allied to those of *V. xanthina, V. ammodytes,* and *V. aspis.*

Symptoms and Physiological Effects—No information is reported.

Treatment and Mortality—Polyvalent antivenins prepared for either Near Eastern, Middle Eastern, or European vipers in general provide appropriate treatment for *V. xanthina* (=*latifii*) envenomation (Latifi 1984, Al-Joufi et al. 1991.)

Remarks

Despite their remote range and rugged habitat, populations of *V. latifii* are threatened. Habitat destruction has reduced populations of *V. latifii* (Tuck 1977; Andrén and Nilson 1979). In the end of the 1970s a major part of the Upper Lar Valley was made a huge water reservoir, much of the known range for *Vipera latifii* was submerged, and microclimate was supposedly altered in the surrounding habitats. There is no information available of the present population's size, but specimens still are brought to the Hessarak Serum Institute in Iran (Nilson own obs.).

Vipera raddei
Boettger 1890

Commonly called the rock viper, Radde's mountain viper, or Armenian mountain viper.

Recognition
(Plate 14.24)

Head—Relatively small and elongate, fairly distinct from neck; supraoculars raised and separated from eye by row of small scales; nostril centered in a large nasal shield which is partially fused with prenasals; loreal present between upper preocular and nasal; supraoculars separated by 6–9 scales at their shortest distance; 7–11 total interocular scales; supraoculars separated by 22–35 scales; supraoculars separated from supranasals by 1–3 canthals; 2–5 apicals bordering rostral; rostral wider than high; 11–24 intercanthals; total scales on head dorsum 35–52; circumorbital scales 12–18, with an incomplete outer ring of 13–18 scales; supralabials 9–10, separated from eye by 2 scale rows; 10–14 sublabials; 2 large anterior chin shields; 4–6 posterior chin shields; 2–4 preventrals.

Body—21–25 midbody dorsal scale rows; 163–181 ventrals; 28–35 subcaudals; anal single.

Size—Maximum for adult males 99 cm, adult females 79 cm; *Vipera r. kurdistanica* maximum female length 70 cm, males 89 cm; tail length approximately 5.4% to 8.2% of total length.

Pattern—Head with 1 dark line from eye to corner of mouth; row of dark blotches along each side of dorsum; venter dark and mottled with lighter shade with scale tips dark edged; throat whitish with dark mottling; back of head with distinct teardrop-shaped black spots; supraoculars noticeably pale.

Body ground color grayish brown, gray, dark gray to blackish; dorsal pattern in northern populations distinct with rounded blotches of yellow, yellowish orange, brownish orange or reddish, sometimes edged with black; dorsal pattern in southern *V. r. kurdistanica* often darker than pattern elements; neck unpatterned or with L-shaped blotches merging to dorsal pattern elements; venter usually dark gray to mottled (Nilson and Andrén 1986; Latifi 1991; U.S. Navy 1991; Leviton et al. 1992).

Sexual dimorphism is subtle, limited to females having a slightly lighter snout (Nilson and Andrén 1986).

Taxonomy and Distribution

Originally described from two specimens taken from Armenia in what is now Anatolia, Turkey, by Boettger (1890). Type specimens in the Georgia Museum in

Tblisi were lost and a topotype has been identified from collections of the British Museum. Two subspecies have been described:
Vipera raddei raddei and
Vipera raddei kurdistanica Nilson and Andrén 1986. *Vipera raddei kurdistanica* is highly polymorphic.

Population distribution covers the region from eastern Turkey and Armenia to Lake Urmia in Iran and to Azerbaijan (*V. r. raddei*). *Vipera r. kurdistanica* occurs in Iran along the border with Turkey and possibly in northern Iraq (Nilson and Andrén 1986; U.S. Navy 1991; Leviton et al. 1992).

Habitat

All forms included in this complex are mountain snakes living in dry sparsely vegetated habitats. *Vipera raddei* inhabits rocky slopes from 1200 to 3000 m. Vegetation, in the form of thin oak or juniper canopy, may be preferred (Nilson and Andrén 1986) although it may be found less commonly in "open, high mountain steppe," (unvegetated rocky areas) (Joger 1984). It occurs in great abundance in volcanic boulder fields below the snow line. The subspecies *V. r. kurdistanica* inhabits relatively open rocky slopes from 2000 to 2500 m (Nilson pers. obs.; Leviton et al. 1992).

In the Lesser Caucasus Mountains of Armenia, *V. raddei* lives in oak forests from 1500 to 1800 m (Darevsky 1966). In juniper scrub and cultivated fields, individuals occupy rock walls and boulder fields. In summer they may shelter in rodent holes, root masses of trees, and large shrubs in the dry soil.

Food and Feeding

Feeds on insects (grasshoppers), lizards, and mice (Latifi 1991). Insects are eaten when snakes emerge from hibernation and orthopteran remains are commonly found in guts of specimens collected in spring (Darevsky 1966). Later in the season individuals switch to small mammals, primarily voles (*Microtus*). Less frequent prey items for adults are lizards and young ground nesting birds. Young feed on insects and small lizards.

Darevsky (1966) gave the minimum number of small mammals consumed per individual snake per year as 100 (stating this was "checked" by unspecified procedures). In captivity, newborns eat pink mice and adults eat laboratory mice. Particularly large individuals eat newly hatched chickens (Kudryavtsev and Mamet 1991).

Behavior

In Armenia, individuals emerge from hibernation by the end of April and remain near the hibernaculum until mid-May. During this period they bask in thin-canopied forest or rocky substrate in masses, with densities as high as 50–60 per hectare. In early spring, basking behavior is limited to the heat of day (Darevsky 1966). In the ensuing weeks, Radde's viper becomes crepuscular. Activity peak times from emergence through June increased from 0900–1300 to 1800–2000 hours. Individuals could be observed on the surface, although in reduced numbers and at reduced activity levels, throughout the day except from 1300 to 1400 hrs when no observations were recorded (Bozhanskii and Kudryavtsev 1986). In summer, individuals disperse during daylight hours making population densities low and individuals difficult to find. In a field experiment in which 63 snakes were marked at the hibernation site in spring, 7 were recaptured at the same location in autumn indicating some degree of site fidelity (Darevsky 1966). In autumn, they return to the hibernacula over an extended period. There is sequential hibernaculum retirement and autumn densities do not exceed 20–50 individuals per hectare (Darevsky 1966).

Radde's viper can be very aggressive when disturbed and exhibits characteristic threat responses (Darevsky 1966). The initial escape effort is a sudden and rapid dive for the nearest cover, accompanied by a distinctive rapidly repeated hissing. If intercepted, individuals adopt a peculiar (for a viper) posture with the forward portion of the body held nearly vertical. From this pose the snake makes repeated strikes at the source of disturbance. Large males are particularly aggressive when irritated.

Reproduction and Development

In the wild, mating occurs upon emergence from hibernation, in association with the initial ecdysis (Darevsky 1966; Bozhanskii and Kudryavtsev 1986). During the spring basking period, males seeking mates are restless and glide around females, shaking their tails, trying to hook the female's tail. If the female is receptive she interlocks her tail and undulates in concert with the male. Females will often break the embrace at this point, only to be pursued by the males. As mating proceeds, the snakes intertwine their bodies, erecting the anterior third. The male pushes the female's head to the ground, sometimes with powerful and violent actions. While intertwined throughout their length, copulation occurs. This may be 2 hours after beginning courtship and copulation lasts up to 30 minutes.

From 3 to 18 young are born per litter in autumn, throughout September in the Lesser Caucasus (Darevsky 1966; Nilson and Andrén 1986). Newborns are relatively large, ranging from 15 cm to over 23 cm in total length and from 3 g to over 12 g per individual.

In captivity, artificial hibernation from November to February is necessary to stimulate courtship (Kudryavtsev and Mamet 1991). Individuals feed 1 day after emergence, mate within a few weeks, and immediately both sexes begin a period of heavy food consumption. Later in spring, pregnant females refuse food until young are born in June or July. Babies born in captivity range from 23 to 35 cm total length and from 8 to 25 g. Feeding is delayed until after the first shed, usually within 8 days of birth (Kudryavtsev and Mamet 1991).

Bite and Venom

Epidemiology—Few bites are reported. This may be due partially to the steep and rugged habitat of these snakes, and partially to the care taken by local people who delay grazing and activities in fields until snakes have dispersed for summer (Darevsky 1966).

Yield—No information is available.

Toxicity—Venom is reputed to be relatively potent (Darevsky 1966).

Content—Little is known regarding venom composition. Gel permeation patterns of Middle Eastern and European species may provide some clues to taxonomic relationships. *Vipera raddei* and *V. latifii* had very similar filtration patterns and these patterns were closely allied to those of *V. xanthina, V. ammodytes,* and *V. aspis.* Neither forms of *V. raddei* exhibited a permeation pattern similar to those for *D. palaestinae,* or *V. berus* (Bernadsky et al. 1986).

Symptoms and Physiological Effects—No information is reported.

Treatment—No information is reported.

Mortality—Human fatalities are known. An adult man died 12 hours after a bite on his shoulder (Darevsky 1966).

Remarks

Despite their remote range and rugged habitat, populations of *V. raddei* are threatened. Overcollecting and habitat destruction have reduced populations of *V. raddei*, occasionally intensified by low local densities or small and highly localized habitats (Bannikov et al. 1978; Honegger 1978b; Dodd 1987).

Vipera wagneri
Nilson and Andrén 1984
Known as the ocellated mountain viper, or Wagner's viper.

Recognition
(Plate 14.25)

Head—Relatively large and elongate, distinct from neck; snout rounded, covered with small keeled scales; large supraoculars in broad contact with eye; interoculars 6–7; 9 supralabials; 1–2 scale rows between eye and supralabials; sublabials usually 12–13; 12–15 circumorbital scales; nostril within a single scale; temporal scales keeled; 2 or 3 apical scales in contact with rostral; usually 1 canthal on each side of head.

Body—23 keeled midbody dorsal scale rows; 161–170 ventrals; 2 to 3 preventrals; 23–31 paired subcaudals; anal single.

Size—Maximum length approximately 70–95 cm.

Pattern—Dorsum of head usually with two black elongated blotches that form a large dark open V marking, without apex; arms of this V end on the neck; there is commonly a dark stripe running from corner of eye to angle of mouth.

Body dorsal ground color grayish; along midline from back of head to tail is a central series of occasionally connected blotches or spots which are light brown to yellowish brown or orange, with black borders on a generally gray ground. Each blotch measures 4 to 8.5 scales wide. (Nilson and Andrén 1986; Joger et al. 1988; Nilson et al. 1988; Bettex 1993; Mulder 1994).

Taxonomy and Distribution

The species was described from a specimen, labeled "Urmia lake" and collected in the regions of the "old Armenian—Persian border" in 1846 (Nilson and Andrén 1984). The known range today is in eastern Turkey, near the borders of Armenia and of Iran. The supposed total range for the species is eastern Turkey and northern Iran (Nilson and Andrén 1986; Nilson et al. 1988, 1990).

Habitat

Although inhabiting dry regions, most populations of subgenus *Montivipera* require some moisture and vegetation. The specific habitats identified for this species include densely vegetated rocky areas, stream valleys, and other moist, vegetated places. It is sympatric but not syntopic with *Vipera raddei* in Turkey. While *V. raddei* prefers rocky, dry, and sun-exposed south-facing slopes *V. wagneri* occur in moist and more shadowy, vegetated north-facing slopes of the same valleys (Nilson pers. obs.).

Food and Feeding

Reports of gut contents or field observations of feeding are not available for *V. wagneri*, but it readily accepts mice in captivity.

Behavior

No information is available.

Reproduction and Development

In 13 females from a *V. wagneri* population, offspring numbered from 4–8 (Bergman and Norstrom 1994).

Bite and Venom

No information is available.

Vipera xanthina (Gray 1849)

Known as the rock viper, coastal viper, Ottoman viper, Turkish viper, and Near East viper.

Recognition
(Plate 14.26)

Head—Relatively large and elongate, distinct from neck; snout rounded, covered with small keeled scales; large supraoculars in broad contact with eye; row of interocular scales between supraoculars 5–8; 9–11 supralabials; 2 scale rows between eye and supralabials; sublabials usually 11–14 (but variable, some specimens as few as 8); 11–14 circumorbital scales; nostril within a single scale; temporal scales keeled; 2 (rarely 3) apical scales in contact with rostral; usually 1 canthal on each side of head.

Body—Usually 23 (rarely 21 or 25) keeled midbody dorsal scale rows; 147–169 ventrals; 1–4 preventrals; 27–38 paired subcaudals; anal single.

Size—Maximum length approximately 130 cm on Greek islands; usual range 70–95 cm. There is no discernible dimorphism (in a series of Turkish *V. xanthina*, both males and females had maximum lengths of approximately 96 cm (Nilson and Andrén 1986); tail length 7.8%–11% of total length.

Pattern—Dorsum of head usually with two large dark oblique spots, sometimes connected with the dorsal zigzag band; there is commonly a dark stripe running from corner of eye to angle of mouth or beyond. Body dorsal ground color often gray, but can be olive, yellowish brown, reddish brown to light brown, or dark gray; along midline from back of head to tail is a central dark dorsal stripe in a wavy pattern colored dark brown or blackish, commonly broken into blotches or saddles. Laterally there is a series of small or larger dark spots or vertical bars along each flank, where these are formed as large bars they fit along indentations in the central stripe; along flanks their pigmentation tends to increase, and many specimens have a darker ground color laterally, leaving a lighter border to the pattern elements; lateral spots may be small and in a double series, one element inserted in each indentation of the central stripe and a corresponding spot parallel to or offset from that element. Venter grayish, yellowish, or whitish gray, finely speckled with darker spots or with larger yellowish or yellow-brown blotches or dark marbling; in many populations the male is brighter and more boldly marked than the female (Steward 1971; Joger 1984; Nilson and Andrén 1986; U.S. Navy 1991; Leviton et al. 1992).

Taxonomy and Distribution

Described by Gray (1849) as *Daboia xanthina* from specimens taken in southwestern Turkey (Xanthus, Kinik). It has remained in the genus *Vipera* since the late 1800s. In general, the *xanthina* complex has a complicated history of species description and is the subject of current controversy in the taxonomic literature (Nilson and Andrén 1986, 1992; Schatti et al. 1991, 1992).

Vipera xanthina is found in small isolated populations over a relatively wide geographic range. The species is confined to the eastern Mediterranean area. It is found in northeastern Greece and Turkish Thrace. Also, it inhabits certain Greek and Turkish Aegean islands. Further, it occurs throughout western and central Turkey (to Kayseri) (Joger 1984; van Wingerde 1986; Nilson and Andrén 1986; Nilson et al. 1990; U.S. Navy 1991).

Habitat

Although inhabiting dry regions, most populations of *V. xanthina* require some moisture and vegetation. They are not found in true desert (Steward 1971; U.S. Navy 1991). The specific habitats identified for this species include densely vegetated rocky areas, rocky scrublands, stream valleys, and other moist vegetated places, talus slopes and rock fields, cedar forests, meadows, and grasslands (Steward 1971; Joger 1984).

The species occurs all the way from sea level to 2500 m in altitude. Frequently, the old ruin towns along the Turkish coast form a very suitable habitat. In eastern Mediterranean and Aegean islands, and possibly in Thrace on the Greek mainland, *V. xanthina* may be found from sea level to lower altitudes of local hills (van Wingerde 1986). In this region, it inhabits dry hills and relatively humid valleys facing the sea, vegetated with oaks, junipers, and olives. It has been captured in cultivated land, both crop fields and olive groves, on both the island and mainland areas of northeast Greece (Dimitropolous 1990). Vegetation cover ranges from sparse grasses and shrubs to cultivated fields and forests. It is often found in stone ruins, on rock walls, and in rocky meadows surrounded by mountain slopes (Nilson and Andrén 1986). Habitat types in which it has been found in alpine systems include relatively humid slopes, sites of ancient cedar forests, and drier, rocky areas (Nilson and Andrén 1986).

Food and Feeding

Reports of gut contents or field observations of feeding are not available for *V. xanthina*. Secondary sources (Steward 1971; Joger 1984) report birds, small mammals, and lizards as primary food sources. Steward (1971) stated that mammals are preferred, some birds are taken, and lizards are "probably eaten," but confirms a lack of direct evidence for the latter.

In captivity, *V. xanthina* feeds readily on mice and adults can be maintained in breeding condition on a ration of one freshly killed laboratory mouse per week (Murphy and Barker 1980). In summer on the Greek islands, *V. xanthina* feeds at night largely on small birds (Dimitropoulos 1990). During late spring and early summer when birds and mammals are actively nesting, *V. xanthina* seeks out and robs nests. Active mammals are generally struck, released, and followed, or held until the venom takes effect (Steward 1971).

Behavior

In warmer portions of the range, it may be crepuscular in spring, becoming strictly nocturnal during the heat of summer (Steward 1971; Dimitropoulos 1990). In high mountain areas in cooler weather, particularly where there are seasonal snows, it becomes diurnal in order to bask (U.S. Navy 1991).

Its movements tend to be sluggish although it is capable of striking rapidly and moving quickly when disturbed. Individuals spend most of their time on the ground or among rocks, ruins, or stone piles. However, it is capable of climbing shrubs and trees, and may bask on low branches (Steward 1971; van Wingerde 1986).

Reproduction and Development

Courtship and mating behavior observations were conducted on two males (91 and 85 cm total length) and a captive female (70 cm) from the Dallas Zoo (Murphy and Barker 1980). Courtship was observed twice between late January and early February. The larger male initiated courtship by head bobbing in front of the female, interspersed with periods of nudging the female's trunk with his snout as she rested in a coil. The male moved his head down the female's body and followed each head bob with a tongue flick. When the female crawled, the male pursued. When the female stopped, the male aligned his tail with hers, pushed his tail beneath, and rotated 180° coming to a vent-to-vent position. At this point the male everted his hemipenes but was unable to insert unless the female raised her tail. This behavioral sequence occurred repeatedly, with many unsuccessful attempts and at least one successful intromission.

For insertion, either hemipenis is used depending on the side of the female's vent on which the male's cloaca rests. During successful copulation which lasted at least 20 minutes, the female crawled forward, dragging the male. The base of the hemipenis is visible during courtship and undergoes a color change from flesh pink to deep purple due to blood engorgement.

In the Dallas Zoo matings, no courtship "dance" was observed. However, courtship displays involving the upper one-third of the male bodies swaying together and engaging in pushing and thrusting have been observed a number of times in captivity (Nilson pers. obs.).

In many areas biennial breeding is likely, and juveniles are encountered on the

Greek coast in October (Dimitropoulos 1990). Young are born in late summer. The number of young per litter commonly varies between 2–10. For *V. xanthina* from southern Turkey, captive mating takes place in March and April. Young are born in late summer, usually from early July to mid-August. Of nine recorded clutches, 2–15 young were born (mean 9). Young averaged between 18 and 20 cm total length and ranged in weight from about 6–9 grams (Nilson and Andrén 1986). In another case, three clutches of 5, 9, and 12 young were born after 96–138 days of gravidity (Radspieler and Schweiger 1990).

Bite and Venom

In older literature many studies on *V. xanthina* venom and evenomation were based on *V. xanthina palaestinae* (=*Daboia palaestinae*), and must be assigned to that species.

Epidemiology—No information is available.

Yield—No information is available.

Toxicity—Possibly dangerous to humans (U.S. Navy 1991).

Content—In general, heavier fractions have little or no proteolytic property although other studies demonstrate some proteolysis. Of six protein fractions separated from whole venom, only one, a relatively high molecular weight fraction, has lethality in itself equivalent to that of whole venom. However, toxicity of this venom results from an important synergistic effect. When one lighter fraction is injected in conjunction with several heavier fractions, lethality is substantially enhanced. This synergistic interaction may be characteristic of snake venoms in general and indicate a property of common evolutionary origin. For example, the protein filtration pattern for *V. bornmuelleri* closely follows *V. xanthina* in general but does not clearly track *D. palaestinae* (Bernadsky et al. 1986).

Six fractions (five anodic, one cathodic) were isolated by Gitter et al. (1963). Of the anodic fractions, one was rich in proteins capable of producing lethal hemorrhages *in vivo*. Four contained neurotoxins, two of which acted as phospholipases, and one delayed coagulation *in vitro*.

Hemorrhagic and neurotoxic components are present in fractionated venom, demonstrated when fractions are tested individually. Zinc is present and is linked to neurotoxic proteins in venom (Gitter et al. 1963). Neurotoxic fractions react with the central nervous system and have some anticoagulant and proteolytic properties (Boquet 1967b).

Symptoms and Physiological Effects—The symptomology follows the general course documented for other European viper bites (Gonzales 1991). Symptoms of *V. xanthina* bite are swelling, and necrosis, and hemorrhage at the bite site. Systemic involvement includes hemolytic and hemorrhagic events leading to hypotension. Injected mice exhibited symptoms of shock. Venom causes red blood cells to undergo spherocytosis (Boquet 1967b).

Kinin release into tissues causing severe shock and local pain is an important symptom of viper bites in general. However, *V. xanthina* venom has relatively low kinin-releasing activity (Al-Joufi et al. 1991).

Treatment and Mortality—Polyvalent antivenins prepared for either Near Eastern, Middle Eastern, or European vipers in general provide appropriate treatment for *V. xanthina* envenomation (Latifi 1984; Al-Joufi et al. 1991). Venom is also detoxified by dihydrothioctic acid (DHTA), a chemical commonly employed to prepare toxoids (Sawai et al. 1967a, b).

Bibliography

Abdalla S, Bilto Y, Disi A. 1992. Effects of the sand viper *Cerastes cerastes* venom on isolated smooth muscle and heart and on hematological and cardiovascular parameters in the guinea pig. Toxicon 30:1247–55.

Abu-Sinna G, Al-Zahaby AS, El-Aal AA, El-Baset AA, Soliman N. 1993. Effect of the viper, *Cerastes cerastes,* venom and venom fractions on carbohydrate metabolism. Toxicon 31(16):791–801.

Abu-Sitta SAM, Whaler BC, Zayat AF. 1978. Cardiotoxicity of Gaboon viper venom. In: Habermehl G, Mebs D, editors. Proceedings of the 3rd Symposium on Plant, Animal and Microbial toxins; 1976 Aug 12, Berlin. Darmstadt: Technische Hochschule. pp 178–85.

Adams Z'S, Gatullo D, Losano G, Marsh NA, Vacca G, Whaler BC. 1981. The effects of *Bitis gabonica* (Gaboon viper) snake venom on blood pressure, stroke volume and coronary circulation in the dog. Toxicon 19(2):263–70.

Agrimi U, Luiselli L. 1992. Feeding strategies of the viper *Vipera ursinii ursinii* in the Apennines. Herpetol J 2:37–42.

Ahuja ML, Veeraraghavan N, Menon IGK. 1946. Action of heparin on the venom of *Echis carinatus*. Nature 158:878.

Akester J. 1979a. Male combat in captive Gaboon vipers. Herpetology 35:124–8.

Akester J. 1979b. Successful mating and reproduction by a Gaboon viper (*Bitis gabonica gabonica*) in captivity. Arnoldia 31(8):1–5.

Akester J. 1983. Male combat and reproductive behavior in captive *Bitis caudalis*. Br J Herpetol 6:329–33.

Akester J. 1984. Further observations on the breeding of the Gaboon viper (*Bitis gabonica*) in captivity. Arnoldia Zimbabwe 9(13):217–22.

Akester J. 1989. Captive breeding of the rhinoceros-horned viper, *Bitis nasicornis*. Br Herpetol Soc Bull 28:31–6.

Al-Badry KS, Nuzhy S. 1983. Hematological and biochemical parameters in active and hibernating sand vipers. Comp Biochem Physiol 74A:137–41.

Alcobendas M, Baud CA, Castanet J. 1991. Structural changes of the periosteocytic area in *Vipera aspis* bone tissue in various physiological conditions. Calcif Tissue Int 49:53–7.

Alcobendas M, Casanet J, Martelly E, Milet C. 1992. Phosphate and calcium level variations in the plasma of the snake *Vipera aspis* during the annual cycle and the reproductive period. Herpetol J 2:42–7.

Al-Joufi A, Bailey G, Reddi K, Smith D. 1991. Neutralization of kinin-releasing enzymes from viperid venoms by antivenin IgG fragments. Toxicon 29:1509–11.

Alloatti G, Camino E, Cedrini L, Losano G, Marsh NA, Whaler BC. 1984. The effects of Gaboon viper (*Bitis gabonica*) venom on the mechanical and electrical activity of the guinea-pig myocardium. J Physiol 163:351–90.

Alloatti G, Camino E, Cedrini L, Losano G, Marsh NA, Whaler BC. 1986. The effects of Gaboon viper (*Bitis gabonica*) venom on the electrical and mechanical activity of the guinea-pig myocardium. Toxicon 24(1):47–61.

Allon N, Kochva E. 1974. The quantities of venom injected into prey of different size by *Vipera palaestinae* in a single bite. Exp Zool 188:71–6.

Al-Sadoon MK. 1991. Metabolic rate-temperature curves of the horned viper, *Cerastes cerastes gasperetti,* the Moila snake, *Malopolon moilensis,* and the adder, *Vipera berus.* Comp Biochem Physiol 99A:119–22.

Amr Z, Amr S. 1983. Snakebites in Jordan. Snake 15:81–5.

Ananjeva NB, Borkin LJ, Darevsky IS, Orlov NL. 1998. Amphibians and Reptiles. Encyclopedia of Nature of Russia. ABF Moscow (In Russian). 574 pp.

Ananjeva NB, Munkhbayar Kh, Orlov NL, Orlova VF, Semenov DV, Terbish Kh. 1997. Amphibians and Reptiles of Mongolia. Russian Academy of Sciences Moscow (In Russian). 416 pp.

Anderson J. 1892. On a small collection of mammals, reptiles, and batrachians from Barbary. Proc zool Soc London 3–24.

Anderson SC. 1963. Amphibians and reptiles from Iran. Pro Calif Acad Sci 31: 417–98.

Andrén C. 1976. The reptile fauna in the lower alpine zone of Aberdare and Mt. Kenya. Brit. J . Herpetology 5(7):566–575.

Andrén C. 1982a. Effect of prey density on reproduction, foraging and other activities in the adder, *Vipera berus*. Amphibia-Reptilia 3:81–96.

Andrén C. 1982b. The role of the vomeronasal organs in the reproductive behavior of the adder *Vipera berus*. Copeia 75:148–57.

Andrén C. 1986. Courtship, mating and agonistic behaviour in a free-living population of adders, *Vipera berus*. Amphibia-Reptilia 7:353–83.

Andrén C, Nilson G. 1979. *Vipera latifii,* an endangered viper from Lar Valley, Iran, and remarks on sympatric herpetofauna. J Herpetol 13:335–41.

Andrén C, Nilson G. 1981. Reproductive success and risk of predation in normal and melanistic color of the adder, *Vipera berus.* Biol J Linn Soc 15:235–46.

Andrén C, Nilson G 1983. Reproductive tactics in an island population of adders, *Vipera berus* (L.), with a fluctuating food resource. Amphibia Reptilia 4:63–79.

Andrén C, Nilson G. 1987. The copulatory plug of the adder, *Vipera berus:* Does it keep sperm in or out? Oikos 49:230–2.

Andrén C, Nilson G, Höggren M, Tegelström H. 1997. Reproductive strategies and sperm competition in the adder, *Vipera berus*. In: Venomous Snakes, Ecology, Evolution and Snakebite (Eds. Thorpe RS, Wüster W, Malhotra A), Symposia of the Zoological Society of London, Oxford University Press, no 70:129–141.

Angel F, Lhote H. 1938. Reptiles et amphibians du Sahara central et du Sudan. Bull Com Et Hist Sci AOF 21:345–84.

Anton T. 1987. General observations on three Soviet snakes, with notes on the Transcausasian-Transcaspian *Vipera*. Bull Chicago Herpetol Soc 22:53–7.

Appleby LG. 1971. British Snakes. London: J Baker. 201 pp.
Arez AP, Laing GD, do Rosario V, Theakston RDG. 1993. Preliminary studies on the characterization of venom from *Vipera latastei latastei* collected in NW Portugal. Portugaliae Zool 2(4):37–42.
Arneil G, MacLaurin J. 1961. Case of adder bite with thrombosis of the saphenous vein. Br Med J 1:1587–8.
Arnold EN, Burton JA. 1978. A Field Guide to the Reptiles and Amphibians of Britain and Europe. London: Collins. 156 pp.
Arora RB, Wig KL, Somani P. 1962. Comparative toxicity. Arch Int Pharmoco 137:299. In: Bucherl W, editor. 1967. Venomous Animals and Their Venoms. Vol. I. Paris: Masson. 347 pp.
Ashe J. 1968. A new bush viper. J E Afr Nat Hist Soc Nat Mus 27:53–9.
Ashe JS, Marx H. 1988. Phylogeny of the viperine snakes. Part 2: Cladistic analysis. Fieldiana Zool 52:1–23.
Ashley BD, Burchfield PM. 1968. Maintenance of a snake colony for the purpose of venom extraction. Toxicon 5:267–75.
Auffenberg W, Rehman H. 1991. Studies on Pakistan reptiles. Part 1: The genus *Echis*. Bull Flor Mus Nat Hist Biol Sci 35(5):263–314.
Augustyn JM, Elliott WB. 1967. Further studies on the production by snake venoms of uncoupling and reverse acceptor control in rat liver mitochondrial preparations. Toxicon 5:135–7.
Baard EHW. 1990. New geographical distribution of *Bitis cornuta cornuta*. J Herpetol Assoc Afr 37:56.
Bailey G, Banks B, Pearce F, Shipolini R. 1975. A comparative study of nerve growth factors from snake venoms. Comp Biochem Physiol 51(B):429–38.
Bajwa SS, Kirakossian H, Reddy KNN, Markland FS. 1982. Thrombin-like and fibrinolytic enzymes in the venoms from the Gaboon viper (*Bitis gabonica*), eastern cottonmouth (*Agkistrodon piscivorus*) and southern copperhead (*Agkistrodon contortrix*) snakes. Toxicon 20(2):427–32.
Balozet L. 1957. La vipere lebetine et son venin. Arch Inst Pasteur Alger 35:220–95.
Balozet L. 1962. Les venins, la sperocytose des hematies et leur sedimentation. Arch Inst Pasteur Alger 40:149–78.
Banks B, Banthorpe D, Berry A, Davies H, Soonan S, Lamont D, Shipolini R, Vernon C. 1968. The preparation of nerve growth factors from snake venoms. Biochem J 108:157–8.
Bannikov A, Darevsky I, Sherbak N. 1978. Amphibia and reptilia. The USSR Red Data Book. Moscow: Lesnaya Promyshlenost. pp 151–72.
Bannister A. 1974. Desert adaptation. Nat Hist NY 83:80–1.
Baran I, Atatür M. 1998. Turkish Herpetofauna (Amphibians and Reptiles). Publication Board of the Ministry of Environment, no: 97/17. 214 pp.
Baran I, Joger U, Kutrup B, Türkozan O. 2001. On new specimens of *Vipera barani* Böhme & Joger, 1983, from northeastern Anatolia, and implications for the validity of *Vipera pontica* Billing, Nilson & Sattler, 1990 (Reptilia, Viperidae). Zoology in the Middle East 23:47–53.

Baran I, Kaya M, Kumlutas Y. 1997. On the herpetofauna of the vicinity of Camlihemsin. Turkish Journal of Zoology 21:409–416.

Barbour T, Loveridge A. 1928. A comparative study of the herpetological faunae of the Uluguru and Usambara mountains, Tanganyika territory with descriptions of new species. Mem Mus Comp Zool 50:136–7.

Baron JP. 1980. Donnees sur l'ecologie de *Vipera ursinii ursinii* au Mont Ventoux. Bull Soc Herpetol Fr 14:26–7.

Baron JP. 1989. Feeding of *Vipera ursinii ursinii* in the Mount Ventoux: Diet and feeding cycle [abstract]. In: First world congress of herpetology; 1988 July 8–14; Cambridge, England.

Barrett R. 1970. The pit organs of snakes. In: Gans C, editor. The Biology of the Reptilia. Vol. 2. New York: Academic Press.

Barton RA. 1966. A record *Causus rhombeatus*. J Herpetol Assoc Afr 2:3.

Barzilay M, de Vries A, Condrea E. 1978. Exposure of human red blood cell membrane phospholipids to snake venom phospholipases A. Part 1: Hydrolysis of substrates by *Vipera palaestinae* phospholipase from within resealed red cells. Toxicon 16:145–52.

Batzri-Izraeli R, Bdolah A. 1982a. A basic phospholipase A—The main toxic component of *Pseudocerastes fieldi* venom. Toxicon 20:203–9.

Batzri-Izraeli R, Bdolah A. 1982b. Isolation and characterization of the main toxic fraction from the venom of the false horned viper (*Pseudocerastes fieldi*). Toxicon 20:867–75.

Bauer AM, Gunther R, Klipfel M. 1995. The Herpetological Contributions of WCH Peters (1815–1883). St. Louis: Society for the Study of Amphibians and Reptiles in cooperation with Deutsche Gesellchaft fur Herpetologie und Terrarienkunde. 51 pp.

Bdolah A. 1979. The venom glands of snakes and venom secretion. In: Lee C-Y, editor. Snake Venoms. New York: Springer-Verlag. pp 41–57.

Bdolah A. 1986. Comparison of venoms from two subspecies of the false horned viper (*Pseudocerastes persicus*). Toxicon 24(7):726–9.

Bea A. 1985. La repartición de las víboras *Vipera aspis* (Linnaeus, 1758) y *Vipera seoanei* (Lataste, 1879) en el Pais Vasco. Cuadernos de sección. Ciencias naturales, 2.

Bea A, Bas S, Brana F, Saint-Girons H. 1984. Morphologie comparée et répartition de *Vipera seoanei* Lataste, 1879, en Espagne. Amphibia-Reptilia 5:395–410.

Bea A, Brana F. 1988. Nota sobre la alimentación de *Vipera latastei,* Boscá, 1878 (Reptilia, Viperidae). Munibe (Ciencias Naturales) 40:121–124.

Bea A, Brana F, Baron JP, Saint-Girons H. 1992. Régimes et cycles alimentaires des vipères européenes (Reptilia, Viperidae). Etude comparée. Ann Biol XXXI(1):25–44.

Beerli P, Billing H, Schätti B. 1986. Taxonomischer Status von *Vipera latastei monticola* Saint Girons, 1953 (Serpentes: Viperidae). Salamandra 22(2/3):101–104.

Bellairs AD'A, Kamal AM. 1981. The chondrocranium and development of the skull. In: Gans C, editor. The Biology of the Reptilia. Vol. 11. New York: Academic Press.

Benbassat J, Shalev O. 1993. Envenomation by *Echis coloratus* (Mid-East saw

scaled viper): A review of the literature and indications for treatment. Isr J Med Sci 29(4):239–50.
Berger-Dell'Mour HAE. 1987. Some new data on the herpetology of southwest Africa. J Herpetol Assoc Afr 33:5–8.
Bergman J, Norstrom M. 1994. Some notes on the genus *Vipera* in Anatolia, Turkey. Litt Serpent 14:166–75.
Bernadsky G, Bdolah A, Kochva E. 1986. Gel permeation patterns of venoms from eleven species of the genus *Vipera*. Toxicon 24:721–5.
Berry PSM. 1963. The Gaboon viper of the plateau forests of northern Rhodesia. Puku 1:71–4.
Bettex, F. 1993. Beobachtungen an *Vipera bulgardaghica*, *Vipera albizona* und *Vipera xanthina* im Freiland und im Terrarium. Herpetofauna 15(86):21–26.
Bhat M, Kasturi S, Gowda T. 1991. Structure-function relationships among neurotoxic phospholipases. Toxicon 29:97–105.
Bhat RN. 1974. Viperine snake bite poisoning. J Indian Med Assoc 63:383–92.
Bicher H, Roth M, Gitter S. 1966. Neurotoxic activity of *Vipera palaestinae* venom. Depression of central autonomic vasoregulatory mechanisms. Med Pharmacol Exp 14:349–59.
Bicher H, Gitter S, Manoach M, Rosen M, de Vries A. 1963. Neurotoxic activity of *Vipera palaestinae* venom depression of cortical and central autonomic vasoregulatory mechanisms. Bull Res Coun Isr 10(E):232–7.
Bicher H, Gitter S, Manoach M, Rosen M, de Vries A. 1964. On the mechanism of *Vipera xanthina palaestinae* induced shock. Isr J Exp Med 11:143–8.
Biella H-J. 1983. Die Sandotter. Die Neue Brehm-Bücherei. A Ziemsen Verlag. Wittenberg Lutherstadt. 84 pp.
Billing H. 1983. Polymorphismus bei *Vipera berus seoanei*. Herpetofauna 21:31–33.
Billing H, Nilson G, Sattler U. 1990. *Vipera pontica* sp.n., a new viper species in the *kaznakovi* group (Reptilia, Viperidae) from northeastern Turkey and adjacent Transcaucasia. Zoologica Scripta 19:227–231.
Biran H, Stern A, Dvilansky A. 1972. Impaired platelet aggregation after *Echis colorata* bite in man. Isr J Med Sci 8:1757.
Biran H, Dvilansky A, Nathan I, Levine A. 1973. Impairment of human platelet aggregation and serotonin release caused in vitro by *Echis colorata* venom. Thromb Diathes Haemorrh 30:191–8.
Blazer GT. 1977. Case history of a horned adder (*Bitis caudalis*) bite in Rhodesia. J Herpetol Assoc Afr 15:23–6.
Bodenheimer F. 1944. Introduction into the knowledge of the amphibia and reptiles of Turkey. Istanbul: Rev Fac Sci Univ.
Boettger O. 1890. Eine neue Viper aus Armenien. Zool Anz 13:62–4.
Boffa M, Boffa G. 1975. Correlation between the enzymatic activities and the factors active on blood coagulation and platelet aggregation from the venom of *Vipera aspis*. Biochim Biophys Acta 354:275–90.
Boffa M, Boffa G, Winchenne J. 1976. A phospholipase A_2 from *Vipera berus* venom and its properties. Biochim Biophys Acta 429:828–38.
Boffa M, Josso F, Boffa G. 1972a. Action of *Vipera aspis* venom on blood clotting factors and platelets. Thromb Diathes Haemorrh 27:8–13.

Boffa M, Delori P, Soulier J. 1972b. Anticoagulant factors from viperidae venoms: Platelet phospholipid inhibitors. Thromb Diathes Haemorrh 28:509–17.

Boffa M, Winchenne J, Lucien N, Boffa G. 1978. The anticoagulant phospholipase from *Vipera berus* venom. In: Rosenberg P, editor. Toxins: Animal, Plant and Microbial. Oxford: Pergamon Press. pp 310–31.

Bogdanovich OP. 1970. Autumn feeding of the saw scaled viper *Echis carinatus*. Zoologicheskii Zhurnal 49(1):409–12.

Bogert CM. 1940. Herpetological results of the Vernay Angola expedition. Bull Am Mus Nat Hist LXXVII:1–107.

Böhme W, Joger U. 1983. Eine neue Art des *Vipera berus*-Komplexes aus der Türkei. Amphibia-Reptilia 4:265–271.

Boltt RE, Ewer RF. 1964. The functional anatomy of the head of the puff adder, *Bitis arietans*. J Morphol 114:83–105.

Bonaparte CL. 1835. *Pelias Chersea*. Rome: Iconogr Fauna Ital. Unnumbered page in Folio 12, 1832–1841.

Bons J, Girot B. 1962. Cole illustree del reptiles du Marc. Trav Ins Ski Cherifen, Ser Zool 26:1–62.

Boquet P. 1964. Venins de serpents. Part I: Physio-pathologie de l'envenimation et properties biologiques des venins. Toxicon 2:4–41.

Boquet P. 1966. Constitution chimique des venins de serpents et immunite antivenimeuse. Toxicon 3:243–79.

Boquet P. 1967a. Chemistry and biochemistry of the snake venoms of Europe and the Mediterranean regions. In: Bucherl W, editor. 1967. Venomous Animals and their Venoms. Vol. I. Paris: Masson. pp 327–39.

Boquet P. 1967b. Pharmacology and toxicology of snake venoms of Europe and the Mediterranean regions. In: Bucherl W, editor. 1967. Venomous Animals and their Venoms. Vol. I. Paris: Masson. pp 340–58.

Boquet P, Izard Y, Jouannet M, Meaume J. 1967. 1st International Symposium on Animal Toxins; 1966 Feb. 7–11; Atlantic City, NJ. New York: Macmillan (Pergamon). 293 pp.

Boshansky AT, Pishchelev VA. 1978. Effect of some forms of economic activity on distribution and number of *Vipera berus*. USSR Ministry of Agriculture. Moscow: Centr Lab Nat Pres.

Botes DP, Viljoen CC. 1974. Purification of phospholipase A from *Bitis gabonica* venom. Toxicon 12:611

Botha AS. 1984. Hatching of snouted night adder (*Causus defilippii*) eggs. J Herpetol Assoc Afr 30:23.

Boulenger GA. 1888. An account of the reptilia obtained in Burma, north of Tenasserim, by ML Fea, of the Genoa Civic Museum. Ann del Mus Civ di Stor Nat Genova 7(Ser 2):602–3.

Boulenger GA. 1893. *Vipera ursinii*. Proc Zool Soc Lond 1893:596.

Boulenger GA. 1896. Catalogue of Snakes of the British Museum. Vol. III. London: Br Mus Nat Hist. 509 pp.

Boulenger GA. 1903. On the geographical variations of the sand-viper. Proc Zool Soc Lond 1903:185–6.

Boulenger GA. 1904. On the sand-viper of Romania. Ann Mag Nat Hist 7:134–5.
Boulenger GA. 1905. New species of *Causus: Causus bilineatus.* Ann Mag Nat Hist 16(7):114.
Boulenger GA. 1913a. On the geographical races of *V. ammodytes.* Ann Mag Nat Hist 8:283–7.
Boulenger GA. 1913b. The Snakes of Europe. London: Methven & Co. Ltd.
Bourret R. 1936. Les serpents de l'Indochine. Tome II. Catalogue systematique et descriptif. Toulouse: Henri Basuyau. pp 440–2.
Boycott R. 1978. Puff adder birth. J Herpetol Assoc Afr 17:21.
Boycott R. 1987. New herpetological distribution records in western Cape Province. J Herpetol Assoc Afr 33:30–1.
Bozhanskii A, Kudryavtsev S. 1986. Ecological observations of the rare vipers of the Caucasus. In: Rocek Z, editor. Studies in Herpetology. Prague: Charles University. pp 495–8.
Brain CK. 1960. Observations on the locomotion of the southwest African adder, *Bitis peringueyi* (Boulenger), with speculations on the origin of sidewinding. Ann Transvaal Mus 24:19–24.
Brana F, Bas S. 1983. *Vipera seoanei cantabrica* ssp. Munibe 35:87–8.
Brana F, Bea A, Saint-Girons H. 1988. (Diet composition and feeding in *Vipera seoanei* Lataste, 1879. Variations in relation to age and reproductive cycle.) (in Spanish) Munibe 40:19–27.
Branch WR. 1977. The venomous snakes of southern Africa. Part 1: Introduction and Viperidae. Bull Maryland Herpetol Soc 313(3):145–69.
Branch WR. 1988. Field Guide to the Snakes and Other Reptiles of Southern Africa. London: New Holland, Ltd. 186 pp.
Branch WR. 1989. A new adder (*Bitis*) from the Cedarberg, and the status of *Bitis inornata:* Preliminary observations. J Herpetol Assoc Afr 36:64–5.
Branch WR. 1992. Field Guide to the Snakes and Other Reptiles of Southern Africa. Cape Town: Struik Publications. 328 pp.
Branch WR. 1997. A new adder (*Bitis*; Viperidae) from the Western Cape Province, South Africa. S Afr J Zool 32(2):37–42.
Branch WR. 1999. Dwarf adders of the Bitis *cornuta-inornata* complex (Serpentes: Viperidae) in South Africa. Kaupia 8:39–63.
Branch WR, Farrell S. 1988. *Bitis arietans arietans,* puff adder: Unusual color pattern. J Herpetol Assoc Afr 34:46.
Breidenbach CH. 1990. Thermal cues influence strikes in pitless vipers. J Herpetol 4:448–50.
Brink S, Steytler JG. 1974. Effects of puff adder venom on coagulation, fibrinolysis and platelet aggregation in the baboon. So Afr Med J 48:1205–13.
Broadley DG. 1960. *Vipera superciliaris,* a rare viper added to the Nyasaland list. J Herpetol Assoc Rhod 12:3–4.
Broadley DG. 1967. The venomous snakes of central and south Africa. In: Bucherl W, editor. Venomous Animals and Their Venoms. Vol. I. Paris: Masson. pp 235–77.
Broadley DG. 1968. The venomous snakes of central and south Africa. In: Bucherl

W, Buckley E, Deulofeu V, editors. Venomous Animals and Their Venoms. Vol. 1. New York: Academic Press. pp 403–35.
Broadley DG. 1971. The reptiles and amphibians of Zambia. Puku 6:1–143.
Broadley DG. 1974a. A puzzling case ophiophagy in Rhodesia. J Herpetol 8(3):247.
Broadley DG. 1974b. Ophiophagy in *Bitis arietans* in Zambia. J Herpetol Assoc Afr 12:30–1.
Broadley DG. 1975. The horned viper *Bitis caudalis* in the central Kalahari. Botswana Notes Rec 4:263–4.
Broadley DG. 1976. The status of herpetology in southern Africa. Zool Afr 11:233–40.
Broadley DG. 1988. A checklist of the reptiles of Zimbabwe, with synoptic keys. Arnoldia Zimbabwe 9(30):369–429.
Broadley DG. 1990. FitzSimons' Snakes of Southern Africa. Parklands (South Africa): J Ball & AD Donker Publishers. 387p.
Broadley DG. 1995a. Distribution record for *Atheris squamigera*. Afr Herpetol News 23:49.
Broadley DG. 1995b. Geographical distribution: *Atheris squamigera squamigera*. Afr Herpetol News 23:49.
Broadley, DG. 1996. A review of the tribe Atherini (Serpentes: Viperidae), with the descriptions of two new genera. Afr. J. Herpetol. 45(2):40–48.
Broadley, DG. 1998. A review of the genus *Atheris,* with the description of a new species from Uganda. Herp. Jour. 8:117–133.
Broadley DG, Cock EV. 1975. Snakes of Rhodesia. Zimbabwe: Longman Zimbabwe Ltd. 97 pp.
Broadley DG, Cock EV. 1993. Snakes of Zimbabwe. Zimbabwe: Longman Zimbabwe Ltd. 126 pp.
Broadley DG, Parker RH. 1976. Natural hybridization between the puff adder and Gaboon viper in Zululand. Durban Mus Novitates 11(3):77–83.
Brodmann P. 1987. Die Giftschlangen Europas und die Gattung *Vipera* in Afrika und Asien. Kümmerly+Frey, Bern 148 pp.
Brown JH. 1973. Toxicology and Pharmacology of Venoms from Poisonous Snakes. London: CC Thomas Publication. 184 pp.
Brown R, Dewar HA. 1965. Heart damage following adder bite in England. Br Heart J 27:144–7.
Brown RS. 1974. Localization of some enzymes in the main venom glands of viperid snakes. J Morphol 143:247–58.
Bruno S. 1968. Sulla *Vipera ammodytes* in Italia. Mem Mus Civico Storia Natur Verona 15:289–336.
Bruno S. 1985. Le Vipere d'Italia e d'Europa. Bologna: Edagricole. 25 pp.
Brushko ZK. 1968. Reproduction of a saw-scaled viper, *Echis carinatus* in natural conditions and in captivity. UZB Biol Zh 12(3):55–9.
Butler J, Reid J. 1986. Habitat preferences of snakes in the southern cross river state, Nigeria. In: Rocek Z, editor. Studies in Herpetology. Proceedings of the

3rd European Herpetological Meeting; 1985 June 3–8; Prague. Prague: Charles University for the Societas Europaea Herpetologica. pp 483–8.

Buys PJ, Buys PJC. 1980. Snakes of South West Africa. Winhoek: Gamsberg Publishers. 64 pp.

Cadle JE. 1988. Phylogenetic relationships among advanced snakes. A molecular perspective. Univ California Publ (Zool) 119:1–77.

Cadle JE. 1992. Phylogenetic relationships among vipers: immunological evidence. Inn: Campbell JA, Brodie ED Jr (eds): Biology of the Pitvipers: 41–48, Tyler Texas (Selva).

Caiger M, Marsh NA, Whaler BC. 1978. Brain hemorrhage following *Bitis gabonica* envenomation. In: Habermehl G, Mebs D, editors. Proceedings of the 3rd Symposium on Plant, Animal and Microbial Toxins; 1977 Mar. 23–27; Frankfurt. Darmstadt: Technische Hochschule. pp 123–31.

Calmette A. 1907. Les venins, les animaux venimeux et la serotherapie antinenimeuse. In: Bucherl W, editor. 1967. Venomous Animals and Their Venoms. Vol. I. Paris: Masson. pp 233.

Cansdale GS. 1948. Field notes on some Gold Coast snakes. Nigerian Field 13:43.

Cansdale GS. 1961. West African Snakes. London: Longman's. 75 pp.

Cattaneo A. 1989. Note erpetologiche sulle isole greche di Serifos, Sifnos e Milos (Cicladi occidentali). Atti della Società Italiana di Scienze Naturali e del Museo Civicio di Storia Naturale di Milano, 130(2):57–76.

Cevese A, Gattullo D, Losano G, Marsh NA, Vacca G, Whaler BC. 1983. The effects of *Bitis gabonica* (Gaboon viper) snake venom on external iliac and mesenteric arterial circulation in the dog. Toxicon 21(1):67–74.

Cevese A, Gattullo D, Losano G, Marsh NA, Vacca G, Whaler BC. 1984. The effects of *Bitis gabonica* (Gaboon viper) snake venom on cardiac stroke work in the anaesthetized rabbit. Life Sci 34:1389–93.

Chadha JS, Ashby DW, Brown JO. 1968. Abnormal electrocardiogram after adder bite. Br Heart J 30:138–40.

Challet E, Pierre J, Reperant J, Ward R, Micelli D. 1991. The serotoninergic system of the brain of the viper *V. aspis*, an immunohistochemical study. Neues Jahrb Geol Palaeontol Abh 4:233–48.

Chapman DS. 1967. The symptomatology, pathology, and treatment of bites of venomous snakes from central and southern Africa. In: Bucherl W, editor. 1967. Venomous Animals and Their Venoms. Vol. I. Paris: Masson. pp 97–137.

Chenaux-Repond R. 1974. Cattle egrets mobbing swimming puff adders. Honeyguide 78:46–7.

Cherlin VA. 1981. The new saw-scaled viper *Echis multisquamatus* sp. nov. from southwestern and middle Asia. Pro Zool Ins Sci USSR 101:92–5.

Cherlin VA. 1983. Dependence of scale pattern in snakes of the genus *Echis* from climatic conditions. Zoologicheskii Zh. 62:252–7.

Cherlin VA. 1984. Morphometric descriptors. Smithsonian Herpetol Inf Serv 61:1

Cheymol J, Boquet P, Detrait J, Roch-Arveller M. 1973. Comparaison des principales proprietes pharmacologiques de differents venins d'*Echis carinatus*. Arch Int Pharmaco 205:293–304.

Chippaux JP, Goyffon M. 1991. Production and use of snake antivenin. In: Tu A, editor. Reptile Venoms and Toxins. New York: Marcel Dekker. pp 529–55.

Chippaux JP, Boche J, Courtois B. 1982. Electrophoretic patterns of the venom from the litter of *Bitis gabonica* snakes. Toxicon 30(2):521–3.

Chippaux JP, Williams V, White J. 1991. Snake venom variability: Methods of study, results and interpretation. Toxicon 29:1279–1303.

Chiszar D, Dickman JD, Colton J. 1986. Sensitivity to thermal stimulation in prairie rattlesnakes (*Crotalus viridis*) after bilateral anesthetization of the facial pits. Behav Neural Biol 45:143–9.

Chopra RN, Chow Han JS, De NN. 1935. An experimental investigation into the action of the venom of *Echis carinatus*. Indian J Med Res 23:391–405.

Christensen PA. 1955. South African Snake Venoms and Antivenins. Johannesburg: South Africa Institute of Medical Research. 35 pp.

Christensen PA. 1967a. The venoms of central and south African snakes. In: Bucherl W, editor. 1967. Venomous Animals and Their Venoms. Vol. I. Paris: Masson. pp 255–85.

Christensen PA. 1967b. Remarks on antivenin potency estimation. Toxicon 5:143–5.

Chugh K, Aikat B, Sharma B, Dash K, Mathew M, Dash K. 1975. Acute renal failure following snakebite. Am J Trop Med Hyg 24:692–7.

Chugh K, Sakhuja V. 1991. Renal diseases caused by snake venom. In: Tu A, editor. Reptile Venoms and Toxins. New York: Marcel Dekker. pp 471–93.

Chugh KS, Mohanthy D, Yash P, Das KC, Ganguly NK, Chakravarty RN. 1981. Hemostatic abnormalities following *Echis carinatus* envenomation in the rhesus monkey. Am J Trop Med Hyg 30(5):1116–20.

Claessen H. 1977. Slangebeten in Europa. Terra Wommelgem 13:41–4.

Clapp JA. 1977. Courtship behavior in the puff adder. E Afr Nat Hist Soc 121:7–8.

Clark R. 1967. Centipede in stomach of young *V. ammodytes*. Copeia 59:224.

Clark RJ, Clark ED, Anderson SC, Leviton AE. 1969. Report on a collection of amphibians and reptiles from Afghanistan. Proc Cal Acad Sci 36:279–315.

Clarkson AR, MacDonald M, Fuster V, Cash J, Robson S. 1970. Glomerular coagulation in acute ischaemic renal failure. Quant J Med 39:585–99.

Cloudsley-Thompson JL. 1988a. The saw-scale viper *Echis carinatus*. Br Herpetol Soc Bull 24:32–3.

Cloudsley-Thompson JL. 1988b. The puff adder (*Bitis arietans*). Br Herpetol Soc Bull 26:23–5.

Coborn J. 1977. Observations on the genus "*Bitis*." Report Cotswold Herpetological Society 15:37–41.

Coborn J. 1991. The Atlas of Snakes of the World. Neptune (NJ): TFH Publications. 204 pp.

Cohen AC, Myres BC. 1970. A function of the horns in the sidewinder rattlesnake *Crotalus cerastes*, with comments on other horned snakes. Copeia 3:574–5.

Cohen S. 1959. Purification and metabolic effects of a nerve growth-promoting protein from snake venoms. J Biol Chem 234:1129–37.

Condrea E, de Vries A. 1965. Venom phospholipase A: A review. Toxicon 2: 261–73.
Condrea E, de Vries A, Mager J. 1962. Action of snake venom phospholipase A on free and lipoprotein bound phospholipids. Biophysica Acta 58:389–92.
Cooper JE. 1971. Surgery on a captive Gaboon viper. Br J Herpetol 4(9):234–5.
Copley A, Banerjee S, Devi A. 1973. Studies of snake venoms on blood coagulation. Part I: The thromboserpentin (thrombin-like) enzyme in the venoms. Thromb Res 2:487–508.
Coppola M, Hogan D. 1992. Venomous snakes of southwest Asia. Am J Emer Med. 10:230–6.
Corbett K. 1989. The Conservation of European Reptiles and Amphibians. London: C Helm. 191 pp.
Corkill NL. 1935. Notes on Sudan snakes. Khartoum: Nat Hist Mus (Publ Sudan Gvt Mus). 3:1–40.
Corkill NL, Cochrane JA. 1965. The snakes of the Arabian Peninsula and Socotra. J Bombay Nat Hist Soc 62:475–506.
Cott HB. 1936. The Zoological Society's expedition to the Zambezi, 1927. Proc Zool Soc Lond 2:923–61.
Coulson IM, Riddell IC. 1988. Growth of an African puff adder (*Bitis arietans arietans*) in captivity: The first year of life. J Herpetol Assoc Afr 35:35.
Cox M. 1991. The Snakes of Thailand and Their Husbandry. Malabar (FL): Krieger Publishing.
Creighton D, Haagner G. 1986. *Causus defilippii* envenomation. J Herpetol Assoc Afr 32:38.
Curry-Lindahl K. 1956. Mambas in combat. Afr Wildl 10:340–1.
Daniel JC. 1983. The book of Indian reptiles. Bombay: Bombay Nat Hist Soc. 119 pp.
Daoud EW, Tu AT, El-Asmar MF. 1986. Mechanism of the anticoagulant, Cerastes F-4, isolated from *Cerastes cerastes* (Egyptian sand viper) venom. Thromb Res 41:791–9.
Daoud EW, Tu AT, El-Asmar MF. 1988. The effect of anticoagulant proteinase (cerastase) from *Cerastes cerastes* venom on the blood coagulation system. Hematology 7:241–52.
Darevsky I. 1966. Ecology of the rock-viper in the natural surroundings of Armenia. Mem Inst Butantan Simp Int 33:81–3.
Darlington J. 1983. Aggressive display by snakes. E Afr Nat Hist Soc Bull May/June 36:35.
Das I. 1996. Biogeography of the Reptiles of South Asia. Malabar (FL): Krieger Publishing. 112 pp.
Daudin F. 1803. Histoire naturelle, generale et partiuliere des reptiles, ouvrage faisant suite, a l'histoire naturelle, generale et particuliere composee par LECLERC DE BUFFON, et redigee par CG Sonnini, Vol. 3. Paris: F. Duffart. 150 pp.
Davenport RC, Budden FH. 1953. Loss of sight following snake-bite. Br J Ophthalmol 57:119.

Dekeyser PL, Derivot J. 1960. Serpents et venins; Morsures et traitements. Notes Africaines. Bull D'information et de correspondance de'l Institut Francais D'Afrique Noire 85:1–34.

Delpierre GR. 1968. Studies on African snake venoms: The proteolytic activities of some African Viperidae venoms. Toxicon 5:233–8.

Delpierre GR, Robertson SSD, Steyn K. 1971. Proteolytic and related enzymes in the venom of African snakes. In: de Vries A, Kochva E, editors. Toxins of Animal and Plant Origin. Vol. I. London: Gordon and Breach Science Publ. 353 pp.

Denson KW. 1969. Coagulant and anti-coagulant action of some snake venoms. Toxicon 7:5–11.

Deoras PJ. 1970. Snakes of India. Delhi: National Book Trust. 133 pp.

Deriyagala PEP. 1951. Some new races of snakes *Eryx, Callophis* and *Echis*. Spolia Zeylan. 26:147–50.

Deriyagala PEP. 1955. A colored Atlas of Some Vertebrates from Ceylon. Vol. 3: Serpentoid Reptilian. Ceylon: Govt. Press. 52 pp.

de Silva A. 1980. Snake Fauna of Sri Lanka. Sri Lanka: National Museum of Sri Lanka. 109 pp.

de Silva A. 1990. Colour Guide to the Snakes of Sri Lanka. Avon (Eng): R&A Publishing, Ltd. 88 pp.

Dessauer HC. 1970. Blood chemistry of reptiles. Physiological and evolutionary aspects. In: Gans C, Parsons T, editors. Biology of the Reptilia. Vol. 3. New York: Academic Press. pp 252–8.

Detrait J, Saint-Girons H. 1979. Communautes antigeniques des venins et systematique des Viperidae. Bijdragen tot de Dierkunde 49:71–80.

Detrait J, Saint-Girons H. 1986. European viper's venoms: Toxicity and immunology. In: Rocek Z, editor. Studies in Herpetology. Prague: Docek. pp 631–6

Detrait J, Bea A, Saint-Girons H. 1982. Etude du venim de *Vipera seoanei* Lataste, 1879 (Reptilia, Viperidae). CR Acad ScI Paris, 295:113–6.

Detrait J, Bea A, Saint-Girons H, Choumet V. 1990. Les variations geographiques du venin de *Vipera seoanei* Lataste (1879). Bull Soc Zool Fr 115(3):277–85.

Devedjiev Y, Atanasov B, Mancheva I, Aleksiev B. 1993. The non-toxic component of vipoxin from the venom of the Bulgarian viper *V. ammodytes*. J Mol Biol 229:1147–9.

Devi A. 1968. The protein and nonprotein constituents of snake venom. In: Bucherl W, Buckley E, Deulofeu V, editors. Venomous Animals and Their Venoms. Vol. I. New York: Academic Press. pp 452–9.

de Vries A, Cohen J. 1969. Hemorrhagic and blood coagulation disturbing action of snake venoms. In: Poller L, editor. Recent Advances in Blood Coagulation. London: J&A Churchill. 277 pp.

de Vries A, Gitter S. 1957. The action of *Vipera palaestinae* venom on blood coagulation in vitro. Brit J Haematol 3:379–86.

de Vries A, Rechnic Y, Moroz CH, Moav B. 1963. Prevention of *Echis colorata* venom induced afibrinogenemia by heparin. Toxicon 1:241–2.

de Vries A, Djaldetti M, Cohen I, Rosin IJ, Bessler H. 1966. Coagulation distur-

bance induced by *Echis coloratus* venom *in vivo*. Thromb Diathes Haemorrh 21(suppl):565–9.
de Witte GF. 1962. Genera Del serpents du Congo Et du Ruanda-Urundi. Ann Mus R Afr Cent 104:1–203.
Dickman JD, Colton JS, Chiszar D, Colton CA. 1987. Trigeminal responses to thermal stimulation of the oral cavity in rattlesnakes (*Crotalus viridis*) before and after anesthetization of the facial pit organs. Brain Res 400(2):365–9.
Dimitropolous A. 1990. A new locality record of the Ottoman viper from the Greek island of Oeousses, N.E. Aegean. Ann Musei Goulandris 8:245–9.
Dimitrov G, Kankokar R. 1968a. Fractionation of *Vipera russelli* venom by gel filtration I. Venom composition and relative fraction function. Toxicon 5:213–21.
Dimitrov G, Kankokar R. 1968b. Fractionation of *Vipera russelli* venom by gel filtration II. Comparative study of yellow and white venoms with special reference to the local necrotizing and lethal actions. Toxicon 5:283–8.
Diniz CR. 1967. Bradykinin formation by snake venoms. In: Bucherl W, editor. Venomous animals and their venoms. Vol. I. Paris: Masson. pp 217–28.
Disi A. 1983. A contribution to the herpetofauna of Jordan. 1. Venomous snakes. Dirasat 10:167–80.
Ditmars R. 1934. Snakes of the World. New York: Macmillan. 286 pp.
Djaldetti M, de Vries A. 1965. Mechanism of thrombocytopenia induced by *Echis coloratus* venom. Thromb Diathes Haemorrh 17:253–7.
Djaldetti M, Cohen I, Joshua H, Bessler H, Lorberbaum O, de Vries A. 1965. *In vivo* and *In vitro* inactivation of fibrin stabilizing factor by *Echis coloratus*. Hemostase 5:121–9.
Djaldetti M, Sandbank U, Dintsman M, Bessler H, Rosin M, de Vries A. 1966. Clot-dissolving action of *Echis colorata* venom in dogs. Thromb Diathes Haemorrh 16:420–9.
Dodd C. 1987. Status, conservation and management. In: Seigel R, Collins JT, Novak SS, editors. Snakes: Ecology and Evolutionary Biology. New York: Macmillan. pp 478–513.
Douglas M. 1981. Notes on the reproduction of the common horned adder *Bitis caudalis*, in the central Namib desert. J Herpetol Assoc Afr 25:5–6.
Dravadamani S. 1989. A statistical report on the Irula snake catcher's co-operative society. Hamadryad 14:20–3.
Dreux P, Saint-Girons H. 1951. Ecologie des viperes: *Vipera ursinii*. Bull Soc Zool Fr 76:47–54.
Drewes RC, Sacherer JM. 1974. A new population of carpet vipers *Echis carinatus aliaborri*, new subspecies from Northern Kenya. J E Afr Nat Hist Soc Nat Mus 145:1–7.
Duff-Mackay A. 1965. Notes on the biology of the carpet viper *Echis carinatus pyramidium* in the northern frontier province of Kenya. J E Afr Nat Hist Soc 25:28–40.
Duguy R. 1963. Biologie de la latence hivernale chez *Vipera aspis*. Vie et Milieu 14:311–443.
Duguy R, Martinez Rica JP, Saint-Girons H. 1979. La répartition des vipères dans

les Pyrénées at les regions voisines du nord de l'Espagne. Bull Soc Hist Nat Toulouse 115:358–77.
Duméril AM, Bibron CG. 1854. Erpetologie general ou histoire naturelle complete des reptiles. Tome Septieme. Deuxieme Partie, comprnant l'histoire des serpents venimeux. Paris: Libraire Encyclopedique de Roret. 1536 pp.
Dvilansky A, Biran H. 1973. Hypofibrinogenemia after *Echis colorata* bite in man. Acta Haematol 49:123–7
Efrati P. 1969. Clinical manifestations and treatment of viper bite in Israel. Toxicon 7:29–31.
Efrati P. 1979. Symptomatology, pathology, and treatment of the bites of viperid snakes. In: Lee C-Y, editor. Handbook of Pharmacology. V.52. Snake Venoms. New York: SprinGer-Verlag. pp 588–99.
Efrati P, Reif L. 1953. Clinical and pathological observations on sixty-five cases of viper bite in Israel. Am J Trop Med Hyg 2:1085–108.
Eiselt J, Baran I. 1970. Ergebnisse zoologischer sammelreisen in der Turkei: Viperidae. Ann Naturhist Mus Wien 74:357–69.
El-Asmar MF, Daoud EW, Tu A. 1985. Purification and characterization of a fibrinogenase activity from the Egyptian sand viper *Cerastes cerastes*. Toxicon 23:562.
El-Asmar MF, Shaban E, Hagag M, Swelam N, Tu A. 1986. Coagulant component in *Cerastes cerastes* (Egyptian sand viper) venom. Toxicon 24:1037–44.
El-Asmar MF, Swaney JB. 1988. Proteolysis *in vitro* of low and high density lipoproteins in human plasma by *Cerastes cerastes* venom. Toxicon 26: 809–16.
El-Hawary MFS, Hassan F. 1974a. Physico-chemical properties of venoms of *Cerastes cerastes* and *Cerastes vipera*. Egy J Physiol Sci 1:9–18.
El-Hawary MFS, Hassan F. 1974b. Proteins and amino acids of *Cerastes cerastes* and *Cerastes vipera* venoms. Egy J Physiol Sci 1:19–37.
El-Hawary MFS, Hassan F. 1976. Cholinesterases and anticholinesterases in venoms of *Cerastes cerastes* and *Cerastes vipera*. Egy J Physiol Sci 3:93–9.
Elkin E. 1979. Some problems in reptilian pathology. J Herpetol 6:15–7.
Ellis CG. 1979. The berg adder: A unique snake. The Practioner 223: 544–7.
Els R. 1988. *Atheris superciliaris* envenomation. J Herpetol Assoc Afr 34:52.
Emmrich D. 1993. Envenomation by *Atheris ceratophorus*. J Herpetol Assoc Afr 42:43.
Emmrich D. 1997. New data for the rare Horned Bush-viper, *Atheris ceratophora* Werner, 1985. Mitt Zool Mus Berl 73(1):131–51.
Ernst CH. 1982. A study of the fangs of Russell's Viper (*Vipera russelli*). J Herpetol 16(1)67–71.
Ernst CH. 1992. Venomous Reptiles of North America. Washington: Smithsonian Institution Press. 326 pp.
Ernst, R. & M.-O. Rödel 2002. A new *Atheris* species (Serpentes: Viperidae), from Taï National Park, Ivory Coast. Herpetological Journal 12(2):55–61.
Esnouf M, Williams W. 1962. Isolation and purification of a bovine-plasma pro-

tein which is substrate for the coagulant fraction of Russell's viper venom. Biochem J 20:62–71.
Ewart J. 1878. Poisonous Snakes of India. London: J & A Churchill. 101 pp.
Fairnaru M, Eisenberg S, Manny N, Hershko C. 1974a. The natural course of defibrination syndrome caused by *Echis colorata* venom in man. Thromb Diathes Haemorrh 31:420.
Fairnaru M, Manny N, Hershko C, Eisenberg S. 1974b. Defibrination following *Echis colorata* bite in man. Isr J Med Sci 6(6):720–5.
Faoro G. 1986. Bemerkungen zur Verbastardierung von *Vipera ammodytes* mit *Vipera aspis*. Herpetofauna 8:6–7 (abstracted in Litt Serp 1987 7:198).
Faoro G. 1987. Report on hybrid breeding *V. aspis X V. ammodytes*. Herpetofauna 8:6–7 and Litt Serpent 7:198.
Filio A, Byron A. 1990. On the legal status concerning the protection of amphibians and reptiles in Greece. Herpetol Rev 21:30–2.
Fisher H. 1982. Horned adders-breeding and keeping. Nyoka News. August.
FitzSimons VFM. 1980. A Field Guide to the Snakes of Southern Africa. London: Collins Publ. 221 pp.
Fletcher AP, Alkaersig N, Sherry S. 1962. Pathogenesis of the coagulation defect developing during pathological plasma proteolytic (fibrinolytic) states. J Clin Invest 41:896.
Flichy B. 1978. L'envenomation viperines. Vipere de France. [Thesis]. Paris: Med Paris.
Flower SS. 1930. On the occurrence of *Pseudocerastes* in Sinai. Ann Mag Nat Hist 6(10):224.
Flower SS. 1933. Notes on the recent reptiles and amphibians of Egypt, with a list of the species recorded from that kingdom. Proc Zool Soc Lond 111:735–851.
Foekema GMM. 1973. *Atheris hispida,* een extremum onder de adders. Aquarium, Den Haag 43:150–5.
Forbes CD, Turpie AGG, Ferguson JC, McNicol GP, Douglas AS. 1969. Effects of Gaboon viper (*Bitis gabonica*) venom on blood coagulation, platelets, and the fibrinolytic enzyme system. J Clin Pathol 22:312–6.
Forbes CD, Turpie AGG, McNicol GP, Douglas AS. 1966. Studies on east African snakes: Mode of action of *Echis carinatus*. Scot Med J 11:168–75.
Fox H. 1976. Urogenital systems of reptiles. In: Gans C, Parsons T, editors. Biology of the Reptilia. Vol. 6. New York: Academic Press. pp 57–119.
Fox W. 1948. Effect of temperature on development of scutellation in the garter snake, *Thamnophis elegans atratus*. Copeia 1948:252–62.
Fox W. 1956. Seminal receptacles of snakes. Anat Rec 124:519–40.
Franza BR, Aronnson DL. 1976. Detection and measurement of low levels of prothrombin. Use of a procoagulant from *Echis carinatus* venom. Thromb Res 8:329.
Franza BR, Aronnson DL, Finlayson JS. 1975. Activation of human prothrombin by a procoagulant fraction from the venom of *Echis carinatus*. J Biol Chem 250:7057
Franzen M, Heckes U. 2000. *Vipera barani* Böhme & Joger, 1983 aus dem

östlichen Pontus-Gebirge, Türkei: Differentialmerkmale, Verbreitung, Habitate. Spixiana 23(1):61–70.

Franzen M, Sigg H. 1989. Bemerkungen zu einegen Schlangen Ostanatoliens. Salamandra 25:203–12.

Fretey J. 1975. Guide des Reptiles et Batraciens de France. Paris: Hatier. 56 pp.

Freyvogel TA. 1965. Quantity and toxicity of *Bitis lachesis* venom. Acta Trop 22:267.

Frieberg M, Walls J. 1984. The World of Venomous Animals. Neptune (NS): TFH Publications. 191 pp.

Friedrich C, Tu AT. 1971. Role of metals in snake venoms for hemorrhagic, esterase, and proteolytic activities. Biochem Pharmacol 20:1549.

Fritsche J, Obst FJ. 1966. *Vipera berus bosniensis* auch in Ungarn. Zool Abh Mus Tierk 28:281–3.

Froesch VP. 1967. *Bitis nasicornis,* ein Problem-Pflegling? Aquar U Terrar Z 20:186–9.

Frommhold E. 1969. Die Kreuzotter. Neue Brehm-Bucherei 332. A Ziemsen Verlag, Wittenberg Lutherstadt, Germany.

Furukawa Y, Matsunaga Y, Hayashi K. 1976. Purification and characterization of a coagulant protein from the venom of Russell's viper. Biochim Biophys Acta 453:48–61.

Gaertner K, Goldblum N, Gitter S, de Vries A. 1962. Venom composition. J Immunol 88:256.

Gans C. 1961. Mimicry in procryptically colored snakes of the genus *Dasypeltis.* Evolution 15:72–91.

Gans C, Elliott WB. 1968. Snake venoms: Production, injection, action. Adv Oral Biol 3:45–81.

Gans C, Kochva E. 1965. The accessory gland in the venom apparatus of viperid snakes. Toxicon 3:61–3.

Gans C, Mendelssohn H. 1971. Sidewinding and jumping progression of vipers. In: de Vries A, Kochva E, editors. Toxins of Animal and Plant Origin. Vol I. London: Gordon and Breach Science Publ. 489 pp.

Gans ZR, Gould RJ, Jacobs JW. 1988. Echistatin: A potent platelet aggregation inhibitor from the venom of the viper *Echis carinatus.* J Biol Chem 263(36): 19827–32.

Gasperetti J. 1988. Fauna of Saudi Arabia. Vol. 9: Snakes of Arabia. Riyadh (Saudia Arabia): Nat. Comm. for Wildlife Conserv. and Development (NCWCD). pp 169–372.

Gattullo D, Marsh NA, Whaler BC. 1983. The cardiovascular effects of Gaboon viper venom. J Physiol 334: 40–1.

Geiger R, Kortmann H. 1977. Esterolytic and proteolytic activities of snake venoms and their inhibition by proteinase inhibitors. Toxicon 15:257–9.

Gennaro J, Callahan W, Lorinez A. 1963. The anatomy and biochemistry of a mucus-secreting cell type present in the poison apparatus of the pit viper *Ancistrodon piscivorus.* Ann NY Acad Sci 106:463–71.

Gennaro JF Jr, Leopold RS, Merriam TW. 1961. Comparative venom biochemistry. In: Bucherl W, editor. 1967. Venomous Animals and Their Venoms. Vol. I. Paris: Masson. pp 349–55.

Geoffroy E, Saint-Hillaire T. 1827. Description del reptiles quo se trouvent en Egypt. In: Audouin V [editor]. Description de l'Egypte, ou recueil observations et Del recherches quo Ott en Egypt pendant l'expedition de l'Armee Franchise. Hist Nat 1(1):121–60.

Ghalayini R, Whaler BC. 1985. Cardiovascular effects of horned viper (*Bitis nasicornis*) venom in the rat. Toxicon 23:567.

Gharpurey K. 1962. Snakes of India and Pakistan. Bombay (India): Popular Prakishan. 79 pp.

Ghosh B, Chaudhuri D. 1968. Chemistry and biochemistry of the venoms of Asiatic snakes. In: Bucherl W, Buckley E, Deulofeu V, editors. Venomous Animals and Their Venoms. Vol. 1. New York: Academic Press. pp 577–610.

Gillingham JC, Carpenter CC, Brecke BJ, Murphy JB. 1977. Courtship and copulatory behavior of the mexican milk snake *Lampropeltis triangulum sinaloae*. Southwest Nat 22:187–94.

Gilon D, Shalev O, Benbassat J. 1989. Treatment of envenomation by *Echis coloratus:* A decision tree. Toxicon 27:1105–12.

Gitter S, Amiel S, Gilat G, Sonnino T, Welwart Y. 1963. Neutron activation analysis of snake venoms: presence of copper. Nature 197:383.

Gitter S, de Vries A. 1967. Symptomatology, pathology and treatment of bites by near Eastern, European, and north African snakes. In: Bucherl W, editor. 1967. Venomous Animals and Their Venoms. Vol. I. Paris: Masson. pp 359–402.

Gitter S, Levi G, Kochva S, de Vries A, Technic J, Casper J. 1960. Studies on the venom of *Echis coloratus*. Am J Trop Med Hyg 9:391–9.

Gitter S, Moroz-Perlmutter C, Boss JC, Livni E, Rechnic J, Goldblum N, de Vries A. 1962. Studies on the snake venoms of the near East: *Walterinnesia aegypti* and *Pseudocerastes fieldii*. Am J Trop Med Hyg 11:861–8.

Glenn JL, Straight R, and Snyder CC. 1973. Surgical technique for isolation of the main venom gland of viperid, crotalid, and elapid snakes. Toxicon 11:231–3.

Golay P, Smith HM, Broadley DG, Dixon JR, McCarthy, Golray P, Schatti J-C, Toriba M. 1993. Endoglyphs and Other Major Venomous Snakes of the World: A Checklist. New York: Springer-Verlag. 393 pp.

Gonzales D. 1991. Snakebite problems in Europe. In: Tu AT, editor. Handbook of Natural Toxins. New York: Marcel Dekker. pp 687–751.

Goode M. 1979a. *Echis colorata* water economy. Herpetol Rev 14(4):120.

Goode M. 1979b. Notes on captive reproduction in *Echis colorata*. Herpetol Rev 10(3):94.

Gopalakrishnakone P. 1985. Light and electron microscopic study of the venom apparatus of the saw-scaled viper *Echis carinatus*. Snake 17:10–4.

Grasset E. 1946. La Vipere du Gabon. Envenimation par *Bitis gabonica*. Son venin et serotherapie antiveneuse specifique. Acta Trop 3(2):97–115

Grasset E, Brechbuler T, Schwartz D, Pongratz E. 1956. In: Venoms. AAAS Publ. No. 44:153.

Grasset E, Goldstein L. 1947. Hyperglycemic action of snake venoms in relation to their toxic antigenic properties. Trans Roy Soc Trop Med Hyg 40:771.

Grasset E, Zoutendyk A. 1936. The antigenic characteristic and relationships of viperine venoms based on the cross neutralizing action of heterologous antivenomous sera. Trans Roy Soc Trop Med Hyg 30:347.

Grasset E, Zoutendyk A. 1937. Preparation d'un serum antivenimeux contre le venin de la vipere du Gabon (*Bitis gabonica*). CR Soc Biol 74:609.

Grasset E, Zoutendyk A. 1938. Studies on the Gaboon viper *Bitis gabonica,* and the preparation of a specific therapeutic antivenene. Trans Roy Soc Trop Med Hyg 21(4):445–50.

Gray J. 1946. The mechanism of locomotion in snakes. J Exp Biol 23(2):101–19.

Gray JE. 1831. Synops. Rept., Animal Kingdom. Appendix 9:1–110.

Gray JE. 1842. Monographic synopsis of the vipers of the family Viperidae. Zool Misc 2:68–71.

Gray JE. 1849. Catalogue of the Specimens of Snakes in the Collection of the British Museum. London: E Newman. 115 pp.

Greene HW. 1988. Antipredator mechanisms in reptiles. In: Gans C, editor. The Biology of the Reptilia. Vol. 16. New York: Academic Press. pp 212–317.

Greenwood BM, Warrell DA, Davidson N, Ormerod LD, Reid HA. 1974. Immunodiagnosis of snake bite. Br Med J 4:743–5.

Groombridge B. 1980. A phyletic analysis of viperine snakes. [Ph-D thesis]. City of London: Polytechnic College. 250 pp.

Groombridge B. 1986. Phyletic relationships among viperine snakes. In: Proceedings of the third European herpetological meeting; 1985 July 5–11; Charles University, Prague. pp 11–17.

Grotto L, Goldblum N, de Vries A. 1966. Studies on snake venom neurotoxins-stability: Enhancement of immunogenicity. Isr J Med Sci 2:245.

Grotto L, Jerushalmy Z, de Vries A. 1969. Effect of purified *Vipera palestine* hemorrhagin on blood coagulation and platelet function. Thromb Diathes Haemorrh 22:482–95.

Grotto L, Moroz C, de Vries A, Goldblum N. 1967. Isolation of *Vipera palaestinae* hemorrhagin and distinction between its hemorrhagic and proteolytic activities. Biochim Biophys Acta 133:356–62.

Gruber U. 1989. Die Schlangen Europas und run ums Mittelmeer. Stuttgart: W Keller & Co. 156 pp.

Gruber U, Fuchs D. 1977. Die herpetofauna des Paros-Archipels. Salamandra 13:60–77.

Grzimek B. 1975. Animal Encyclopedia. Vol. 6: Reptiles. New York: Van Nostrand Reinhold. 457 pp.

Gubensek F, Sket D, Lebez D. 1974. Fractionation of *V. ammodytes* venom and seasonal variation of its composition. Toxicon 12:167–74.

Gulden J. 1988. Hibernation and breeding of *V. ammodytes ammodytes*. Litt Serp 8:168–72.

Gumaa KA, Osman OH, Kertesz G. 1974. Distribution of I^{125}-labelled *Bitis arietans* venom in the rat. Toxicon 12:565–8.

Gumprecht A, Lauten U. 1997. Zur fortpflanzung und Haltung der Levante-Otter *Macrovipera lebetina lebetina* (Linnaeus, 1758). Sauria, Berlin 19(1):39–43.
Gunther A. 1878. On reptiles from Midian collected by Major Burton. Proc Zool Soc Lond 1878:977–8.
Guttman-Friedman A. 1956. Blindness after snake-bite. Br J Opthalmol 40:57.
Haacke WD. 1975. Description of a new adder (Viperidae, Reptilia) from Southern Africa, with discussion of related forms. Cimbebasia 4:115–28.
Haagner G. 1986a. *Causus defilippii* reproduction. J Herpetol Assoc Afr 32:38.
Haagner G. 1986b. Notes on the diet of the Gaboon viper. Lammergeyer 37:56.
Haagner G. 1987. *Bitis caudalis* reproduction. J Herpetol Assoc Afr 33:38.
Haagner G. 1988. Gluttony causes death in juvenile puff adder *Bitis arietans*. Koedoe 31:246.
Haagner G. 1990a. The maintenance of puff adders (*Bitis a. arietans*) in captivity with notes on captive breeding. J Herpetol Assoc Afr 38:53–5.
Haagner G. 1990b. *Bitis atropos*, Berg adder reproduction and sperm retention. J Herpetol Assoc Afr 37:48.
Haagner G, Hurter J. 1988. Additional distribution records of the berg adder *Bitis atropos* in the south-eastern Transvaal and Swaziland. Koedoe 31:71–7.
Haas G. 1951. On the present state of our knowledge of the herpetofauna of Palestine. Bull Res Counc 3(B):67–94.
Haas G. 1957. Some amphibians and reptiles from Arabia. Proc Calif Acad Sci, 4th Series, 29:47–86.
Haas G. 1973. Jaw muscles in Rhynchocephalia and Squamata. In: Gans C, editor. The Biology of the Reptilia. Vol. 4. New York: Academic Press. pp 1–113.
Habermann E. 1961. Zuordnung pharmakologischer und enzymatischer Wirkungen von Kallikrein und Shlangengiften mittels diisoprophylfluorophosphat und elektrophorese. Naunyn-Schmiedebergs Arch Exp Pharmakol 240:552–72.
Hadar H, Gitter S. 1959. The results of treatment with Pasteur antiserum in cases of snake bite. Harefuah 56:1–4.
Hagstrom T. 1994. Een succesvolle kweek met de Bosadder (*A. squamiger*). Lacerta 52:101–3.
Hall L. 1962. Investigations in a case of snake bite. E Afr Med J 22:174.
Harding K, Welch K. 1980. Venomous Snakes of the World: A Checklist. Malabar (FL): Krieger Publishing. 120 pp.
Harper JB. 1963. Captive snakes in Ghana. Br J Herpetol 3(4):71–4.
Harrison JR. 1992. Envenomation by a desert horned viper, *Cerastes cerastes:* a case history. In: Contributions to Herpetology. Gr Cincinn Herpetol Soc. pp 15–7.
Hassan F, El-Hawary MFS. 1975. Immunological properties of antivenins. Part I: Bivalent *Cerastes cerastes* and *Cerastes vipera* antivenin. Am J Trop Med Hyg 24(6):1031–4.
Hassan F, El-Hawary MFS. 1977. Fractionation of the snake venoms of *Cerastes cerastes* and *Cerastes vipera*. Toxicon 15:170–3.
Hati R, Mandal M, Hati A. 1993. Active immunization of rabbit with gamma irradiated Russell's viper venom toxoid. Toxicon 28:895–902.

Hattingh J, Willemse GT. 1976a. Hematological observations on the puff adder, *Bitis arietans*. Herpetologica 32:245–7.
Hattingh J, Willemse GT. 1976b. Osmotic fragility of puff adder blood. S Afr J Sci 72:181.
Hawgood B, Bon C. 1991. Snake venom presynaptic toxins. In: Tu A, editor. Reptile venoms and toxins. New York: Marcel Dekker. pp 3–52.
Heatwole H, Davison E. 1976. A review of caudal luring in snakes with notes on its occurrence in the Saharan viper *Cerastes vipera*. Herpetologica 32:332–6.
Hedges NG. 1983. Reptiles and amphibians of East Africa. Nairobi: Kenya Literature Bureau. 65 pp.
Heise PJ, Maxson LR, Dowling HG, Hedges SB. 1995. Higher level snake phylogeny inferred from mitochondrial-DNA sequences of 12S ribosomal RNA and 16S ribosomal RNA genes. Mol Biol Evol 12(2):259–65.
Hellmich W. 1962. Reptiles and Amphibians of Europe. London: Blandford Press. 107 pp.
Herprint International. 1994. Catalogue of Valid Species and Synonyms. Pretoria (South Africa): Herprint International. 280 pp.
Herrmann H-W, Joger U. 1997. Evolution of viperine snakes. In: Thorpe RS. Wüster W, Malhotra A (eds.): Venomous snakes: ecology, evolution and snakebite: 43–61; Symp. Zool. Soc. Lond., 70; Oxford (Oxford University Press).
Herrmann H-W, Joger U, Nilson G. 1992. Phylogeny and systematics of viperinae snakes III: Resurrection of the genus *Macrovipera* (Reuss, 1927) as suggested by biochemical evidence. Amphibia-Reptilia 13: 375–392.
Hillaby J. 1964. A plague of vipers. New Scientist 410:780–3.
Hoffman LAC. 1988. Note on the ecology of the horned adder *Bitis caudalis* from Gobabeb, Namib-Naukluft Park. J Herpetol Assoc Afr 35:33–4.
Hofstetter H, Lott Stolz G. 1982. Nierenveran derungen beim hund infolge Vipernbiss. Schwiz Arch Tierheilk 124:625–9.
Höggren M. 1995. Mating strategies and sperm competition in the adder (*Vipera berus*). Acta Univ Ups, Comprehensive Summaries of Uppsala Dissertations from the Faculty of Science and Technology 163. 27 pp. Uppsala. ISBN 91-554-3630-7.
Höggren M, Nilson G, Andrén C, Orlov NL, Tuniyev BS. 1993. Vipers of the Caucasus: Natural History and Systematic Review. Herpetological Natural History 1(2):11–19.
Höggren M, Tegelström H. 1996. Does long-term storage of spermatozoa occur in the adder (*Vipera berus*)? J Zool Lond 240:501–510.
Hogue-Angeletti R, Bradshaw R. 1979. Nerve growth factors in snake venoms. In: Lee C, editor. Snake Venoms. New York: Springer-Verlag. pp 276–94.
Holzberger H. 1980. Erganzende zur Haltung und Zucht von *Vipera ammodytes*. Herpetofauna 2:32 (summarized in Litt Serpent 1983 3:37–8).
Holzberger H. 1981. Zur Haltung der einheimischen Kreuzotter. Herpetofauna 3:6–9.
Honegger RE. 1978a. Red Data Book. Vol. 3: Amphibia and Reptilia. Switzerland: IUCN. 215 pp.

Honegger RE. 1978b. Threatened amphibians and reptiles in Europe. Report for the Council of Europe.

Honegger RE. 1981. Threatened amphibians and reptiles in Europe. Wiesbaden: Akademische Verlagsgesellschaft.

Hopkins PW. 1957. Additions to the literature on Devonshire amphibians and reptiles. Br J Herpetol 2:90–2.

Huang H. 1984. Effects of phospholipases A from *Vipera russelli* snake venom on blood pressure, plasma prostacyclin level and renin activity in rats. Toxicon 22:253–64.

Huang H, Lee C. 1984. Isolation and pharmacological properties of phospholipases A from *Vipera russelli* snake venom. Toxicon 22:207–17.

Huang S, Perez J. 1980. Comparative study on hemorrhagic and proteolytic activities of snake venoms. Toxicon 18:421–6.

Huffman TN. 1974. Reproduction of a Gaboon viper, *Bitis gabonica gabonica*, in captivity. Arnoldia Rhodesia 6:1–7.

Hughes B. 1968. An unusual rhinoceros viper, *Bitis nascicornis*, from Ghana, west Africa. Zoologische Mededelingen 43:107.

Hughes B. 1976. Notes on African carpet vipers, *Echis carinatus, E. leucogaster,* and *E. ocellatus*. Rev Suisse Zool 83(2):359–71.

Hughes B. 1977. Latitudinal clines and ecogeography of the west African night adder, *Causus maculatus* (Hallowell 1842). Bull de l'Institut Fondamental d'Afrique Noire 39:359–84.

Hughes B, Barry DH. 1968. The snakes of Ghana: A checklist and key. Bull IFAN 31:1004–41.

Hulselmans JLJ, de Roy A, de Vree F. 1970. Contribution a l'herpetologie de la Republique du Togo. 1: List preliminaire Des serpents recoltes par la premiere Mission Zoologique Bulge au Togo. Rev Zool Bot Afr 81:193–6.

Hulselmans JLJ, de Vree F, Van Del Straeten E. 1971. Contribution a l'herpetologie de la Republique du Togo. 3. Liste preliminaire des serpents recoltes par la troisieme Mission Belge au Togo. Rev Zool Bot Afr 83:47–9.

Hurrell DP. 1981. Namaqua dwarf adder bite. S Afr Med J 59:491–2.

Hurter J. 1986. *Bitis atropos* envenomation. J Herpetol Assoc Afr 32:33.

Hurwitz BJ, Hull PR. 1971. Berg adder bite. S Afr Med J 45:969–71.

Hyslop SNA, Marsh NA. 1991. Comparison of the physiological effects in rabbits of Gaboon viper (*Bitis gabonica*) venoms from different sources. Toxicon 29(10):1235–50.

Iddon D, Theakston RDG, Hommel M. 1985. A monoclonal antibody active against the haemorrhagin of Nigerian *Echis carinatus* venom. Toxicon 23: 576.

Ionides CJP, Pitman CRS. 1965a. Notes on two east African venomous snake populations—*Echis carinatus* pyramidum (Geoffroy), Egyptian saw-scaled viper and *Vipera hindii* Boulenger, Montane viper. J E Afr Nat Hist Soc 2(111):116–21.

Ionides CJP, Pitman CRS. 1965b. Notes on three east African venomous snake populations. Puku 3:87–95.

Isemonger RM. 1962. Snakes of Africa. London: T Nelson & Sons. 153 pp.

Ishunin GI. 1964. Feeding of *Echis carinatus* in Uzbekistan and Turkmeniya. Leningrad Prize State Univ: Pro Herr Con 12–14.
Iwanaga S, Suzuki T. 1979. Enzymes in snake venom. In: Lee C-Y, editor. Snake Venoms. New York: Springer-Verlag. pp 61–8.
Izard Y, Boquet P. 1958. Toxic fraction activity in separated venoms. In: Bucherl W, editor. 1967. Venomous Animals and Their Venoms. Vol. I. Paris: Masson. pp 352–71.
Jackson C, Gordon J, Hanahan D. 1971. Separation of the tosyl arginine esterase activity from the factor X activating enzyme of Russell's viper venom. Biochim Biophys Acta 252:255–61.
Jacobs DM, Belcher AD. 1983. Notes on captive reproduction in the west African Gaboon viper (*Bitis gabonica rhinoceros*) at the Rio Grande Zoological Park. Proceedings of a Reptile Symposium; 1982 July 8–12; San Antonio, Texas. Capt Prob Husb 7:103–7.
Jacobson ER. 1985. Use of a polyvalent autogenous bacterin for treatment of mixed gram-negative bacterial osteomyelitis in a rhinoceros viper. JAVMA 187(11):1224–5.
Jacobson ER, Spencer CP. 1983. Colono-uterine fistula in a rhinoceros viper. JAVMA 183(11):1309–10.
Jacobsen NHG. 1986a. Growth of puff adders (*Bitis arietans*) in captivity. J Herpetol Assoc Afr 32:19–23.
Jacobsen NHG. 1986b. Life history notes: *Bitis atropos and Bitis caudalis reproduction*. J Herpetol Assoc Afr 32:38.
Janisch M. 1993. A mutant form of the meadow viper (*V. u. rakosiensis*) from Hungary with a peculiar color pattern. Misc Zool Hung 8:45–9.
Janssen M, Freyvogel TA, Meier J. 1990. Antigenic relationship between the venom of the night adder *Causus maculatus* and venoms from other viperids. Toxicon 28(8):975–83.
Janssen M, Meier J, Freyvogel TA. 1992. Purification and characterization of an antithrombin III inactivating enzyme from the venom of the African night adder (*Causus rhombeatus*). Toxicon 30:985–99.
Jayanthi G, Kasturi S, Gowda T. 1989. Dissociation of catalytic activity and neurotoxicity of a basic phospholipase A2 from Russell's viper venom. Toxicon 27:875–85.
Jena I, Sarangi A. 1993. Snakes of Medical Importance and Snake-bite Treatment. New Delhi: SB Nangia, Ashish Publ House. 293 pp.
Jimenez-Porras JM. 1967. Differentiation between *Bothrops nummifer* and *Bothrops picadoli* by means of the biochemical properties of their venoms. In: Russell FE, Saunders PR, editors. Animal toxins. Oxford: Pergamon Press. pp 196–213.
Jobin F, Esnouf MP. 1966. Coagulant activity of tiger snake (*Notechis scutatus scutatus*) venom. Nature 211:873–5.
Joger U. 1984. The Venomous Snakes of the Near and Middle East. Wiesbaden: Dr. Ludwig Reichert Verlag. 175 pp.
Joger U, Herrmann H-W, Nilson G. 1992. Molecular Phylogeny and Systematics

of Viperine Snakes II: A revision of the *Vipera ursinii* Complex. Proc Sixth Ord Gen meet SEH, Budapest 1991, 239–244.
Joger U, Lenk P, Baran I, Böhme W, Ziegler T, Heidrich P, Wink M. 1997. The phylogenetic position of *Vipera barani* and of *Vipera nikolskii* within the *Vipera* berus complex.
Joger, U., Teynié, A., Fuchs, D. 1988. Morphological characterization of *Vipera wagneri* Nilson & Andrén, 1984 (Reptilia: Viperidae), with first description of the males. Bonn. Zool. Beieträge 39(2/3):221–228.
Johnson BD. 1968. Selected Crotalidae venom properties as a source of taxonomic criteria. Toxicon 6:5–10.
Jooris R. 1986. Breeding *Vipera aspis aspis*. Litt Serpent 6:117.
Joshua H, de Vries A. 1965. Mechanism of thrombocytopenia induced by *Echis colorata* venom *in vivo*. Thromb Diathes Haemorrh 17:253–7.
Joshua H, Djaldetti M, Oskan E, Bessler H, Rosen M, de Vries A. 1964. Mechanism of thrombocytopenia in the dog and guinea pig following *Echis colorata* inoculation. Hemostase 4:333–40.
Kaiser E, Raab W. 1967. Collagenolytic activity of snake and spider venoms. Toxicon 4:251–5.
Kamiguti AS, Theakston RD, Tomy SC. 1988. An investigation of the coagulant activity of the venom of the saw-scale viper *Echis carinatus* from Saudi Arabia. Ann Trop Med Parasitol 82(5):503–9.
Kardong KV. 1986a. The predatory strike of the rattlesnake: when things go amiss. Copeia 1986(3):816–20.
Kardong KV. 1986b. Observations on live *Azemiops feae,* Fea's viper. Herpetol Rev 17(4):81–2.
Karstens R. 1986. Breeding *V. aspis x V. ammodytes*. Litt Serpent 6:203–4.
Kellaway CH, Williams FE. 1933. Bite incidences of elapid snakes. Aus J Expl Biol Med Sci 11:84.
Khalaf KT. 1959. Reptiles of Iraq with notes on the Amphibians. Baghdad: Ar-Rabitta Press. 80 pp.
Khan MS. 1983. Venomous terrestrial snakes of Pakistan. The Snake 15:101–5.
Khan MS. 1990. Venomous terrestrial snakes of Pakistan and the snake bite problem. In: Gopalalakrishnakone P, Chou LM, editors. Snakes of Medical Importance (Asia-Pacific Region). Singapore: Venom and Toxin Research Group, Nat Univ of Singapore. pp 135–58.
Khole V. 1991. Toxicities of snake venoms and their components. In: Tu A, editor. Reptile Venoms and Toxins. New York: Marcel Dekker. pp 405–70.
Kini RM, Evans HJ. 1990. Effects of snake venom on blood platelets. Toxicon 28:387.
Kirk R, Corkill NL. 1946. Venom of the rhinoceros viper, *Bitis nasicornis*. J Trop Med Hyg 49:9–14.
Klauber LM. 1972. Rattlesnakes. Berkeley (CA): Univ Cal Press. 350 pp.
Klemmer K. 1967a. Methods of classification of venomous snakes. In: Bucherl W, editor. 1967. Venomous Animals and Their Venoms. Vol. I. Paris: Masson. pp 275–8.

Klemmer K. 1967b. Classification and distribution of European, north African, and north and west Asiatic venomous snakes. In: Bucherl W, editor. 1967. Venomous Animals and Their Venoms. Vol. I. Paris: Masson. pp 309–26.

Klemmer K. 1973. List Der rezenten Giftschlangen. In: Die Giftschlangen der Erde. 255–449. Behringwerke Mitteil., Sonderband., Elwert Unix. Buchhandl., Mar burg/Lahn.

Knight A, Mindell DP. 1993. Substitution bias, weighting of DNA-sequence evolution, and the phylogenetic position of Fea's viper. Sys Biol 42(1):18–31.

Knoepffler L. 1965. Autoobservation d'envenimation par morsure d'*Atheris* sp. Toxicon 2:275–6.

Knoepffler P, Sochurek E. 1955. Neues uber die Rassen der wiesenotter (*Vipera ursinii*). Burgenl Heimatbl 17:185–8.

Kocholaty WF, Ashley B. 1966. Detoxification of Russell's viper and water moccasin venom by photooxidation in the presence of methylene blue. Toxicon 3:187–94.

Kocholaty WF, Ledford EB, Daly JG, Billings TA, Kochva E. 1960. Venom activity in viperid snakes. Am J Trop Med Hyg 9:381.

Kochva E. 1962. On the lateral jaw musculature of the Solenoglypha with remarks on some other snakes. J Morphol 110:227–71.

Kochva E. 1978. Oral glands of the Reptilia. In: Gans C, Gans M, editors. Biology of the reptilia. Vol. 8. New York: Academic Press. pp 85–156.

Kochva E. 1987. The origin of snakes and evolution of the venom apparatus. Toxicon 25:65–106.

Kochva E, Gans C. 1964. Snake venoms: production, injection, action. Anat Rec 148:302.

Kochva E, Gans C. 1965. The venom gland of *Vipera palaestinae* with comments on the glands of some other viperines. Acta Anat 62:365–401.

Kochva E, Gans C. 1966. Histology and biochemistry of the venom gland of some crotaline snakes. Copeia 3:506.

Kochva E, Gans C. 1967. The structure of the venom gland and secretion of venom in viperid snakes. 1st International Symposium on Animal Toxins; 1966 Nov. 11–15; Atlantic City, New Jersey. New York: Macmillan (Pergamon). pp 290–321.

Kochva E, Gans C. 1970. Salivary glands of snakes. Clin Tox 3:363–73.

Kochva E, Gans C. 1971. Venom glands of the Viperidae. In: Minton S Jr, editor. Snake Venoms and Envenomation. New York: Marcel Dekker. pp 40–89.

Kochva E, Shayer-Wollberg M, Sobol R. 1967. The special pattern of the venom gland in *Atractaspis* and its bearing on the taxonomic status of the genus. Copeia 59:763–8.

Kochwa S, Gitter S, Strauss A, de Vries A, Leffkowitz M. 1959a. Immunologic study of *Vipera xanthina palaestinae* venom and preparation of potent antivenin in rabbits. J Immunol 82:107–15.

Kochwa S, Izard Y, Boquet P, Gitter S. 1959b. Sur preparation d'un immun-serum equin antivenimeux au moyen des fractions neurotoxiques isolees du venin de *Vipera xanthina palaestinae*. Ann Inst Pasteur 97:370–6.

Kochwa S, Perlmutter C, Gitter S, Rechnic J, de Vries A. 1960. Studies on *Vipera palaestinae* venom. Fractionation by ion exchange chromatography. Am J Trop Med Hyg 33:374–9.

Kornalik F, Blomback B. 1975. Prothrombin activation induced Ecarin, a prothrombin converting enzyme from *Echis carinatus*. Thromb Res 6:53–63.

Kornalik F, Hladovec J. 1975. The effect of ecarin-defibrinating enzyme isolated from *Echis carinatus* on experimental thrombosis. Thromb Res 7:611.

Kornalik F, Pudlak P. 1971. A prolonged defibrination caused by *Echis carinatus* venom. Life Sci 10(2):309–14.

Kornalik F, Pudlak P. 1973. Coagulation defect following non-toxic doses of *Echis* viper venom. Experimentia 18:381–2.

Kornalik F, Taborska E. 1973. Enzyme activity in *Echis* venom. In: Kaiser I, editor. Animals and plant toxins. Munich: Goldmann. pp 161–6.

Kotenko TI. 1989. *Vipera ursinii renardi* in Ukraine. [abstract]. First World Congress of Herpetology; 1988 Nov. 6–9; Cambridge, England. 57 pp.

Kramer E. 1961a. Uber zwei afrikanische Zwergpuffottern, *Bitis hindii* und *Bitis superciliaris*. Vjschr Naturf Ges Zurich 106:419–23.

Kramer E. 1961b. Variation, sexual dimorphism, description and taxonomy of *Vipera ursinii* and *Vipera kaznokovi*. Rev Suisse de Zool 68:627–726.

Kramer E. 1971. Revalidierte und neue Rassen der europaischen Schlangenfauna. Lavori Soc Ital Biogeogr NS 1:667–76.

Kramer E, Schnurrenberger H. 1958. Zur Schlangenfauna von Libyen. Die Aquarien- und Terrarien-Zeitschr XI.2., 1.2.:57–9.

Kramer E, Schnurrenberger H. 1963. Systematik, verbreitung undo okologie Der Libyschen schlangen. Rev Suisse Zool 70:453–568.

Krause PB. 1982. Differences in the musculature of the venom gland among river jack vipers, *Bitis nasicornis*. J Herpetol 16(1):87–9.

Kudryavtsev S, Mamet S. 1991. Husbandry and propagation of the Radde's viper. Herpetol Rev 22:96.

Kudryavtsev S, Mamet S, Proutkina M. 1993. Keeping and breeding in captivity snakes of Russia and adjacent countries. Part II. The Snake 25:121–30.

Kuntz R. 1963. The snakes of Taiwan. Taipei: US Navy Medical Research Unit No. 2. 43 pp.

Labib RS, Awad ER, Farag NW. 1981b. Proteases of *Cerastes cerastes* and *Cerastes vipera* snake venoms. Toxicon 19:73–83.

Labib RS, Azab MH, Farag NW. 1981a. Effects of *Cerastes cerastes* (Egyptian sand viper) and *Cerastes vipera* (Sahara sand viper) snake venoms on blood coagulation: separation of coagulant and anticoagulant factors and their correlation with arginineesterase and protease activities. Toxicon 19: 85–94.

Labib RS, Halim HY, Farag NW. 1979. Fractionation of *Cerastes cerastes* and *Cerastes vipera* snake venoms by gel filtration and identification of some enzymatic and biological activities. Toxicon 17:337–45.

Lake AR, Trevor-Jones TR. 1987. Formation of the poison fang canal of the puff adder *Bitis arietans*. S Afr J Sci 83:668–9.

Lake AR, Trevor-Jones TR, le Roux CDJ, Hattingh J. 1988. Histology of the venom apparatus of the puff adder *Bitis arietans*. S Afr J Sci 84:150–2.
Lambiris AJ. 1966. Observations on Rhodesian reptiles. J Herpetol Assoc Afr 2: 33–4.
Lanceau M-T, Lanceau Y. 1974. A la Decouverte des Reptiles. Paris: Fleurus. 89 pp.
Lane M. 1963. Life with Ionides. London: Hamish-Hamilton. 157 pp.
Lanoie L, Branch W. 1991. *Atheris squamiger:* fatal envenomation. J Herpetol Assoc Afr 39:29.
Lasfargues E, Di Fine-Lasfargues J. 1951. Vipers venimeux. Ann Inst Pasteur 81:642–51.
Latifi M. 1973. Comparative venom yield in five species of viperid snakes. In: Proceedings of the 9th International Congress on Tropical Medicine and Malaria; 1972 June 3–11; Bilbao, Spain. London: Blandford Press. Abstract 107.
Latifi M. 1984. Variation in yield and lethality of venoms from Iranian snakes. Toxicon 22:373–80.
Latifi M. 1991. The snakes of Iran. Published by the Department of the Environment and the Society for the Study of Amphibians and Reptiles, 2nd ed. 156 pp.
Latifi M, Farzanpay R, Tabatabai M. 1973. Comparative studies of Iranian snake venoms by gel diffusion and neutralization tests. In: Kaiser E, editor. Symposium on Animal and Plant Toxins. Munich: Golman. pp 345–57.
Laurent RF. 1955. Diagnoses preliminaires de quelques serpents venimeux. Rev Zool Bot Afr 51:127–39.
Laurent RF. 1956a. Contribution a l'herpetologie de la region Des Grands Lacs de l'Afrique Central. Parts 1–3. Ann Mus Congo Belg 48:1–390.
Laurent RF. 1956b. Notes herpetologiques africaines. I. Rev Zool Bot Afr Vol. LIII, Fasc. 3–4.
Laurent RF. 1958. Notes herpetologiques africaines. II. Rev Zool Bot Afr 58: 115–28.
Laurent RF. 1960. Notes complementaires sur les Cheloniens et les Ophidiens du Congo Oriental. Ann Mus Roy Congo-Belge Serie No. 8, Sci Zool 84.
Laurenti JN. 1798. Specimen medicum, exhibens synopsin reptilium emendatum cum experimentis circa venena et antidota reptilium austracorum, quod authoritate et consensus. Vienna: Joan Thomae. 217 pp.
Lawson, DP 1993. The reptiles and amphibians of the Korup National Park Project, Cameroon. Herpetological Natural History. 1(2):27–90.
Lawson, DP 1999. A new species of arboreal viper (Serpentes: Viperidae: *Atheris*) from Cameroon, Africa. Proc. Biol. Soc. Washington. 112(4):793–803.
Lawson DP, Noonan BP, Ustach PC 2001. *Atheris subocularis* (Serpentes: Viperidae) Revisited: Molecular and Morphological Evidence for the Resurrection of an Enigmatic Taxon. Copeia. 2001(3):737–744.
Lawson, DP, Ustach PC 2000. A redescription of *Atheris squamigera* (Serpentes: Viperidae) with comments on the validity of *Atheris anisolepis*. J. Herpetol. 34(3):386–389.
Lee CY, Ho CL, Botes DP. 1983. Site of action of caudoxin, a neurotoxic phos-

pholipase A2 from the horned puff adder (*Bitis caudalis*) venom. Toxicon 20(3):637–47.
Leighton G. 1901. The Life History of British Serpents and Their Local Distribution in the British Isles. London: Blackwell & Son. 112 pp.
Lenk P, Herrmann H-W, Joger U, Wink M. 1999. Phylogeny and taxonomic subdivisions of *Bitis* (Reptilia: Viperidae) based on molecular evidence. Kaupia 8:31–38
Lenk P, Kalyabina S, Wink M, Joger U. 2001. Evolutionary relatonships among the true vipers (Reptilia: Viperidae) inferred from mitochondrial DNA sequences. Molecular Phylogenetics and Evolution 19(1):94–104.
Leston D. 1970a. The activity pattern of *Causus rhombeatus* in Ghana. Br J Herpetol 4(6):139–40.
Leston D. 1970b. Some snakes from the forest zone of Ghana. Br J Herpetol 4(6):141–4.
Leston D, Hughes B. 1968. The snakes of Tafo, a forest cocoa-farm in Ghana. Bull de IFAN 30(2):737–58.
Leviton AE. 1967. The venomous terrestrial snakes of east Asia, India, Malaya, and Indonesia. In: Bucherl W, editor. 1967. Venomous Animals and Their Venoms. Vol. I. Paris: Masson. pp 529–76.
Leviton AE, Anderson SC. 1967. Survey of the reptiles of the Sheikdom of Abu Dhabi, Arabian Peninsula. Part II. Systematic account of the collection of reptiles made in the Sheikdom of Abe Dhabi by John Gasperetti. Pro Calif Acad Sci 35:157–92.
Leviton A, Anderson SC, Adler K, Minton S. 1992. Handbook to Middle East Amphibians and Reptiles. Society for the Study of Amphibians and Reptiles, St. Louis University, St. Louis, USA. 160 pp.
Li Y, Liu K, Wang Q, Ran R, Tu G. 1985. A platelet function inhibitor purified from *Vipera russelli siamensis* snake venom. Toxicon 23:895–903.
Liem KF, Marx H, Rabb GB. 1971. The viperid snake *Azemiops:* Its comparative cephalic anatomy and phylogenetic position in relation to Viperinae and Crotalinae. Field Zool 59:67–126.
Lloyd CNV. 1977. Report on a Berg adder bite. J Herpetol Assoc Afr 16:8–10.
Lombardi G, Bianco F. 1974. La Vipera. Florence (Italy): Nardini Editore. 384 pp.
Lorberbaum O, Cohen I, Joshua H, de Vries A. 1966. Action of *Echis colorata* venom on intrinsic thromboplastin generation. Isr J Med Sci 2(2):248–9.
Lotze H. 1973. Die Schlangen der Erimonissia in den Kykladen. Salamandra 9:58–70.
Louw GN. 1972. The role of advective fog in the water economy of certain Namib desert animals. Symp Zool Soc Lond 31:297–314.
Loveridge A. 1933. Report on the scientific results of an expedition to the southwestern highlands of Tanganyika Territory. Bull Mus Comp Zool 74:195–423.
Loveridge A. 1946. A guide to the snakes of the Nairobi District. J E Afr Nat Hist Soc 18(3–4):97–115.
Loveridge A. 1953. Zoological results of a fifth expedition to East Africa. III. Reptiles from Nyasaland and Tete. Bull Mus Comp Zool 110:143–322.

Loveridge A. 1959. On a fourth collection of reptiles, mostly taken in Tanganyika territory by Mr. CJP Ionides. Proc Zool Soc Lond 133(1):29–44.

Luiselli L, Akani GC; Angelici FM. 2000. Arboreal habits and viper biology in the African rainforest: The ecology of *Atheris squamiger*. Israel Journal of Zoology 46(4): 273–286.

Luiselli L, Rugiero L. 1990. On habitat selection in six species of snakes in Canale Monterano, including data on reproduction and feeding in *Vipera aspis franciscired i*. Herpetozoa 2:107–15.

Lynn WG. 1931. The structure and function of the facial pit of the pit vipers. Am J Anat 49:97.

Mackay A. 1980. An adventuresome puff adder. E Afr Nat Hist Soc Bull 88:53.

Mackay N, Ferguson JC, McNichol GP. 1970. Effects of the venom of the rhinoceros horned viper (*Bitis nasicornis*) on blood coagulation, platelet aggregation, and fibrinolysis. J Clin Pathol 23:789–96.

Mahendra BC. 1984. Handbook of the snakes of India, Ceylon, Burma, Bangladesh, and Pakistan. Ann Zool 22.

Manacas S. 1981–1982. Ofideos venenosos da Guiné, S. Tomé, Angola e Mocambique. Garcia de Orta, Ser Zool Lisboa 10(1–2):13–46.

Manjunatha R, Evans HJ. 1990. Effects of snake venom proteins on blood platelets. Toxicon 28(12):1387–422.

Mara WP. 1993. Venomous Snakes of the World. Neptune (NS): TFH Publications. 275 pp.

Marais J. 1981. Case history of snouted night adder bite. J Herpetol Assoc Afr 26:6–7.

Marais J. 1992. A Complete Guide to the Snakes of Southern Africa. Malabar (FL): Krieger Publishing. 284 pp.

Marian M. 1956. Adatok a keresztes vipera (*Vipera b. berus*) somogyi elterjedesi viszonyaihoz. Budapest (Hungary): Ann Hist Nat Mus Nat Hung pp 115–9.

Marian M. 1963. Nehany adat a keresztes vipera (*Vipera b. berus*) szaporadas biologiajahoz. Budapest: Vertebr Hung. 158 pp.

Markland FS, Pirkle H. 1977. Biological activities and biochemical properties of thrombin-like enzymes from snake venoms. In: Lundbard R, Fenton JW, Mann KG, editors. Chemistry and biology of thrombin. Ann Arbor (MI): Ann Arbor Science. pp 250–60.

Marsh NE. 1975. Gaboon viper venom: a comparative study of the coagulant, proteolytic and toxic properties of four commercially dried preparations and freshly collected venom. Toxicon 13:171–5.

Marsh NE, Glatston A. 1974. Some observations on the venom of the rhinoceros horned viper, *Bitis nasicornis*. Toxicon 12:621–8.

Marsh NE, Smith ICH, Whaler BC. 1979. Effects of envenomation on cardiac cell permeability. J Phys 291:72–3.

Marsh NE, Whaler BC. 1974. Separation and partial characterization of a coagulant enzyme from *Bitis gabonica* venom. Br J Haematol 26:295–306.

Marsh NE, Whaler BC. 1984. The Gaboon viper (*Bitis gabonica*) its biology, venom components and toxinology. Toxicon 22(5):669–94.

Marx H. 1956. Keys to the Lizards and Snakes of Egypt. Cairo: US Navy Research Report. 47 pp.
Marx H. 1958. Sexual dimorphism in coloration in the viper *Cerastes vipera*. Nat Hist Misc 164:1–2.
Marx H, Olechowski TS. 1970. Fea's viper and the common gray shrew: a distribution note on predator and prey. J Mammol 51:205.
Marx H, Rabb GB. 1965. Relationships and zoogeography of the viperine snakes (Family Viperidae). Field Zool 44:161–206.
Marx H, Rabb GB. 1972. Phyletic analysis of fifty characters of advanced snakes. Field Zool 63:1–321.
Master RWP, Rao SS. 1961. Identification of enzymes and toxins in venoms of Indian cobra and Russell's viper after starch gel electrophoresis. J Biol Chem 236:1986–90.
Master RWP, Rao SS. 1963. Enzyme content of elapid and viperid venoms. In: Bucherl W, editor. 1967. Venomous Animals and Their Venoms. Vol. I. Paris: Masson. pp 88–135.
Matthews JG. 1968. The rhinoceros viper: a note on envenomation and capture of prey. Uganda J 32:81–2.
Mattison C. 1982. The Care of Reptiles and Amphibians in Captivity. Poole (Eng): Blandford Press. 160 pp.
Mattison S. 1986. Snakes of the World. Poole (Eng): Blandford Press. 145 pp.
Maurer FW Jr. 1975. Observation of a mating pair of African puff adders, *Bitis arietans arietans*. Botsw Not Rec 7:198–201.
McDiarmid, Campbell JA, Touré T'SA 1999. Snake Species of the World. A Taxonomic and Geographical Reference. Vol. 1. The Herpetologists' League. 511 pp.
McDonald LJ. 1962. Food and feeding in *Bitis arietans*. Lammergeyer 2:70.
McKay D, Moroz C, de Vries A, Csavossy J, Cruse V. 1970. The action of hemorrhagin and phospholipase derived from *Vipera palaestinae* venom on the microcirculation. Lab Invest 22:387–99.
McMahon M. 1990. *Vipera ammodytes meridionalis* envenomation. J Herpetol Assoc Afr 37:60.
Meaume J. 1956. Les venins de serpents agents modificateurs de la coagulation sanguine. Toxicon 4:25–58.
Mebs D. 1970a. Biochemistry of kinin-releasing enzymes in the venom of the viper *Bitis gabonica* and of the lizard *Heloderma suspectum*. In: Peters T, editor. Bradikinin and Related Kinins. New York: Plenum Press. pp 107.
Mebs D. 1970b. A comparative study of enzyme activities in snake venoms. Int J Biochem 1:4335.
Mebs D. 1978. Pharmacology of reptile venom. In: Gans C, editor. The Biology of the Reptilia. Vol. 8. New York: Academic Press. pp 1–255.
Mebs D, Kuch U, Meier J. 1994. Studies on venom and venom apparatus of Fea's viper, *Azemiops feae*. Toxicon 32(10):1275–8.
Mehely L. 1911. Systematisch-phylogenetische studien and Viperiden. Budapest: Ann Hist Nat Mus Nat Hung 9:186–243.

Mehrtens JM. 1987. Living Snakes of the World in Color. New York: Sterling Publishers. 256 pp.
Meier J, Stocker KF. 1991. Snake venom protein C activators. In: Tu A, editor. Reptile Venoms and Toxins. New York: Marcel Dekker. pp 265–79.
Meier J, Stocker K, Svendson L, Brogli M. 1985. Chromogenic proteinase substrates as possible tools in the characterization of crotalid and viperid snake venoms. Toxicon 23:393–7.
Melichercik J. 1993. The viper *Ammodytes meridionalis* and its breeding in the terrarium. Akvarium Terarium 36:36–7.
Mellanby J. 1909. The coagulation of blood. Part II. The actions of snake venom, peptone and leech extract. J Physiol 38:441.
Mendelssohn H. 1963. On the biology of venomous snakes of Israel. Part I. Isr J Zool 12:143–70.
Mendelssohn H. 1965. On the biology of venomous snakes of Israel. Part II. Isr J Zool 14:185–212.
Mendelssohn H, Golani I, Marder U. 1971. Agricultural development and the distribution of venomous snakes and snake bite in Israel. In: de Vries A, Kochva E, editors. Toxins of animal and plant origin. Vol 1. New York: Gordon and Breach Science Publ. pp 3–16.
Mermod C. 1970. Domaine vital et deplacements chez *Cerastes vipera* et *Cerastes cerastes*. Rev Suisse De Zool 77:555–61.
Merrem B. 1820. Versuch eines Systems der Amphibien I. (Tentamen Systematis Amphibiorum.) Marburg: JC Kriegeri. 191 pp.
Mertens R. 1947. Die Lurche und Kriechtiere des Rhein-Main Gebietes. Verlag Dr. Waldemar Kramer, Frankfurt am Main, Germany.
Mertens R. 1950. Über reptilienbastarde. Senckenbergiana 31.10.1950, Frankfurt am Main, Germany.
Mertens R. 1952. Amphibien und Reptilien aus der Türkei. Rev Fac Sci Univ Istanbul 1:41–75
Mertens R. 1964. Über Reptilienbastarde III. Senckenbergiana 13.3. Frankfurt am Main, Germany.
Mertens R. 1969. Die Amphibien und Reptilien West Pakistan. Stuttgarter Beitr Z Naturkunde 197:1–96.
Mertens R, Darevsky I, Klemmer K. 1967. *Vipera latifii,* eine neue Giftschlange aus dem Iran. Senckenb Biol 48:161–8.
Mertens R, Müller L. 1928. Liste der amphibien und reptilien Europas. Abh Senckenb Naturf Ges 45:1–62.
Mierte D. 1992. Cles de determination des serpents d'Afrique. Ann Sci Zool Mus Roy L'Afrique Centr Tervuren Belgique 267:1–152.
Minnich JE. 1982. The use of water. In: Gans C, editor. The Biology of the Reptilia. Vol. 12. New York: Academic Press. pp 86–207.
Minton SA Jr. 1966. A contribution to the herpetology of west Pakistan. Bull Am Mus Nat Hist 134:27–184.
Minton SA Jr. 1967a. Snakebite. In: Beeson PB, McDermott W, editors. Cecil and Loeb Textbook of Medicine. Philadelphia: Saunders. 420 pp.

Minton SA Jr. 1967b. Observations on the toxicity and antigenic makeup of venoms from juvenile snakes. Toxicon 4:294–301.
Minton SA Jr. 1968. Antigenic relationships of the venom of *Atractaspis microlepidota* to that of other snakes. Toxicon 6:59–65.
Minton SA Jr. 1974. Venom Diseases. Springfield (IL): CC Thomas Publ. 386 pp.
Minton SA Jr. 1975. A note on the venom of an aged rattlesnake. Toxicon 13:73.
Minton SA Jr. 1976. Neutralization of old world viper venoms by American pit viper antivenin. Toxicon 14:146–8.
Minton SA Jr. 1979. Common antigens in snake venoms. In: Lee CY, editor. Handbook of Experimental Pharmacology. Berlin: Springer-Verlag. 52:847.
Minton SA Jr. 1987. Present tests for detection of snake venom: clinical applications. Ann Emer Med 15:77–88.
Minton SA Jr. 1990. Neurotoxic snake envenoming. Sem Neurol 10(1):52–61.
Minton SA Jr. 1994. Clinical hemosatic disorders caused by venoms. In: Ratnoff MC, Forbes, editors. Disorders of Hemostasis. Philadelphia: Saunders. pp 518–31.
Minton SA Jr, Minton MR. 1969. Venomous Reptiles. New York: Charles Scribner's Sons. 215 pp.
Moav B, Moroz CH, de Vries A. 1963. Activation of the fibrinolytic system of the guinea pig following inoculation of *Echis* venom. Toxicon 1:109–12.
Mohamed AH, Abdel-Baset A, Hassan A. 1980. Immunological studies on monovalent and bivalent *Cerastes* antivenin. Toxicon 18:384–7.
Mohamed AH, Bakr IA, Kamel A. 1966. Egyptian polyvalent antisnakebite serum technic of preparation. Toxicon 4:69–72.
Mohamed AH, Darwish MA, Hani-Ayobe M. 1973. Immunological studies on Egyptian polyvalent antivenins. Toxicon 11:457–60.
Mohamed AH, Darwish MA, Hani-Ayobe M. 1974a. Immunological studies on an Egyptian bivalent *Naja* antivenin. Toxicon 12:321–3.
Mohamed AH, Darwish MA, Hani-Ayobe M. 1974b. Studies on Egyptian *Cerastes cerastes* antivenin. Toxicon 12:599–601.
Mohamed AH, El-Damarawy NA. 1974. The role of the fibrinolytic enzyme system in the haemostatic defects following snake envenomation. Toxicon 12:467–75.
Mohamed AH, El-Serougi IS, Hannah MM. 1969. Observations on the effects of *Echis carinatus* venom on blood clotting. Toxicon 6:215–9.
Mohamed AH, El-Serougi M, Khaled LZ. 1969. Effects of *Cerastes cerastes* venom on blood coagulation mechanisms. Toxicon 7:181–4.
Mohamed AH, El-Serougi MS, Kamel A. 1963. Effects of *Echis carinatus* venom on blood glucose and liver and muscle glycogen concentrations. Toxicon 1:243–4.
Mohamed AH, Fawzia K, Khalil FK, Baset AA. 1977. Immunological studies on polyvalent and monovalent snake antivenins. Toxicon 15:271–4.
Mohamed AH, Fouad S, Abbas F, Abdel-Aal A, Abdel-Baset A, Hassan A, Abbas N, Zahran F. 1980. Metabolic studies of Egyptian and allied African snake venoms. Toxicon 18:381–3.

Mohamed AH, Fouad S, Abdel-Aal A, Abdel-Baset A, Hassan AA, Abbas N, Zahran F, Abbas F. 1980. Effect of some Egyptian and African snake venoms on blood levels of sodium and potassium. Toxicon 18:479.

Mohamed AH, Kamel A, Ayobe MH. 1969a. Studies of phospholipase A and B activities of Egyptian snake venoms and a scorpion toxin. Toxicon 6:293–8.

Mohamed AH, Kamel A, Ayobe MH. 1969b. Some enzymatic activities of Egyptian snake venoms and a scorpion venom. Toxicon 7:185–8.

Mohamed AH, Khaled LZ. 1966. Effect of the venom of *Cerastes cerastes* on nerve tissue and skeletal muscle. Toxicon 3:223–4.

Mohamed AH, Khaled LZ. 1969. Effects of *Cerastes cerastes* venom on blood and tissue histamine and on arterial blood pressure. Toxicon 6:221–3.

Mohamed AH, Khaled LZ, Abdel-Rehim MS. 1969. Effects of different Egyptian venoms on the oxygen consumption of isolated tissue slices. Toxicon 7:251–4.

Mohamed AH, Saleh AM, Ahmed S, El-Maghraby M. 1977. Effects of *Cerastes vipera* snake venom on blood and bone marrow cells. Toxicon 15:35–40.

Mohamed AH, Saleh AM, Ahmed S, El-Maghraby M, Allam HN. 1977. Effect of *Cerastes vipera* snake venom on muscle spindles, spinal ganglia and spinal cord. Toxicon 15:235–45.

Mohamed AH, Saleh AM, El-Maghraby M. 1975. Histopathological and histochemical changes in skeletal muscles after *Bitis gabonica* envenomation. Toxicon 13:165–9.

Mole RH, Everard A. 1947. Snake bite by *Echis carinatus*. Quart J Med 16: 291–302.

Monney J-C. 1989. An ecological study of the asp viper in prealpine environment. [Abstract]. 1st World Congress of Herpetology; 1988 June 2–6; Paris. London: Blackwell. 99 pp.

Moore RW. 1968. *Vipera berus* bite: a case study. Adv Oral Biol 3:22–23.

Morgan D. 1988. The lowland viper (*A. superciliaris*) in captivity: a preliminary perspective. J Herpetol Assoc Afr 34:38–9.

Morita T, Iwanaga S, Suzuki T. 1976. The mechanism of activation of bovine prothrombin by an activator isolated from *Echis carinatus* venom and characterization of the new active intermediates. J Biochem 79:1089.

Moroz C, de Vries A, Sela M. 1966a. Isolation and characterization of a neurotoxin from *Vipera palaestinae*. Biochim Biophys Acta 124:136–46.

Moroz C, Grotto L, Goldblum N, de Vries A. 1966b. Enhancement of immunogenicity of snake venom neurotoxins. In: Russell FE, Saunders PR, editors. Animal Toxins. Oxford: Pergamon Press. pp 68–75.

Morton CB. 1960. Adder bites in Cornwall. Br Med J 1960 1:373–6.

Mulder, J. 1994. Additional information on *Vipera albizona* (Reptilia, Serpentes, Viperidae). Deinsea—Annual of the Natural History Museum of Rotterdam 1:77–83.

Murphy JB. 1974. Case history of snakebite due to the horned puff adder *Bitis caudalis*. J Herpetol Assoc Afr 12:22–4.

Murphy JB, Barker DG. 1980. Courtship and copulation of the Ottoman viper with special reference to use of the hemipenes. Herpetologica 36:165–70.

Murphy JB, Joy JE. 1973. Removal of a deficient fang mechanism in a captive puff adder. Br J Herpetol 4:314–6.

Murthy TSN, Venkateswarlu T. 1974. Notes on the occurrence of the saw-scaled viper, *Echis carinatus* around Madras and its bites. Indian J Zool 13(3):117–8.

Nadjafov J, Iskenderov T. 1994. The peculiarities of reproduction of levantine viper. Zool Zhurn 73:79–84.

Nahas L, Denson KWE, MacFarlane RG. 1964. A study of the coagulant action of eight venomous snakes. Thromb Diathes Haemorrh 12:355–67.

Naulleau G. 1970. Le reproduction de *Vipera aspis* en captivite dans des conditions artificielles. J Herpetol 4:113–21.

Naulleau G. 1973a. Rearing the asp viper in captivity. In: Edwards T, editor. International Zoo Yearbook 13. Cambridge: Cambridge Press. (copy not paginated, reported by other authors to be 108–111).

Naulleau G. 1973b. Reproduction twice in one year in a captive viper. Br J Herpetol 5:353–7.

Naulleau G. 1973c. Contribution a l'etude d'une population malanique de *Vipera aspis,* dans les Alpes Suisses. Bull Soc Sci Nat Ouest de la France 71:15–21.

Naulleau G. 1979. Etude biotelemetrique de la thermoregulation chez *Vipera aspis* elevee en conditions artificielles. J Herpetol 13:203–8.

Naulleau G. 1983a. Action de la temperature sur la digestion chez cinq espece europeennes du genre *Vipera*. pp Cent Pir Biol Expl 13:89–94.

Naulleau G. 1983b. The effects of temperature on digestion in *V. aspis*. J Herpetol 17:166–70.

Naulleau G. 1983c. Teratologie chez *Natrix natrix et Vipera aspis*. Angers: C R Ier Coll Intern Pathologie Reptiles et Amphibiens. pp 245–9.

Naulleau G. 1986. Effects of temperature on gestation in *V. aspis* and *V. berus*. In: Rocek Z, editor. Studies in Herpetology. Proceedings of European Herpetological Symposium. Prague: Charles University, Prague for the Societas Europea Herpetologica. pp 489–94.

Naulleau G, Marques M. 1973. Etude biotelemetrique preliminaire de la thermoregulation de la digestion chez *Vipera aspis*. C R Acad Sci 276, Serie D:3433–6.

Naulleau G, van den Brule B. 1980. Captive reproduction of *Vipera russelli*. Herpetol Rev 11:110–2.

Neelin JM. 1963. Enzyme fractionation of viper venoms. Can J Biochem Physiol 41:1073–9.

Nikolsky AM. 1913. [Reptiles and Amphibians of the Caucasus]. The Caucasus Museum Press, Tiflis, Georgia. 272 pp. (In Russian).

Nikolsky AM. 1916. Fauna of Russia and adjacent countries. Volume II: Ophidia. Petrograd. Translation from the Israel Program for Scientific Translations, Jerusalem, 1964, 247 pp.

Nilson G. 1976. The reproductive cycle of *Vipera berus* in SW Sweden. Norw J Zool 24:233–4.

Nilson G. 1980. Male reproductive cycle of the European adder, *Vipera berus,* and its relation to annual activity periods. Copeia 75:727–37.

Nilson G. 1981. Ovarian cycle and reproductive dynamics in the female adder, *Vipera berus*. Amphibia-Reptilia 2:63–82.
Nilson G. 2002. Eurasian vipers and the systematics of the *Vipera ursinii* complex. In: Kovács T, Korsós Z, Rehák I, Corbett K, Miller PS (eds). Population and Habitat Viability Assessment for the Hungarian Meadow Viper (*Vipera ursinii rakosiensis*). Workshop Report. Apple Valley MN: IUCN/(SSC Conservation Breeding Specialist Group. 65.
Nilson G, Andrén C. 1982. Function of renal sex secretion and male hierarchy in the adder, *Vipera berus*, during reproduction. Horm Behav 16:404–413
Nilson G, Andrén C. 1984. Systematics of the *Vipera xanthina* complex. II. An overlooked viper in the *xanthina* species-group in Iran. Bonn Zool Beitr 35(3–4) :175–184.
Nilson G, Andrén C. 1985a. Systematics of the *Vipera xanthina* complex. I. A new Iranian viper in the *raddei* species-group. Amphibia-Reptilia 6:207–14.
Nilson G, Andrén C. 1985b. Systematics of the *Vipera xanthina* complex. III. Taxonomic status of the Bulgar Dagh viper in south Turkey. J Herpetol 19(2): 276–283.
Nilson G, Andrén C. 1986. The mountain vipers of the middle east: The *Vipera xanthina* complex. Bonner Zoologische Monographien 20.
Nilson G, Andrén C. 1987. Morphological and phylogenetical considerations of alpine European and Asiatic *Vipera ursinii* populations. In: Timon A, editor. Proceedings of the 4th Ordinary General Meeting of Societas Europea Herpetologica; 1986 Mar 2–6; Nijmegen. pp 61–6.
Nilson G, Andrén C. 1988a. A new subspecies of the subalpine meadow viper, *Vipera ursinii* from Greece. Zool Scrip 17:311–4.
Nilson G, Andrén C. 1988b. *Vipera lebetina transmediterranea*, a new subspecies of viper from North Africa, with remarks on the taxonomy of *Vipera lebetina* and *Vipera mauritanica* (Reptilia: Viperidae). Bonn Zool Beitr 39(4):371–379.
Nilson G, Andrén C. 1992. The species concept in the *Vipera xanthina* complex: reflecting evolutionary history or hiding biological diversity? Amphibia-Reptilia 13:421–4.
Nilson G, Andrén C. 1997a. Evolution, systematics and biogeography of Palearctic vipers. In: Thorpe RS. Wüster W, Malhotra A (eds.): Venomous snakes: ecology, evolution and snakebite 31–42; Symp. Zool. Soc. Lond., 70; Oxford (Oxford University Press).
Nilson G, Andrén C. 1997b. *Vipera nikolskii* Vedmederja, Grubant and Rudayeva 1986. (In: Gasc et al. (eds.): Atlas of Amphibians and Reptiles in Europe. Societas Europaea Herpetologica & Muséum National d'Histoire naturelle, Paris 496 pp.) 396–397.
Nilson G, Andrén C. 2001: The Meadow and Steppe Vipers of Europe and Asia, the *Vipera ursinii* complex. Acta Zoologica Academiae Scientiarum Hungaricae 47(2–3):87–267 .
Nilson G, Andrén C, Flärdh B. 1988. Die vipern der Turkei. Salamandra 24:215–47.
Nilson G, Andrén C, Flärdh B. 1990. *Vipera albizona*, a new mountain viper from

central Turkey, with comments on isolating effects of the Anatolian "diagonal." Amphibia-Reptilia 11:285–94.
Nilson G, Andrén C, Joger U. 1993. A reevaluation of the taxonomic status of the Moldavian steppe viper based on immunological investigations, with a discussion of the hypothesis of secondary intergradiation between *Vipera ursinii rakosiensis* and *Vipera (ursinii) renardi*. Amphibia Reptilia 14(1):45–57.
Nilson G, Höggren M, Tuniyev BS, Orlov NL, Andrén C. 1994. Phylogeny of the vipers of the Caucasus (Reptilia, Viperidae). Zoological Scripta 23(4):353–360
Nilson G, Sundberg P. 1981. The taxonomic status of the *Vipera xanthina* complex. J Herpetol 15:379–81.
Nilson G, Tuniyev BS, Andrén C, Orlov NL, Joger U, Hermann H-W. 1999. Taxonomic position of the *Vipera xanthina* complex. Kaupia (Darmstadt) 8:99–102.
Nilson G, Tuniyev BS, Orlov NL, Höggren M, Andrén C. 1995. Systematics of the vipers of the Caucasus: Polymorphism or sibling species? Asiatic Herpetological Research 6:1-26.
Novak V, Sket D, Gubensek F. 1973. Arterial blood pressure depressing proteins in *Vipera ammodytes* venom. In: Kaiser E, editor. Animal and Plant Toxins. Munich: Wilhelm Goldmann. pp 159–62.
Obst F. 1983. Zur Kenntnis der Schlangengattung *Vipera*. Zool Abh Staatl Mus Tierk Dresden 38:229–35.
Oram S, Ross G, Pell L, Winteler J. 1963. Snake bite incidence in West Africa. Br Med J 2:1647.
Orel B, Ritonja A, Gubensek F. 1983. Raman spectra of two phospholipases A2 from *Vipera ammodytes* venom. Per Biol 85:123–6.
Orlov N. 1995. Rare snakes of the mountainous forests of northern Indochina. Russian Journal of Herpetology 2(2): 179–183.
Orlov N. 1997. Viperid snakes (Viperidae Bonaparte, 1840) of Tam-Dao mountain range. Russian Journal of Herpetology 4(1):67–74.
Orlov NL, Tuniyev BS. 1986. [The recent areas, their possible genesis and the phylogeny of three viper species of Eurosiberian group of the *Vipera kaznakowi* complex in the Caucasus]. In: N. Ananjeva and L. Borkin (eds.), Systematics and Ecology of Amphibians and Reptiles. Proceedings of the Zoological Institute, Leningrad, 157: 107–135. (In Russian).
Orlov NL, Tuniyev BS. 1990. Three species in the *Vipera kaznakowi* complex (Eurosiberian Group) in the Caucasus: Their present distribution, possible genesis, and phylogeny. Asiatic Herpetological Research 3:1–36.
Oshaka A, Omori-Satoh T, Kondo H, Kondo S, Murata R. 1966. Biochemical and pathological aspects of hemorrhagic principles in snake venoms with special reference to Habu (*Trimeresurus flavoviridis*) venom. Mem Inst Butantan 33:193.
Oshaka A, Suzuki K, Ohashi M. 1975. The spurting of erythrocytes through junctions of the vascular endothelium treated with snake venom. Microvasc Res 10:208–13.
Oshima G, Iwanaga S. 1969. Occurrence of glycoproteins in various snake venoms. Toxicon 7:235–8.

Oshima G, Sato-Ohmori T, Suzuki T. 1969. Proteinase, arginine ester hydrolase and a kinin releasing enzyme in snake venoms. Toxicon 7:229–33.

Osman H, El-Sir I, El-Sir NT. 1988. The snakes of the Sudan. 1: The snakes of Khartoum Province. The Snake 20:74–9.

Ostrovskikh 1997. Different forms of melanism and its age development in the populations of steppe viper *Vipera renardi* (Christoph, 1861). Russian Journal of Herpetology 4(2):186–191.

Otis VS. 1973. Hemocytological and serum chemistry parameters of the African puff adder, *Bitis arietans*. Herpetologica 29(2):110–6.

Ouyang C, Teng CM, Huang TF. 1992. Characterization of snake venom components acting on blood coagulation and platelet function. Toxicon 30(9):945–66.

Ovadia M. 1978a. Isolation and characterization of three hemorrhagic factors from the venom of *Vipera palaestinae*. Toxicon 16:479–87.

Ovadia M. 1978b. Purification and characterization of an anti-hemorrhagic factor from the serum of the snake *Vipera palaestinae*. Toxicon 16:661–72.

Ovadia M, Kochva E. 1977. Neutralization of Viperidae and Elapidae snake venoms by sera of different animals. Toxicon 15:541–7.

Packard GC, Packard MJ. 1988. The physiological ecology of reptilian eggs and embryos. In: Gans C, editor. Biology of the Reptilia. Vol. 16. New York: Academic Press. pp 119–256.

Paget D, Cock EV. 1979. Case history of a Berg adder bite. Centr Afr J Med 25(2):30–3.

Paine MJI, Desmond HP, Theakston RDG, Crampton JM. 1992. Gene expression in *Echis carinatus* (carpet viper) venom glands following milking. Toxicon 30:379–86.

Palmer NG. 1986. *Bitis atropos* envenomation. J Herpetol Assoc Afr 32:33.

Parellada X. 1995a. About the apparent inexistence of a spring mating in the Catalan population of *Vipera latasti* (Reptilia: Viperidae), and note about the reproductive success. In: Llorente GA, Montori A, Santos X, Carretero MA, editors. Scientia Herpetologica. 7th Ordinary General Meeting of Societas Europaea Herpetologica; 1993 Sep 15–19; Barcelona. Asociacion Herpetologica Espanola, Barcelona. 1995:1–385.

Parellada X. 1995b. Status of *Vipera aspis* and *Vipera latasti* (Viperidae, Reptilia) in Catalonia (NE Spain). In: Llorente GA, Montori A, Santos X, Carretero MA, editors. Scientia Herpetologica. 7th Ordinary General Meeting of Societas Europaea Herpetologica; 1993 Sep 15–19; Barcelona. Asociacion Herpetologica Espanola, Barcelona. 1995:328–34.

Parker HW. 1949. The snakes of Somaliland and the Sokotra Islands. Zoo Verh Rijksmus Nat Hist Leiden 6:1–115.

Parker HW. 1977. Snakes: A natural history. Second edition. New York: Cornell University Press. 188 pp.

Parry CR. 1975. Notes on the eye movement in *Bitis gabonica*. J Herpetol Assoc Afr 13:14.

Pe L, Cho K. 1986. Amount of venom injected by Russell's Viper. Toxicon 24: 730–3.

Penzes B. 1974. Die Weisenotter—Eine den schonsten Schlanger Europes. Aquar Terrar 7:236–7.
Peters JA. 1967. The scientific name of the African puff adder. Copeia 1967 (4): 864–5.
Peters WCH. 1854. Diagnosen neuer Batrachier, welche zusammen mit der fruher gegebenen Ubersicht der Schlangen und Eidechsen mitgetheilt werden. Ber Bekanntmach Geeignet Verhandl Konigl-Preuss. Akad Wiss, Berlin. November:614–28.
Petkovic D, Javanovic T, Micevic D, Unkovic-Cvetkovic N, Cvetkovic M. 1979. Action of *Vipera ammodytes* venom and its fractionation on the isolated rat heart. Toxicon 17:639–44.
Petkovic D, Pavlovic M, Matejevic D, Unkovic-Cvetkovic N, Jovanovic T, Aleksic N, Cvetkovic M, Colovic J, Stamenovic B. 1983. Influence of calcium on the action of *Vipera ammodytes* snake venom on the myocardium. Toxicon 21:887–92.
Phelps T. 1981. Poisonous snakes. Poole, Dorset (Eng): Blandford Press. 138 pp.
Phelps T. 1989. Poisonous snakes. 2nd edition. Poole, Dorset (Eng): Blandford Press. 156 pp.
Phillips LL, Weiss HJ, Christy NP. 1973. Effects of puff adder venom on the coagulation mechanism II. *In vitro*. Thromb Diathes Haemorrh 30:499–508.
Phillips LL, Weiss HJ, Pessar L, Christy NP. 1973. Effects of puff adder venom on the coagulation mechanism I. *In vivo*. Toxicon 11:423–31.
Phisalix M. 1940. Viperes de France. Paris: Stock. 135 pp.
Pielowski Z. 1962. Untersuchungen uber die Okologie der Kreuzotter (*Vipera berus*). Zool Jb Syst 89:479–500.
Pienaar U de V. 1966. The reptiles of the Kruger National Park. Koedoe 1:214–21.
Pienaar U de V. 1978. The reptile fauna of the Kruger National Park. National Parks Board of South Africa. pp 19.
Pirkle H. 1988. Gabonase: Haemostasis and animal venoms. Hematology 7: 117–9.
Pitman CRS. 1938. A Guide to the Snakes of Uganda. Uganda: Uganda Society Publishers. 27 pp.
Pitman CRS. 1961. Jumping snakes. J Bombay Nat Hist Soc 58:809.
Pitman CRS. 1966. More snake and lizard predators of birds. Bull BOC 82: 33–40.
Pitman CRS. 1973. The saw scaled viper or carpet viper, (*Echis carinatus*) in Africa and its bite. J Herpetol Assoc Afr 9:6–34.
Pitman CRS. 1974. A Guide to the Snakes of Uganda. London: Codicote, Wheldon & Wesley, Ltd. 102 pp.
Pomianowska I. 1972. Metabolic rate in the adder (*Vipera berus*). Bull Acad Pol Sci Ser Sci Biol 20:143–6.
Pomianowska-Pilipiuk I. 1974. Energy balance and food requirements of the adult vipers *Vipera berus*. Ekol Pol 22:195–211.
Porath A, Gilon D, Schulchynska-Castel H, Shalev O, Keynan A, Benbassat J. 1992. Risk indicators after envenomation in humans by *Echis coloratus* (Mideast saw scaled viper). Toxicon 30:25–32.

Pozzi A. 1966. Geonemia e catalogo ragionato degli Anfibi e Rettili della Jugoslavia. Natura 51:1–55.
Prestt I. 1971. An ecological study of the viper *V. berus berus* in southern England. J Zool 164:373–418.
Price RM. 1982. Dorsal snake scale microdermatoglyphics: ecological indicator or taxonomic tool. J Herpetol 16: 294–306.
Price RM. 1987. Microdermatoglyphics: Suggested taxonomic affinities of the viperid genera *Azemiops* and *Pseudocerastes*. The Snake 19:47–50.
Pugh RNH, Theakston RDG. 1980. Incidence and mortality of snake bite in savanna Nigeria. The Lancet 2(8205):1181–3.
Pukrittayakamee S, Ratcliffe P, McMichael A, Warrell D, Bunnag D. 1987. A competitive radioimmunoassay using a monoclonal antibody to detect the Factor X activator of Russell's viper venom. Toxicon 25:721–9.
Radspieler C, Schweiger M. 1990. Die Bergotter *Daboia* (Synonym *Vipera*) *xanthina* Gray, 1849. Herpetofauna 12(66):11–20.
Rahmy T, Tu AT, El-Banhawey A, El-Asmar MF, Hassan FM. 1991. Electron microscopic study of the effect of Egyptian sand viper (*Cerastes cerastes*) venom and its hemmorhagic toxin on muscle. J Wild Med 2:7–14.
Rao SS, Kadaba IS, Rao SS. 1959. Pit viper venoms. In: Bucherl W, editor. 1967. Venomous animals and their venoms. Vol. I. Paris: Masson. pp 325–49.
Rechnic J, de Vries A, Moroz-Perlmutter C, Levi C, Kochwa S, Gitter S. 1967. Comparative enzymology. In: Bucherl W, editor. 1967. Venomous animals and their venoms. Vol. I. Paris: Masson. pp 111–9.
Rechnic J, Trachtenberg P, Casper J, Moroz CH, de Vries A. 1962. Afibrinogenemia and thrombocytopenia in guinea pigs following injection of *Echis colorata* venom. Blood 20:735–49.
Reid HA. 1967. Defibrination by *Agkistrodon rhodosoma* venom. In: Russell FE, Saunders PR, editors. Animal Toxins. Oxford: Pergamon Press. pp 323–6.
Reid HA. 1968a. Symptomology, pathology, and treatment of land snake bite in India and southeast Asia. In: Bucherl W, Buckley E, Deulofeu V, editors. Venomous Animals and Their Venoms. Vol. 1. New York: Academic Press. pp 611–42.
Reid HA. 1968b. The paradox in therapeutic defibrination. The Lancet 1968:7541–2.
Reid HA. 1972. Snake bite Sri Lanka. Trop Doct 2:159.
Reid HA. 1974. Daily patterns of snake bite in the Asian tropics. Abstract and comment No. 166. Trop Dis Bull 71:80–1.
Reid HA. 1977. Prolonged defibrination syndrome after bites by the carpet viper *Echis carinatus*. Br Med J 2:1326.
Reid HA, Thean PC, Martin NJ. 1963. Specific antivenine and prednisone in viper bite poisoning: controlled trial. Br Med J 2:1378.
Reymond A. 1956. Contribution a l'etude de l'action du venin de *Vipera lebetina*. Trav Inst Sci. Cherifien, Serie Zool 9.
Rivers IL, Koenig HF. 1981. Analysis of Berg adder bite (*Bitis atropos*) venom. J Herpetol Assoc Afr 26:1–2.
Robinson MD, Hughes DA. 1978. Observations on the natural history of Peringuey's adder, *Bitis peringueyi* (Boulenger). Ann Transv Mus 31(16):189–93.

Roman B. 1980. Serpentes De Haute-Volta. C.N.R.S.T. Ouagadougou, Haute-Voltera. 189 pp.

Roman E. 1972. Deux sous-especes de la viper *Echis carinatus* Dan Leo territoires de Haute-Volta et du Niger: *Echis carinatus ocellatus* (Stemmer). 42 pp.

Rosenberg P. 1965. Effects of venoms on the squid giant axon. Toxicon 3:125–31.

Rosenfeld G, Nahas L, Kelen EMA. 1967. Coagulant, proteolytic, and hemolytic properties of some snake venoms. In: Bucherl W, editor. 1967. Venomous Animals and Their Venoms. Vol. I. Paris: Masson. pp 229–74.

Rosing J, Tans G. 1992. Structural and functional properties of snake venom prothrombin activators. Toxicon 30(12):1515–27.

Roux-Esteve J. 1965. Les Serpents de la region la Maboke-Boukoko. La Maboke 3(1):51–92.

Russell FE. 1983. Snake Venom Poisoning. New York: Scholium International Inc. 211 pp.

Russell FE, Emery J. 1961. Incision and suction following injection of rattlesnake venom. Am J Med Sci 241:160–6.

Russell FE, Lauritzen L. 1966. Antivenin. Trans Roy Soc Trop Med Hyg 60(6): 797–800.

Saha BK. 1989. Snakebite cases at Baruipur, West Bengal treated by 'Mangta' medicine men. Hamadryad 14:23–6.

Saint-Girons H. 1947. Ecologie des Viperes. I: *Vipera aspis*. Bull Soc Zool Fr 72:158–69.

Saint-Girons H. 1953. Une vipère naine: *Vipera latastei montana*. Bull Soc Zool Fr 78:24–28.

Saint-Girons H. 1956. Les serpents du Maroc. Var Sci Soc Sci Nat Psyc Maro 8:1–29.

Saint-Girons H. 1957. Le cycle sexuel chez *Vipera aspis* dans l'Ouest de la France. Bull Biol Fr et Belg 91:284–350.

Saint-Girons H. 1977. Systématique de *Vipera latastei latastei* Bosca, 1878 et description de *Vipera latastei gaditana*, subsp. n. (Reptilia, Viperidae). Rev Suisse Zool 84:599–607.

Saint-Girons H. 1978. Morphologie externe comparee et systematique des Vipere d'Europe. Rev Suisse Zool 85:565–95.

Saint-Girons H. 1980. Biogéographie et évolution des vipères européenes. CR Soc Biogéeogr 496:146–72.

Saint-Girons H. 1986. Comparative data on lepidosaurian reproduction and some time tables. In: Gans C, editor. The Biology of the Reptilia. Vol. 15. New York: Academic Press. pp 75–150.

Saint-Girons H, Bea A, Brana F. 1986. La distribución de los diferentes fenotipos de *Vipera seoanei* Lataste, 1879, en la región de los Picos de Europa (Norte de la Península Ibérica). Ciencias Naturales 38:121–8.

Saint-Girons H, Duguy R. 1976. Ecoplogie et position systématique de *Vipera seoanei* Lataste, 1879. Bull Soc Zool Fr 101:325–339.

Saint-Girons H, Duguy R, Detrait J. 1983. Les vipères du Sud du Massif Central: Morphologgie externe et venin. Bull Soc Hist Nat Toulouse 119:81–6.

Saint-Girons H, Kramer E. 1963. Le cycle sexuel chez *Vipera berus* (L.) en montagne. Reveu Suisse Zool. 70:191–221.
Saint-Girons H, Naulleau G. 1981. Poids des noveau-nes et strategies reproducrices des viperes europeenes. Rev Ecol (Terre et Vie) 35:597–616.
Samel M, Siigur E, Siigur J. 1987. Purification and characterization of two arginine ester hydrolases from *Vipera berus* venom. Toxicon 25:379–88.
Sandbank J, Jerushalmy Z, Ben-David I, de Vries A. 1974. Effect of *Echis coloratus* venom on brain vessels. Toxicon 12:267–71.
Sandbank U, Djaldetti M. 1966. Effect of *Echis colorata* venom inoculation on the nervous system of the dog and guinea pig. Acta Neuropathol 6:61–9.
Sanguanrungsirikul S, Chomdej B, Suwanpraser K, Wattanavaha P. 1989. Acute effect of Russell's viper venom on renal hemodynamics and autoregulation of blood flow in dogs. Toxicon 27:1199–207.
Sant SM. 1978. Pathogenesis of viperine envenomation. Toxicon 16:136.
Sarkar NK, Devi A. 1967. Enzymes in snake venoms. In: Bucherl W, editor. 1967. Venomous Animals and Their Venoms. Vol. I. Paris: Masson. pp 167–217.
Sawai Y, Kawamura Y, Fukuyama T, Keegan H. 1967a. Studies on the toxoids against the venoms of certain Asian snakes. Toxicon 7:19–24.
Sawai Y, Kawamura Y, Fukuyama T, Keegan H. 1967b. Studies on the inactivation of snake venom by dihydrothioctic acid. Jap J Exp Med 37:121–8.
Sawai Y, Makino M, Tateno I, Okonogi T, Mitsuhashi S. 1962. Studies on the improvement of treatment of Habu snake bite. 3: Clinical analysis and medical treatment of Habu snake bite on the Amami Islands. Jap J Exp Med 32: 117–38.
Schaeffer RC Jr. 1987. Hetereogeneity of *Echis* venoms from different sources. Toxicon 25(12):1343–6.
Schaeffer RC Jr, Barnhart MI, Carlson RW. 1987. Pulmonary fibrin deposition and increased permeability to protein following microthromboembolism in dogs: a structure-function relationship. Microvasc Res 33:327.
Schatti B. 2001. A new species of *Coluber* (sensu lato) from the Dahlak Islands, Eritrea, with a review of the herpetofauna of the archipelago. Russian Journal of Herpetology 8(2):139–148.
Schatti B, Baran I, Sigg H. 1991. Rediscovery of the Bolkar viper: Morphological variation and systematic implications on the "*Vipera xanthina* complex." Amphibia-Reptilia 12:305–27.
Schatti B, Baran I, Sigg H. 1992. The *Vipera xanthina* complex: A reply to Nilson and Andrén. Amphibia-Reptilia 13:425.
Schleich HH, Kästle W, Kabisch K. 1996. Amphibians and Reptiles of North Africa. Koeltz Scientific Books, Koenigstein, Germany 630 pp.
Schmidt C. 1939. Reptiles and amphibians from Southwestern Asia. Field Mus Nat Hist Zool Ser 24:49–92.
Schmidt KP. 1923. Contributions to the herpetology of the Belgian Congo. Bull Am Mus Nat Hist 49:1–146.
Schmidt KP. 1955. Amphibians and reptiles from Iran. Vidensk Med Dansk Nat Foren Kbh 117:193–207.

Schmidt KP, Marx H. 1956. The herpetology of the Sinai. Field Zool 39(4):21–40.
Schneider JG. 1801. Historic Amphiorum naturalia et litterariae. Fasc 2:1–364.
Schnurrenberger H. 1957. Het levenvan *Aspis cerastes* (Linné) in de vrije natuur. Lacerta 16:58–60.
Schnurrenberger H. 1959. Observations on behavior in two Libyan species of viperine snake. Herpetologica 15:70–2.
Schulchynska-Castel H, Dvilansky A, Keyman A. 1986. *Echis colorata* bites: clinical evaluation of 42 patients. A retrospective study. Isr J Med Sci 22:880–4.
Schwarz, E. (1936) Untersuchungen über Systematik und Verbreitung der europäischen und mediterranen Ottern. Behringwerke-Mitteilungen 7:159–262.
Schwick G, Dichgeisser F. 1963. Snake venom classification. In: Bucherl W, editor. 1967. Venomous Animals and Their Venoms. Vol. I. Paris: Masson. pp 315–23.
Sebela M. 1978. Contribution to the knowledge of the common viper's diet (*Vipera berus*) in the Ceskomoravska vysocina. Cas Mor Nusea Sci Nat 58:213–6.
Seigel RA, Collins JT. 1993. Snakes: Ecology and Behavior. New York: McGraw-Hill, Inc. 190 pp.
Seigel RA, Collins JT, Novak SS. 1987. Snakes: Ecology and Evolutionary Biology. New York: Macmillan. 310 pp.
Shabo-Shina R, Bdolah A. 1985. Effects of the neurotoxic complex from *Pseudocerastes fieldi* venom on rat brain synaptosomes. Toxicon 23:610.
Shabo-Shina R, Bdolah A. 1987. Interactions of the neurotoxic complex from the venom of the false horned viper (*Pseudocerastes fieldi*) with rat striatal synaptosomes. Toxicon 25(3):253–66.
Sharma RC, Vazirani TG. 1977. Food and feeding habits of some reptiles of Rajasthan. Rec Zool Surv India 73:77–93.
Shaw CE. 1959. Longevity of snakes in the United States as of January 1, 1959. Copeia 1959(4):336–7
Shaw CE, Campbell S. 1974. Snakes of the American West. In: Findlay FE, editor. 1983. Snake Venom Poisoning. Frankfurt: Scholium International, Inc. pp 54–82.
Shaw CJ. 1924. Notes on the effect of the bite of McMahon's viper (*Eristocophis macmahoni*). J Bombay Nat Hist Soc 30:485–6.
Shaw G, Nodder F. 1797. The Naturalist's Miscellany. Volume 8. London: Nodder and Co. 65 pp.
Shayer-Wollberg M, Kochva E. 1967. Embryonic development of the venom apparatus in *Causus rhombeatus*. Herpetologica 23(4):249–59.
Shine R. 1986. Evolutionary origins of viviparity in squamate reptiles. In: Gans C, editor. Biology of the Reptilia. Vol. 15. New York: Academic Press. pp 1–119.
Siddiqi AR, Persson B, Zaidi ZH, Jornvall H. 1992. Characterization of two platelet-aggregation inhibitor-like polypeptides from viper venom. Peptides 13(6):1033–7.
Siddiqi AR, Zaidi ZH, Jornvall H. 1991. Purification and characterization of two

highly different group II phospholipase A2 isozymes from a single viperid (*Eristocophis macmahoni*) venom. Eur J Biochem 201:675–9.

Simpson JW, Taylor JC, Levy BM. 1972. Collagenolytic activity in some snake venoms. Comp Biochem Physiol 39:963.

Sitprija V, Benyajati C, Boonpucknoaviq V. 1974. Further observations of renal insufficiency in snakebite. Nephron 13:396–403.

Sitprija V, Sribhibhadh R, Benyajati C, Tangchai P. 1971. Acute renal failure in snakebite. In: de Vries A, Kochva E, editors. Toxins of Animal and Plant Origin. Vol. 1. New York: Gordon and Breach Science Publishers. Paper VI-5.

Sket D, Gubensek F. 1976. Pharmacological study of phospholipase A from *Vipera ammodytes* venom. Toxicon 14:393–6.

Smetsers P. 1990. Sensitive to stress or not, that is the question. Litt Serpent 10:158–60.

Smith M. 1969. The British Reptiles and Amphibians. The New Naturalist. London: Collins. 120 pp.

Sochurek E. 1953. Irrtumer um *Vipera berus bosniensis*. Carinthia II V. 63:150–2. Mit Naturw. Ver Karnten, Klagenfurt, Germany.

Spawls S. 1979. An unusually heavy puff adder. E Afr Nat Hist Soc 25:18–20.

Spawls S, Branch B. 1995. The Dangerous Snakes of Africa. Ralph Curtis Books. Dubai: Oriental Press. 192 pp.

Spawls S, Branch B. 1998. Dangerous snakes of Africa. London: Ralph Curtis. 211 pp.

Spawls S, Howell K, Drewes R, Ashe J. 2002. A Field Guide to the Reptiles of East Africa. Academic Press. 543 pp.

Steehouder T. 1989. De gewone padadder (*Causus rhombeatus*) in het terrarium. Lacerta 47(3):87–90.

Stemmler O. 1969. Die Sandrasselotter aus Pakistan: *Echis carinatus sockureki* subspp. Nov Aquater 6(10):118–25.

Stemmler O. 1970a. Uber den Geburtsvorgang bei einer Sandrasselotter (*Echis carinatus*). Salamandra 6:18–25.

Stemmler O. 1970b. Die Sandrasselotter aus Westafrika: *Echis carinatus ocellatus* subsp. Nov Rev Suisse De Zool 77(2):273–82.

Stemmler O. 1971a. Die Reptilien der Schwiez mit besonderer Berucksichtigung der Basler Region. Veroff Nat Mus Basel, Number 5.

Stemmler O. 1971b. Zur Haltung von *Carinatus leakeyi* und *Echis carinatus sockureki* subspp. Nov Aquater 6(10):118–25.

Stemmler O, Sochurek E. 1969a. Der Kenya Sandrasselotter. Zool Garten NF Leipzig 40:200–10.

Stemmler O, Sochurek E. 1969b. Die Sandrasselotter von Kenya: *Echis carinatus leakeyi* subsp. Nov Aquater 6:89–94.

Sterer Y. 1992. A mixed litter of horned and hornless vipers *Cerastes cerastes*. Isr J Zool 37:247–9.

Stettler PH. 1993. Bemerkunge zur postembryonalen Entwicklung der Waldsteppenotter *Vipera nikolskii*—ein Vergleich mit Schwärzlingen der Kreuzotter

Vipera berus. Zusammenfassungen Jahrestagung der Deuthschen Gesellschaft für Herpetologie und Terrarienkunde, Idar-Oberstein, 7.

Stevens RA. 1973. A report on the lowland viper, *Atheris superciliaris* from the Lake Chilwa floodplain of Malawi. Arnoldia (Rhodesia) 22:1–22.

Steward JW. 1971. The Snakes of Europe. London: David & Charles, Newton Abbot. 191 pp.

Stewart M, Wilson VJ. 1966. Herpetofauna of the Nyika Plateau (Malawi and Zambia). Ann Nat Mus 18:287–314.

Steyn K, Delpierre GR. 1973. The determination of proteolytic activity of snake venom by means of a chromogenic substrate. Toxicon 11:103–5.

Stille B, Madsen T, Niklasson M. 1986. Multiple paternity in the adder, *Vipera berus.* Oikos 47:173–5.

Stille B, Niklasson M, Madsen T. 1987. Within season multiple paternity in the adder, *Vipera berus:* A reply. Oikos 49:232–3.

Strauch A. 1869. Synopsis der Viperiden nebst Bemerkungen uber die geographische Verbreitung dieser Giftshlangen-familie. Mem Acad Sci St. Petersburg 14.

Street D. 1979. The reptiles of northern and central Europe. London: BT Batsford LTD 185 pp.

Strydom DJ, Joubert FJ, Howard NL. 1986. Chemical studies on protease A of *Bitis arietans* venom. Toxicon 24(3):247–57.

Stucki-Stern MC. 1979. Snake Report 721: A comparative study of the herpetological fauna of the former West Cameroon/Africa. Switzerland: Herpeto-Verlag. 98 pp.

Sweeney RCH. 1961. Snakes of Nyasaland. Zomba, Nyasaland: The Nyasaland Society and Nyasaland Government. 74 pp.

Swiecicki AW. 1965. Snakes and snake bite in the western region, Ghana. J Trop Med Hyg 68:300.

Szyndlar Z. 1988.Two new extinct species of the genera *Malpolon* and *Vipera* (Reptilia, Serpentes) from the Pliocene of Layna (Spain). Acta Zool Cracov 31:687–706.

Szyndlar Z, Rage J-C. 1999. Oldest fossil vipers (Serpentes: Viperidae) from the Old world. Kaupia (Darmstadt) 8:9–20.

Taborska E. 1971. Intraspecies variability of the venom of *Echis carinatus.* Physiol Bohemoslov 20:307.

Takahashi H, Iwanaga S, Kitagawa T, Hokama Y, Suzuki T. 1974. Snake venom proteinase inhibitors II. Chemical structure of inhibitor II isolated from the venom of Russell's viper. J Biochem 76:721–3.

Takahashi H, Iwanaga S, Suzuki S. 1974a. Distribution of proteinase inhibitors in snake venoms. Toxicon 12:193–7.

Takahashi H, Iwanaga S, Suzuki T. 1974b. Snake venom protein inhibitors I. Isolation and properties of two inhibitors of kallikrein, trypsin, plasmin, and alpha-chymotrypsin from the venom of Russell's viper. J Biochem 76:709–19.

Takahashi WY, Tu AT. 1970. Puff adder snake bite. JAMA 211(11):1857.

Tan NH, Ponnudurai G. 1988. A comparative study of cobra (*Naja*) venom enzymes. Comp Biochem Physiol 90(B):745–50.
Tan NH, Ponnudurai G. 1990a. A comparative study of the biological properties of venoms from snakes of the genus *Vipera* (true adders). Comp Biochem Physiol 96(B):683–8.
Tan NH, Ponnudurai G. 1990b. A comparative study of the biological activities of venoms from snakes of the genus *Agkistrodon* (Moccasins and Copperheads). Comp Biochem Physiol 95(B):577–82.
Tan NH, Ponnudurai G. 1990c. A comparative study of the biological properties of Krait (genus *Bungarus*) venoms. Comp Biochem Physiol 95(C):105–9.
Tan NH, Ponnudurai G. 1992. A comparative study of the biological properties of venoms from some old world vipers (Subfamily Viperinae). Int J Biochem 24:331–6.
Taub AM, Elliott WB. 1964. Some effects of snake venoms on mitochondria. Toxicon 2:87–92.
Taub RG, Pugh FH. 1979. The effect of rattlesnake venom on digestion of prey. Toxicon 17:221.
Taylor D, Iddon D, Sells P, Theakston RDG. 1986. An investigation of venom secretion by the venom glands of the carpet viper (*Echis carinatus*). Toxicon 24(7):651–9.
Taylor J, Mallick SMK. 1935. Observation on the neutralization of the hemorrhaging of certain viper venoms by antivenins. Indian J Med Res 23:121–30.
Tchorbanov B, Grishin E, Aleksiev B, Ovchinnikov Y. 1978. A neurotoxic complex from the venom of the Bulgarian viper (*V. ammodytes ammodytes*) and a partial amino acid sequence of the toxic phospholipase A_2. Toxicon 16:37–44.
Teng C, Chen Y, Ouyang C. 1984. Purification and properties of the main coagulant and anticoagulant principles of *Vipera russelli* snake venom. Biochim Biophys Acta 786:204–12.
Teng C, Wang J, Huang T, Liau M. 1989. Effects of venom proteases on peptide chromogenic substrates and bovine prothrombin. Toxicon 27:161–7.
Terentyev P, Chernov S. 1949. Key to amphibians and reptiles. 3rd edition, translated through Jerusalem, Isr. 88 pp.
Than T, Sein M, Hla-Pe U. 1985. Variations in Russell's viper venom composition with different batches of collection. Toxicon 23. pp 17–24.
Theakston RDG, Reid HA. 1982. Epidemiology of snake bite in West Africa. Toxicon 20:364.
Theakston RDG, Warrell DA. 1991. Antivenins: A review. Toxicon 29:1419.
Theakston RDG, Zumbuehl O, New PRC. 1985. Use of liposomes for protective immunization in sheep against *Echis carinatus* snake venom. Toxicon 23: 921–9255.
Theodor O. 1955. On Poisonous Snakes and Snake Bite in Israel. The Israeli Scientific Press. 86 pp.
Thireau M. 1967. Contribution a l'etude de la morphologie caudale, de l'anatomie vertebrale et costale des genres *Atheris, Atractaspis et Causus*. Bull du Mus Nat D'Hist Nat 39:454–70.

Thomas E. 1972. *Bitis arietans* (Viperidae). Kommentkampf der Mannchen. Publ Wiss Film Sekt Biol 5:291–9.

Thomas RG, Pugh FH. 1979. The effect of rattlesnake venom on digestion of prey. Toxicon 17:221.

Thwin M, Mee-Mee K, Kyin M, Than T. 1988. Kinetics of envenomation with Russell's viper venom and of antivenin use in mice. Toxicon 26:373–8.

Thwin M, Than T, Hla-Pe U. 1985. Relationship of administered dose to blood venom levels in mice following experimental envenomation by Russell's viper venom. Toxicon 23:43–52.

Tilmisany AK, Mustafa AA, Aziz AA, Osman OH. 1986a. Evidence for the presence of histamine in gaboon viper (*Bitis gabonica*) venom. Toxicon 24(11–12): 1159–61.

Tilmisany AK, Najjar TA. 1982. Effects of the venom of the snake *Cerastes cerastes* on the response of the rat vas deferens to field stimulation. Toxicon 20:505–8.

Tilmisany AK, Osman OH, Aziz AA, Mustafa AA. 1986b. Effects of venom from *Bitis nasicornis* (rhinoceros horned viper) on isolated aortic strips. Toxicon 24(3):309–12.

Tsai MC, Lee CY, Bdolah A. 1983. Mode of neuromuscular action of a toxic phospholipase A_2 from *Pseudocerastes fieldi* (Field's horned viper) snake venom. Toxicon 21(4):527–34.

Tu AT. 1969. Effects of snake venoms on mammalian cells in tissue culture. Toxicon 6:277–80.

Tu AT. 1991. Handbook of Natural Toxins. New York: Marcel Dekker. 556 pp.

Tu AT, Gordon PJ, Chua A. 1965. Some biochemical evidence in support of the classification of venomous snakes. Toxicon 3:5–8.

Tu AT, Homma M, Hong B. 1969. Hemorrhagic, myonecrotic, thrombotic and proteolytic activities of viper venoms. Toxicon 6:175–8.

Tu AT, Toom PM. 1967. The presence of L-leucyl *b*-napthlamide hydrolyzing enzyme in snake venoms. Experimentia 23:439.

Tu AT, Toom PM. 1968. Hydrolysis of peptides by Crotalidae and Viperidae venoms. Toxicon 5:201–5.

Tu AT, Toom PM, Murdock DS. 1966. Chemical differences in the venoms of genetically different snakes. In: Russell FE, Saunder PR, editors. Animal Toxins. New York: Pergamon Press. pp 280–307.

Tuck R. 1977. Iranian reptiles and amphibians believed to merit designation as "rare" and/or "endangered" species. Report to DOE Committee for Rare and Endangered Species, Tehran, Iran.

Tuniyev BS, Ostrovskikh SV. 2001. Two new species of vipers of "*kaznakovi*" complex (Ophidia, Viperinae) from the western Caucasus. Russian Journal of Herpetology 8(2):117–126.

Turk V, Ritonja A, Gubensek F. 1979. The isolation and characterization of trypsin and chymotrypsin inhibitors from *Vipera ammodytes* venom. Toxicon 17:195.

Turner RM. 1972. Snake bite treatment. Black Lechwe 10(3):24–33.

Tweedie MWF. 1983. The Snakes of Malaya. Singapore: Singapore National Printers, Ltd. 105 pp.
Underwood G. 1967. A contribution to the classification of snakes. England: Staples Printers LTD. 135 pp.
Underwood G. 1968. On the status of some south African vipers: A new bush viper. J E Afr Nat Hist Soc Nat Mus 27:53–9.
Underwood G. 1970. The eye. In: Gans C, editor. Biology of the Reptilia. Vol. 2. New York: Academic Press. pp 9–140.
U.S. Navy. 1991. Poisonous Snakes of the World. New York: Dover Books. (Reprint of US Govt. Printing Office, Washington, DC.) 133 pp.
van Damme R, Castilla AM. 1996. Chemosensory predator recognition in the lizard *Podarcis hispanica:* effects of predation pressure relaxation. J Chem Ecol 22(1):13–22.
van de Beek H. 1991. Breeding V. *ammodytes ammodytes.* Litt Serpent 11:144.
van der Rijst H. 1990. Thermoregulation of the adder. Litt Serpent 10:62–70.
van der Walt SJ, Joubert FJ. 1971. Studies on puff adder (*Bitis arietans*) venom. I: Purification and properties of protease A. Toxicon 9:153–61.
van der Walt SJ, Joubert FJ. 1972a. Studies on puff adder (*Bitis arietans*) venom. II: Specificity of protease A. Toxicon 10:341–9.
van der Walt SJ. Joubert FJ. 1972b. Studies on puff adder (*Bitis arietans*) venom. III: Ultracentrifuge and ORD studies on protease A. Toxicon 10:351–6.
van Wingerde J. 1986. The distribution of *Vipera xanthina* on the east Aegean islands and in Thrace. Litt Serpent 6:131–9.
Vedmederja VI, Grubandt VN, Rudaeva AV. 1986. On the question of the name (Nomenclature) of the black viper from the forest steppe zone of the European part of the USSR. Vestnik Charkov Univ 288:83–85. (In Russian).
Vedmederja VL, Orlov NL, Tuniyev BS. 1986. [On the taxonomy of the three viper species of the *Vipera kaznakowi* complex]. In: N. Ananjeva and L. Borkin (eds.), Systematics and Ecology of Amphibians and Reptiles. Proceedings of the Zoological Institute, Leningrad, 157: 55–61. (In Russian).
Vesey-Fitzgerald D. 1958. The snakes of northern Rhodesia and the Tanganyika borderlands. Proc Trans Rhodesia Sci Assoc 46(XLVI)17–102.
Vest DK. 1986. Preliminary studies on the venom of the Chinese snake *Azemiops feae.* Toxicon 24(5):510–3.
Vick JA, Ciuchta HP, Manthie JH. 1966. Pathophysiological studies of 10 snake venoms. In: Russell FE, Saunder PR, editors. Animal Toxins. Oxford: Pergamon Press. pp 160–77.
Vick J, Ciuchta H, Polley E. 1964. Effect of snake venom and endotoxin on cortical electrical activity. Nature 203:1387.
Viitanen P. 1967. Hibernation and seasonal movements of the viper *Vipera berus berus* in southern Finland. Ann Zoo Fennici 4:472–546.
Viljoen CC, Botes DP, Kruger H. 1982. Isolation and characterization of the amino acid sequence of caudoxin, a presynaptic acting toxic phospholipase A2 from the venom of the horned puff adder (*Bitis caudalis*). Toxicon 20(4):715–37.

Villiers A. 1950a. Contribution a l'etude du peuplement de la Mauritania. Bull IFAN 27:1192–5.
Villiers A. 1950b. La collection de serpents de L'I. F.A.N. Institut Francais D'Afrique Noire. Catalogues VI.
Villiers A. 1975. Les Serpents de L'Ouest Africain. Initiations et Etudes Africaines No. II. Universite de Dakar.
Vishwanath B, Manjunatha R, Gowda T. 1988. Purification and partial biochemical characterization of an edema inducing phospholipase A2 from *Vipera russelli* snake venom. Toxicon 26:713–20.
Visser J. 1981. Additional records and corrections for the Cape Province. J Herpetol Assoc Afr 25:6–7.
Visser J, Carpenter G. 1977. Notes on a Gaboon adder bite. J Herpetol Assoc Afr 15:21–2.
Visser J, Chapman DS. 1978. Snakes and Snakebite: Venomous Snakes and Management of Snakebite in Southern Africa. South Africa: Purnell and Sons Ltd. 152 pp.
Vit Z. 1977. The Russell's Viper. Prezgl Zool 21:185–8.
Vogel Z. 1964. Reptiles and Amphibians: Their Care and Behaviour. London: Studio Vista. 185 pp.
Wakeman BN. 1966. Uganda's poisonous snakes: Further observations on feeding habits. Uganda J 30(1):101–3.
Walker CW. 1945. Notes on adder bite (England and Wales). Br Med J 2:13–4.
Wall F. 1906. The breeding of Russell's viper. J Bombay Nat Hist Soc 16: 292–312.
Wall F. 1913. The poisonous snakes of India. Trans Bombay Med Phys Soc 3:57–9.
Wall, F. 1921. The Snakes of Ceylon. Ceylon: HR Cottle 71 pp.
Warburg MR. 1964. Observations on microclimate in habitats of some desert vipers in the Negev, Arava and dead Sea regions. Vie Milleu 15:1017–41.
Warrell D. 1985. Tropical snakebite: Clinical studies in south east Asia. Toxicon 23:543.
Warrell D. 1986. Tropical snake bite: Clinical studies in Southeast Asia. In: Harris J, editor. Natural Toxins, Animal, Plant and Microbial. Oxford: Oxford Scientific Publications. pp 25–52.
Warrell DA, Arnett TP. 1976. The importance of bites by the saw scaled or carpet viper (*Echis carinatus*) epidemiological studies in Nigeria and a review of the world literature. Acta Tropica 23:309–41.
Warrell DA, Davidson N, Greenwood BM, Ormerod LD, Pope HM, Watkins BJ, Prentice CRM. 1977. Poisoning by bites of the saw scaled or carpet viper (*Echis carinatus*) in Nigeria. Quart J Med 46:33–62.
Warrell DA, Ormerod LD, Davidson N. 1975. Bites by puff-adder (*Bitis arietans*) in Nigeria, and value of antivenin. Br Med J 4:697–700.
Warrell DA, Ormerod LD, Davidson N. 1976. Bites by the night adder (*Causus maculatus*) and burrowing vipers (*Atractaspis*) in Nigeria. Am J Trop Med Hyg 25(3):517–24.
Warrell DA, Pope I, Prentice CRM. 1976. Disseminated intravascular coagulation

caused by the carpet viper (*Echis carinatus*): Trial of heparin. Br J Haematol 33:335–12.
Watkins BJ, Greenwood BM, Reid HA. 1974. Bites by the saw-scaled or carpet viper (*Echis carinatus*): Trial of two specific antivenins. Br Med J 4:437–40.
Weima A. 1990. Breeding *Vipera mauretanica*. Litt Serp 10:161–5.
Weinstein S, Minton SA. 1984. Lethal potencies and immunoelectrophoretic profiles of venoms of *Vipera bornmuelleri* and *Vipera latifii*. Toxicon 4:625–9.
Weiss HJ, Phillips LJ, Hopewell WS, Phillips G, Christy NP, Nitti JF. 1973. Heparin therapy in a patient bit by a saw scaled viper *Echis carinatus*, a snake whose venom activates prothrombin. Am J Med 51:653–62.
Welch KRG. 1983. Herpetology of Europe and Southwest Asia. A Checklist of the Orders Amphisbaenia, Sauria and Serpentes. Malabar (FL): Krieger Publishing 185 pp.
Weldon P, Demeter B, Walsh T, Kleister J. 1992. Chemoreception in the feeding behavior of reptiles: considerations for maintenance and management. In: Murphy J, Adler, K, Collins J, editors. Captive Management and Conservation of Amphibians and Reptiles. St. Louis (MO): Society for the Study of Amphibians and Reptiles, St. Louis University. pp 61–70.
Werman SD. 1986. Phylogenetic relationships of the true viper *Eristocophis macmahoni*, based on parsimony analysis of allozyme characters. Copeia 1986(4):1014–20.
Werner, F. 1898. Über einige neue Reptilien und einen neuen frosch aus dem cilicischen Taurus. Zool. Anz. 21:217–223.
Werner, F. 1922. Synopsis den Schlangenfamilien der Amblycephalidae und Viperidae. Arch. Nat. Gesch. 88A:185–244.
Werner F. 1938. Eine verkannte Viper (*Vipera palaestinae* n. sp.). Zool Anz 122: 313–8.
Werner W. 1895. Die Wiessenotter. Verb Zool-Bot Ges Wien 45:194.
Werner YL. 1967. Dark adaptation of the vertical pupil in a snake. Herpetologica 23(1):62–3.
Werner YL 1995. A guide to the reptiles and amphibians of Israel. Jerusalem (Nature Reserves Authority) (In Hebrew) 86 pp.
Werner YL, Avital E. 1980. The herpetofauna of Mt. Hermon and its altitudinal distribution. Isr J Zool 29:192–3.
Werner YL, Sivan N. 1992. Systematics and zoogeography of *Cerastes* (Ophidia: Viperidae) in the Levant: 2, Taxonomy, ecology, and zoogeography. The Snake 24:34–49.
Werner YL, Le Verdier A, Rosenman D, Sivan N. 1991. Systematics and zoogeography of *Cerastes* (Ophidia: Viperidae) in the Levant: 1, Distinguishing Arabian from African "*Cerastes cerastes*." The Snake 23:90–100.
Werner YL, Sivan N, Kushnir V, Motro U. 1999. A statistic approach to variation in *Cerastes* (Ophidia: Viperidae), with the description of two endemic subspecies. Kaupia 8:83–97.
Whaler BC. 1971. Venom yields from captive Gaboon vipers (*Bitis gabonica*). Uganda J 35(2):195–206.
Whaler BC. 1972. Gaboon viper venom and its effects. J Physiol 222:61–2.

Whaler BC. 1975. Cardiovascular and respiratory effects of Gaboon viper venom. Gen Pharm 6:35.
Whitaker R. 1973a. Climbing response of two snake species during rain (*Echis carinatus* and *Vipera russelli*). J Bombay Nat Hist Soc 70(2):387–9.
Whitaker R. 1973b. *Echis* collection in Ratnagiri District. J Bombay Nat Hist Soc 71(3):617–20.
Whitaker R. 1975. *Echis* in Tamil Nadu. J Bombay Nat Hist Soc 72(2):563.
Whitaker R. 1978. Common Indian Snakes. New Delhi (India): Macmillan. 85 pp.
Willemse GT, Hattingh J, Coetzee N. 1979. Precipitation of human blood clotting factors by puff adder (*Bitis arietans*) venom. Toxicon 7(5):331–5.
Willemse GT, Hattingh J, Karlsson RM, Levy S, Parker C. 1979. Changes in composition and protein concentration of Puff adder (*Bitis arietans*) venom due to frequent milking. Toxicon 17:37–42.
Wilson VJ. 1965. The snakes of the eastern province of Zambia. Puku 3:149–69.
Woodhams B, Wilson S, Xin B, Hutton R. 1990. Differences between the venoms of two sub-species of Russell's viper: *Vipera russelli pulchella* and *Vipera russelli siamensis*. Toxicon 28:427–33.
Woodward SF. 1933. A few notes on the persistence of active spermatozoa in the African night adder *Causus rhombeatus*. Proc Zool Soc Lond 82:189–90.
Wright D. 1982. The maintenance of *Vipera berus* in captivity. The Herptile 7: 2–4.
Wüster W. 1992. Cobras and other herps in south-east Asia. Br Herpetol Soc Bull 39:19–24.
Wüster W. 1998. The genus *Daboia* (Serpentes: Viperidae): Russell's viper. Hamadryad 23(1):33–40.
Wüster W, Otsuka S, Malhotra A, Thorpe R. 1992. Population systematics of Russell's viper: a multivariate study. Biol J Linn Soc 47:97–113.
Yatziv S, Manny N, Ritchie J, Russell A. 1974. The induction of afibrinogenemia by *Echis colorata* snake bite. J Trop Med Hyg 77:136–43.
Yu P, Tang A, Liang J, Huang Q, Mo F. 1989. An investigation on Wuzhou's snakebite epidemiology during the 1973–1984 period. Curr Herpetol E Asia: 489–492.
Yu P, Tang Z, Yue S, Xie R, Shen B, Lu X, Yu Z, and Li L. 1989. Survey of epidemiology of snake-bite in two cities, six counties, 54 towns, and 153 villages. Curr Herpetol E Asia 493–7.
Zaki OA, Hanna NM, Beskharoun MA, Kajubi S, Petkovit D. 1976. Circulatory effect of *Bitis gabonica* venom. Ain Shams Med J 27:135.
Zeller EZ. 1948. Enzymes of snake venoms and their biological significance. Adv Enzymol 8:459.
Zeller EZ. 1951. Enzymes as essential components of bacterial and animal toxins. In: Sunner JF, Myback K, editors. The Enzymes. Vol I. New York: Academic Press. pp 40–53.
Zhao R, Er-Mi TH, Zhao G. 1981. Notes on Fea's viper from China. Acta Herpetol Sinica 5(11):66–71.
Zuffi MAL, Bonnet X. 1999. Italian subspecies of the asp viper, *Vipera aspis:* patterns of variability and distribution. Ital J Zool 66:78–95.

Glossary

acetylcholine — a chemical neurotransmitter produced in the presynaptic membrane that carries a nerve impulse across a synapse.
acetylcholinesterase — an enzyme which breaks down acetylcholine at the postsynaptic membrane to prevent continual nerve firing.
adrenal corticoid activity — action of hormones produced by adrenal glands that produce many effects including dilation and constriction of blood vessels, emotional state adjustments, and metabolism of various materials and systems.
adjuvant mixture — addition of a chemical which enhances the effects of the principal ingredient.
afibrinogenemia — disorder in blood characterized by absence of fibrinogen in the plasma causing incoagulability.
albumin — a simple, water-soluble protein widely distributed in tissues and fluids of animals. Found in blood as serum albumin.
albuminuria — presence of detectable amounts of serum albumin in urine. Usually a sign of renal impairment.
allopatric — living in different areas.
amino acids — nitrogen-containing organic acids, the building blocks of protein.
anal — scale in front of and covering the cloacal opening in snakes.
anaphylaxis — severe hypersensitivity reaction which can cause circulatory, respiratory, and neurological distress. May be fatal if untreated.
anemia — condition of reduced numbers of circulating red blood cells and/or hemoglobin.
anticoagulant — a substance that inhibits normal coagulation of blood.
antivenin — concentrated and purified serum antibodies obtained from an animal hyperimmunized to a venom. Usually obtained from horses, rabbits, and recently sheep.
apnea — cessation of breathing, usually temporary; often a result of stimuli reduction to the respiratory center.
arboreal — living in trees or bushes.
arenicolous — organisms associated with sandy habitats.
asymptomatic — without symptoms.
atrophy — the wasting away of a body part due to disease, injury, lack of nutrition, etc.
autonomic — part of the nervous system concerned with controlling involuntary body functions, such as glands, smooth muscle, and the heart.
A-V (atrio-ventricular) — bundle of modified cardiac muscle fibers that form part of the heart's impulse conductance system.
axilla — pertaining to the area of the armpit.

azotemia—presence of nitrogen, especially urea, in increased amounts in the blood.
basement membrane—thin layer of delicate noncellular material underlying epithelium.
beta-globulin—globulin proteins in blood plasma associated with certain antibodies. See globulins.
bilirubin—orange-yellow pigment in bile produced from red blood cell hemoglobin.
bleb—irregular elevation of epidermis containing body fluid.
bradykinin—a polypeptide found in plasma capable of influencing smooth muscle contraction; inducing hypotension; increasing blood flow in, and permeability of capillaries.
canaliculated—traversed by a small tubular passage or channel.
canthus—angle between the flat crown of the head and the side, between the snout and eye.
cardiotoxin—a toxin that interferes with heart muscle contraction or rhythm.
cerebral-spinal fluid—extracellular fluid located within the meninges and ventricles of the central nervous system.
cerebral-ventrical—one of the cavities inside the brain.
chin shields—paired enlarged scales near the ventral midline of the lower jaw in snakes.
cholinergic—pertaining to nerve endings which liberate acetylcholine.
clonic—alternating contraction and relaxation of muscles.
clotting time—time necessary for whole blood to clot, normally 4–8 minutes.
coagulation—the process of changing from a liquid to a solid or semi-solid formation of a blood clot from liquid blood.
coagulation defect—see coagulopathy.
coagulopathy—the state of impaired blood coagulation.
collagenolytic—breakdown of fibrous proteins commonly found in skin, bone, ligaments, and cartilage.
compliment$_3$-C$_9$ sequence—a pathway of protein activation involved in the immune response.
corticoid—steroid materials produced in the cortex of the adrenal glands.
crepuscular—activity occurring at dawn and dusk.
cristae—highly folded inner membrane of the mitochondria; site of energy producing reactions.
cross-reactivity—the effect of an antivenin prepared from the venom of one type of snake on the venom of a different type of snake.
cyanotic—slight bluish coloration due to abnormal amounts of reduced hemoglobin in the blood.
cystine—a sulfur containing amino acid important in supplying sulfur in metabolism, and in maintaining protein structure.
cytolysis—dissolution or destruction of living cells.
cytotoxin—venom that affects walls and contents of cells.
debridement—removal of necrotic tissue.
defibrination—process of removal or destruction of fibrin.
denervation—a condition in which the nerve supply is blocked or cut off.

GLOSSARY

diastolic—measurement of blood pressure which records the lower pressure in the arterial vascular system.
diplopic—the perception of double images; "double vision."
DIC—disseminated intravascular coagulopathy, an abnormal coagulation state that uses up stored blood clotting factors and platelets, causing incoagulable blood and risk of severe bleeding.
diurnal—normally being active during the day.
dorsals—rows of scales that cover the top surface of a snake's body.
dyspnea—shortness of breath, usually in response to low blood oxygen levels.
ecarin—a prothrombin-converting enzyme isolated from *Echis carinatus* venom.
ecchymosis—discoloration of skin resembling a bruise; caused by extravasation of blood forming dispersed blood clots.
ECG or EKG (electrocardiogram)—graph showing electrical activity of the heart over time.
edema—presence of excessive fluid in intercellular spaces; causes swelling of soft tissues.
electrophoresis—movement of charged colloidal particles through a medium in which they are suspended; movement is the result of electrical potential.
electrolyte—substances in solution that conduct electrical currents; salts, acids, and bases are examples.
endemic—confined to a limited habitat or country.
endothelium—flat cellular layer that lines the inner wall of blood vessels or gut.
envenomation—injection of venom into tissues.
enzymes—proteins which mediate chemical reactions by a catalytic action.
epistaxis—hemorrhage from the nose (nosebleed).
erythema—redness of the skin.
erythrocytic—pertaining to red blood cells.
erythropoiesis—formation of red blood cells.
exopeptidases—enzymes which break peptide bond of terminal amino acids.
extravasation—escape of fluids into surrounding tissues; usually in reference to blood leaking from vessels.
exudate—any fluid that leaks out of a tissue.
facilitated diffusion—transport of materials through the cell membrane using membrane proteins or other means to facilitate movement.
fascia—a band or sheath of tough fibrous connective tissue that covers groups of muscles.
fasciculation—involuntary repetitive contraction of the muscles.
fibrin—filamentous protein formed by the action of thrombin on fibrinogen; this forms the basis of blood clot formation.
fibrinogen—protein in blood plasma converted to fibrin through action of thrombin and calcium.
fibrinogenemia—the amount of fibrinogen in the blood.
fibrinolysis—breakdown of fibrin caused by action of a proteolytic enzyme system.
fibrinolysin—substance, also called "plasminogen," which dissolves fibrin, breaking down formed clots; promotes bleeding.

fibrinogenolysis—destruction of fibrinogen.
FDP—"Fibrinogen degradation products" are substances in blood that indicate fibrinogen breakdown has occurred.
flaccid paralysis—paralysis involving a loss of muscle tone.
frontal—single enlarged median scute on the crown between supraoculars and behind the prefrontals.
gangrene—necrosis or death of tissues usually due to insufficient blood supply.
gingival sulcus—slight depression at base of tooth where it enters the gums.
globulins—proteins in blood plasma with several functions, including speeding conversion of prothrombin to thrombin, and antibody activity.
glomerulus—the functional subunits of the kidneys (plural glomeruli).
glomerular capillaries—bed of capillaries supplying each glomerulus with blood.
glomerulitis—inflammation of the glomeruli.
glycogen—the stored form of carbohydrate in animals.
hematocrit—separation of cells from plasma in the blood to determine erythrocyte volume in a given volume of blood.
hematological—pertaining to the blood.
hematoma—swelling or mass of blood (usually clotted) confined to an organ, tissue or space; caused by a break in a blood vessel.
hematuria—presence of blood in urine.
hemoglobin—a protein in red blood cells that carries oxygen and gives blood its red color.
hemoglobinuria—presence of hemoglobin in urine.
hemolysis—destruction of red blood cells causing liberated hemoglobin to diffuse into surrounding fluids.
hemoperitoneum—escaping of blood into the peritoneal cavity.
hemorrhagin—a toxic protein that damages the walls of blood vessels and causes bleeding.
hemostatic disruption—failure of hemorrhage to be stopped.
hemotoxin—a toxin that destroys blood cells and/or causes coagulopathy.
heparin—a substance that inhibits coagulation by preventing conversion of prothrombin to thrombin.
hepatotoxicity—malfunctioning of the liver due to a toxin.
heterologous antivenin—refers to antivenin prepared from sources other than the species causing the envenomation.
histamine—an amine derived from histidine that participates in many allergic reactions.
histidine—an amino acid obtained from the hydrolysis of tissue protein.
hyaluronidase—an enzyme that promotes the local spread of toxins by lysis of intercellular connective tissue that supports organs.
hyperacousia—abnormal sensitivity to sound.
hypercoagulability—coagulation of blood at an accelerated rate resulting in excessive clotting.
hyperglycemia—increase in blood sugar.
hyperplasia—excessive proliferation of normal cells in the normal tissue arrangement of an organ.

GLOSSARY

hypofibrinogenemia—having lower than normal levels of fibrinogen resulting in a coagulation defect.
hypoglycemia—deficiency of sugar in the blood.
hypotension—low blood pressure.
iliac—pertaining to the iliac bone, the widest and most superior part of the pelvis.
i.m.—intramuscular; into the muscles, usually with regards to injection mode.
imbricate—condition of overlapping scales, as in roof tiles.
incoagulable—blood that does not form clots.
induration—an area of hardened tissue.
internasals—scutes of snakes (usually paired) on the crown just behind the nasal.
intracapillary—within the capillaries.
intragluteally—into the gluteus (buttock) muscle.
intravascular—within the blood vessels.
i.p.—intraperitoneal; into the abdominal cavity, usually with regards to injection mode.
ischemia—local and temporary anemia due to obstruction of circulation to a body part.
i.v.—intravenous; into a vein, usually with regards to injection mode.
keel—in herpetology, a prominent ridge on a scale.
labial—pertaining to the lip area.
l-amino-acid oxidase—an enzyme that produces hydrogen peroxide, a powerful oxidizing agent.
lateral—pertaining to the sides of the body.
LD_{50}—the dose that will kill half of the animals it is injected into, usually white mice; a measure of toxicity.
leukocytosis—increase in number of leukocytes (white blood cells).
leukopenia—abnormal decrease in white blood cells.
loreal—a scale on the side of a snakes head, between the eye and nostral, but touching neither.
lymph—plasmalike fluid lacking red blood cells that bathes tissues.
lymphadenitis—inflammation of the lymph nodes.
lymphadenopathy—enlargement of lymph nodes, may be tender if inflamed.
lyophilization—process of quick freezing and dehydration of venom under a high vacuum.
mediastinum—space between the pleural cavities which encloses the heart, trachea, and esophagus.
meizothrombin—an intermediary compound in the formation of thrombin.
mental—in herpetology, the triangular scale at the lower jaw symphasis, corresponds to the rostral of the upper jaw.
mesenteric—pertaining to the mesentery, a peritoneal fold encircling the intestines, attaching it to the abdominal wall,
microangiopathic—small diseased changes of blood vessels.
microemboli—small, undissolved matter present in blood or lymphatic vessels.
microvasculature—referring to tiny vessels such as capillaries.
mitochondria—structures inside the cell responsible for energy production.
MLD_{50}—minimal lethal dose that will kill a test animal.

monovalent—an antivenin made of antigens produced from venom injection of one species.
motor end-plate—flat end point of a motor nerve fiber where it connects near a muscle fiber.
mucosal—pertains to mucus secreting membranes lining passages and cavities.
myocardial—pertaining to the muscles of the heart.
myocarditis—inflammation of cardiac muscle tissue.
myocardium—middle layers of the walls of the heart, comprising the cardiac muscle.
myofibrils—small fibers found in, and making up muscle tissue.
myonecrotic—destruction or death of muscle tissue.
myotoxic—venoms that affect muscle tissue.
nasal—in reptiles, the scale enclosing the nostril.
nasorostral—in reptiles, the enlarged, usually paired, scale just behind the rostral scale.
necrosis—death of a portion of a tissue.
nephritis—inflammation of the kidneys.
neurological—pertaining to the nervous system.
neutrophil—a type of leukocyte, the staining of which indicates a neutral pH.
neuromuscular junction—the space (synapse) between a neuron and a muscle fiber across which nerve impulses are transmitted.
neurotoxin—a toxin that has a marked effect on nerve tissue.
nocturnal—the behavior of being active at night.
nodal rhythm—cardiac rhythm with origin at the atrioventricular node.
nystagmic—condition of involuntary eyeball movements.
occipitals—in reptiles, the paired enlarged scutes lying immediately behind the parietals.
oliguria—reduced amount of urine formation.
oligemic—deficient amount of blood volume in the body.
opthalmoplegia—paralysis of the eye muscles.
oviparous—the reproductive method of producing external hard shelled eggs.
paralytic ileus—paralysis of the small intestine.
parapatric—living in neighboring (bordering) areas.
paresis—a slight or incomplete paralysis.
paraesthesia—an abnormal sensation, as a pricking numbness or burning.
parasympathetic—part of the autonomic nervous system, distributed to the heart, smooth muscles, and glands of the head and neck, and the thoracic, abdominal, and pelvic viscera.
parietals—in reptiles, the large paired scutes at the posterior end of the crown, immediately behind the frontal and supraoculars.
paroxism—a sudden spasm or convulsion.
passive diffusion—movement of molecules through the cell membrane using energy.
PCV—packed cell volume, in reference to blood cell volume.
pelvicalyceal system—pertaining to the kidney, pelvis, and calyx.

peptide—compound formed by the bonding of two or more amino acids.
peritoneum—membrane that lines the abdominal cavity.
perivascular hemorrhage—bleeding around the blood vessels.
petechial hemorrhages—small, purplish hemorrhagic spots on the skin.
phospholipase—an enzyme which breaks down phospholipids, a major building block of cell membranes.
plasma—the liquid, noncellular portion of blood.
plasminogen—a protein found in many tissues and body fluids.
platelets—disk-shaped, nonnucleated elements of blood, involved in blood coagulation.
polymerization—process of making a large compound from many smaller, similar subunit compounds.
polypeptide—compound formed by the bonding of many peptides.
polyvalent—a serum containing antibodies against the toxins of several different organisms.
postoculars—in reptiles, the scale(s) immediately behind and in contact with the eye.
post-synaptic membrane—muscle membrane that receives the stimulus to start contraction in response to a nerve impulse.
prefrontals—in reptiles, the enlarged scales just behind the internasals, or that area if it is covered with small scales.
prehensile—adapted for grabbing, as in the tails of certain snakes.
preocular—in reptiles, the scale lying immediately in front of and in contact with the eye.
preoccupied—the rule of priority (according to the code of zoological nomenclature). The first use of a name has priority. The same name cannot be used again for another species.
presynaptic membrane—membrane of the nerve ending from which a neurotransmitter passes into synapse en route to the post-synapse.
procoagulant—tending to favor the action of coagulation.
protease—a protein-splitting enzyme.
proteinuria—protein, usually albumin, in the urine.
proteolytic—action of enzymes which hastens the hydrolysis of proteins.
prothrombin—a blood clotting factor which forms thrombin in the coagulation cascade.
prothrombin time—a test determining clotting time after calcium and thromboplastin are added to decalcified plasma.
ptosis—paralysis of the eye muscles that keep the eyes open, causes drooping eyelids.
pulmonary—concerning or involving the lungs.
putrefactive—causing malodorous decomposition of animal tissue.
Q_{10}—a measure of the amount of oxygen used for every 10°C increase in ambient temperature.
recalcification time—time required for clotting to occur in uncoagulated blood following the addition of calcium.

renal—pertaining to the kidneys.
retroperitoneal—located behind the peritoneum.
rostral—pertaining to the rostrum (nose), a scale at the front of the nose of a snake.
sarcolemma—a delicate membrane surrounding each striated muscle fiber.
sarcoplasmic reticulum—network of ultramicroscopic filaments between fibrils of muscle tissue.
s.c.—subcutaneous; located between the skin and muscle layer.
scutes—overlapping or adjacent scales that cover the body of a reptile.
serine—an amino acid found in many proteins.
serous fluid—liquids of the body, similar to blood serum, which are in part secreted by serous membranes.
shock—a disturbance of oxygen supply to tissues and return of blood to the heart, often a result of hypotension.
serotonin—a chemical present in platelets, a potent vasoconstrictor and stimulator of smooth muscles.
s. l.—from the Latin, *sensu lato,* or in its broad sense.
spherocytosis—condition of red blood cells taking on a spherical shape.
s. str.—from the Latin, *sensu stricto,* or in its strict sense.
stroke volume—amount of blood ejected from the left ventricle at each beat.
subarachnoid—located below the arachnoid meninge (one of three membranes surrounding the central nervous system.
subcaudals—scales beneath the tail of a snake.
subconjunctival—below the mucus membranes which line the eyelids (conjunctiva).
sublabials—in reptiles, the scales (usually enlarged) along the border of the lower lip behind the mental.
sublethal—dose of toxin less than the amount required to kill.
sublingual—located below the tongue.
suboculars—in reptiles, the scales immediately below and in contact with the eye.
suffusion—spreading of body fluid into surrounding tissues.
supralabials—in reptiles, the scales (usually enlarged) along the border of the upper lip behind the rostral.
supranasal—located above the nasal.
supraocular—in reptiles, the enlarged scales or scutes on the crown directly above each eye.
systolic pressure—maximum blood pressure measured during the contraction of the ventricles.
sympatric—living in the same area.
symptomatology—a complex of symptoms.
synergistic—two or more components combined having a greater effect than each did separately.
synapse—space between two neurons.
tachycardia—rapid heart beat.
tachyphylaxis—rapid decrease in response to a toxic dose of a substance by previously injecting small doses of the same substance.

GLOSSARY

temporals—scales on the side of the head between the parietals and the supralabials, and behind the postoculars.

thrombin—an enzyme formed in shed blood from prothrombin that reacts with fibrinogen to form blood clots.

thrombocytopenia—abnormal decrease in the number of blood platelets.

thromboplastin—a blood coagulation factor (Factor III) that accelerates clotting time.

thrombocyte—synonymous term for blood platelet.

toxicoglomerulitis—inflammation of the nephron capillary beds due to the presence of a toxin.

toxicity—the deleterious effects of a toxin or other substance.

toxin—a naturally occurring poisonous substance.

toxinology—the study of the clinical, physiological, pharmacological, and pathological effects of biological toxins.

transaminase—an enzyme that transfers an amino acid from one compound to another, or the transposition of an amino acid within the same compound.

tripeptide—a proteinlike substance comprising three peptides bonded together.

urticaria—vascular reaction of the skin characterized by brief elevated eruptions on the skin with severe itching, commonly called "hives."

vascular bed—a high concentration of blood vessels in a given area.

vacuolization—process of forming vacuoles, or spaces in tissues.

vasculitis—inflammation of a vessel.

vasodilator—causing the relaxation of the blood vessels.

ventrals—the enlarged scales that extend own the underside of the body.

viscous—sticky, resistant to flow.

viviparous—reproduction by giving birth to live young that develop inside the mother's body.

vitellogenesis—production of egg yolk.